Fundamental Laboratory Approaches for Biochemistry and Biotechnology

Alexander J. Ninfa, Ph.D.
David P. Ballou, Ph.D.

University of Michigan

Fitzgerald Science Press, Inc.
Bethesda, Maryland

Fitzgerald Science Press, Inc.
Editorial Offices:
7307 MacArthur Blvd, Suite 214
Bethesda, MD 20816

fitzscipress@msn.com

Fitzgerald Science Press, Inc.
Orders and Fulfillment:
P.O. Box 605
Herndon, VA 20172
(800) 869–4409
(703) 661–1577

To the users of this book:
We have constructed a web site

http://www.umich.edu/~aninfa/book/

for this text, which will provide frequent updates, including additional experiments, refinements of experimental protocols, guidance for setting up experiments, and worked out answers to problems in the book. This site will also serve as a forum for other instructors to communicate improvements to protocols and new experiments and teaching ideas.

Library of Congress Cataloging-in-Publication Data

Ninfa, Alexander J.
      Fundamental laboratory approaches for biochemistry and biotechnology: a text with experiments / Alexander J. Ninfa and David P. Ballou.
            p.   cm.
      Includes bibliographical references and index
      ISBN 1-891786-00-8 (alk. paper)
      1. Biochemistry—Laboratory manuals.   2. Biotechnology—Laboratory manuals.   3. Molecular biology—Laboratory manuals.   I. Ballou, David P.   II. Title.
QP519.N55   1998
      572—dc21
                                                                                          98-9662
                                                                                          CIP

Production Management: Susan Graham
Production Service: Dennis Burke, The Publishing Management Connection
Cover Design: Susan Schmidler
Cover Illustration: Lisa Thomas
Printer/Binder: United Book Press

to Jean and Kathleen

Contents

Preface

Fundamental Laboratory Approaches for Biochemistry and Biotechnology was written as a biochemistry laboratory text for advanced undergraduate and first-year graduate students in biochemistry and other life sciences. Our goal was to provide a logical framework for training students in how to approach research problems and conduct and evaluate scientific research. Each chapter provides extensive background on the principles underlying methods used in research, followed by experiments designed to illustrate some of those principles. These experiments were designed for a 13-week semester during which the laboratory meets twice weekly for 3 hours/session. Since students taking such courses are often from diverse areas of study, we have tried to include many different areas of biochemistry in the text; for example, experiments ranging from enzyme kinetics and ligand binding to enzyme purification, recombinant DNA, and PCR are included.

The 25 experiments described have all been successfully lab tested in our class and found to be safe and easy to conduct. Some of them constitute "classical" experiments used in biochemistry courses at The University of Michigan for more than 20 years. Early chapters provide detailed protocols, while later chapters require students to formulate their own protocols as they develop experience during the course. To ensure year-to-year consistency and to streamline laboratory preparations, a list of all equipment needed and instructions for the preparation of laboratory materials (including possible vendors and catalogue numbers for each reagent) are provided at the end of each chapter.

We have constructed a web site

http://www.umich.edu/~aninfa/book/

for this text, which will provide frequent updates, including additional experiments, refinements of experimental protocols, guidance for setting up experiments, and worked out answers to problems in the book. This site will also serve as a forum for other instructors to communicate improvements to protocols and new experiments and teaching ideas.

Because this text provides considerable background and references for each of the areas covered, we feel that it will be suitable as a text to complement a set of "homegrown" experiments used in typical biochemistry laboratory courses at the advanced undergraduate or beginning graduate levels. Also, if resources and time are available for more advanced experiments, these experiments can be easily incorporated or substituted into the course, as much of the background material needed will already be present. Indeed, in Chapter 1 we present a list of questions for each chapter that may serve as extensions of the experiments. We encourage students to ask questions that take them beyond the formal exercises and to design protocols to test hypotheses.

Several factors have influenced our choices of experiments. Foremost is their potential for teaching the principles of biochemical research. The constraints are cost, space, safety, and time. Fortunately, there are many wonderful biochemistry experiments that are illustrative of general principles and useful for nurturing the development of skills for carrying out biochemical research, and which can be performed both inexpensively and safely with simple instrumentation and common reagents. The time factor is perhaps the most difficult problem to overcome because many students are likely to have scheduled classes or other commitments for the period immediately following the lab session and will be unable to stay late. Furthermore, student access to teaching laboratories after hours is often limited by the other responsibilities of the instructors. Thus, we have chosen experiments that can be performed in the allotted lab sessions. This often requires that the instructors perform the initial steps or some follow-up steps of a procedure, so that the students can conduct the most interesting and instructive measurements during the lab period.

Given the inexperience of the students and the potential for serious accidents, we have decided to conduct the course completely without use of radioactive materials. This is somewhat unfortunate, since a great range of biochemical research requires the use of radioactivity. However, we feel that the general principles of experimental design and good laboratory practices can be taught effectively without radioactive materials. Specialized training for the use and disposal of radioactive materials probably should involve instruction by trained university radiation control personnel so that they adhere to practices consistent with NRC regulations.

All of the experiments have been designed for performance without the use of animals. Although the sacrifice of animals in research is often necessary, the principles and most techniques that are important for carrying out biochemical research can be taught using plant or microbial starting materials.

Four topics important for research, but not often found in laboratory textbooks, have been included here:

1. A brief introduction to several ethical issues commonly encountered in research is in Chapter 1. The intent is to bring awareness and stimulate classroom discussion of these issues.

2. A three-pronged approach is used to encourage students to think like a research scientist. First, in Chapter 1 we introduce the basic intellectual tool of the scientific method, the hypothesis, and its use as a framework for developing conclusive experiments. Second, the laboratory experiments are designed to teach increasingly complex techniques and the analysis of data. Third, gradually, we require the students to devise their own protocols. Obviously, in the beginning of the course we must start with some "cookbook" procedures. However, we try to wean the students from detailed procedures and get them to design their own protocols to

accomplish the aims of the session. For example, in Chapters 9, 10, and 13, the students are required to design experiments to answer questions.

3. We have included some examples of how biochemistry is used and has altered our daily lives. Chapter 12 introduces some of the principles of forensic DNA analysis, and the students test whether their instructors can be excluded as suspects in a hypothetical crime. Chapter 9 involves the development of an assay for an enzyme used in clinical diagnoses. A variation of this assay is used in the clinic to distinguish between heart attacks and other medical problems presenting similar symptoms. Also in Chapter 9, an assay for ethanol is presented. We have found that these real-life examples stimulate student interest and provide a good framework for extension of the discussion of ethical and social issues related to these techniques and other techniques.

4. Because modern biological research requires the use of the wealth of information and software tools available on the Internet, Chapter 13 provides an introduction to these resources. Literature searches, protein and nucleotide database searches, and the use of RasMol, a popular molecular graphics visualization program, are presented. A challenging homework problem requires the student to use all of these resources. Numerous useful web sites are listed.

Finally, we want to emphasize that research is fun and satisfying. We hope that our love of biochemical research is evident throughout this text and that students might share in the excitement of "doing" modern biochemical and molecular biology research once they have been exposed to the ideas and techniques covered in this text.

We would like to acknowledge our lab colleagues for discussions, figures, and review of the manuscript. We thank the teaching assistants and students of the course, who have hunted down and helped us correct omissions and errors in the text, as well as developed improvements in the experimental protocols. We also thank the external reviewers who have provided insightful comments for improvement of the original versions of this manuscript.

Alexander J. Ninfa
David P. Ballou
Ann Arbor, Michigan 1998

1 Basic Practices and Techniques in the Biochemistry Laboratory

LABORATORY SAFETY

Safety in the laboratory depends primarily on the awareness and sense of responsibility of everyone involved. When you work in a laboratory, pay close attention to your surroundings, including your laboratory mates. Plan ahead and think through all of the operations of your experiments. These are some of the most important criteria for carrying out safe (as well as successful) experiments. Most laboratory accidents occur because common sense and planning are not used. We will try to avoid using dangerous materials or instruments in the teaching laboratory environment; nevertheless, many of the materials and instruments used can be harmful if proper care is not exercised. It is essential that you follow careful laboratory safety practices at all times.

You must take personal responsibility for safety. Rules do not make the laboratory safe; only people can minimize danger in the laboratory. Having stated this, a few basic safety considerations are listed below.

1. Do not eat, drink, smoke, or apply make-up in the laboratory.

2. Do not pipette by mouth; use pipetting devices.

3. Wear disposable gloves while handling harmful chemicals and be aware that the gloves are a source of contamination. Replace these gloves as appropriate to avoid spreading contamination. Never touch your face with your gloves.

4. Wear safety goggles to protect your eyes when your instructor tells you to do so. Wear safety goggles that block UV light when using a powerful UV source such as the transilluminator. Wearing contact lenses in the laboratory can be dangerous, especially if a reagent accidentally sprays into your eyes.

5. Carry out your operations in a fume hood if you are using a foul-smelling compound, .

6. Exercise caution when using instruments such as centrifuges and electrophoresis power supplies that have potential for serious accidents. If you are uncertain about the proper procedures, ask your instructor.

7. Inform your instructor immediately if an accident should occur.

8. Inform your instructors of any dangerous situation that you notice.

9. Immediately clean up any spills on bench tops, around instruments, or on the floor. Spills on the floor can be slippery and dangerous.

10. Clearly label samples, reagent bottles, and stock solutions with the chemical identity of contents, your name, and date of preparation.

11. Be aware of the toxicity of the chemicals and the biological hazards of materials used in the laboratory. Your instructors should keep a guide book that describes the safe handling of toxic substances and provides emergency first aid information.

Make safety part of your laboratory culture. Well-planned safe experiments will lead to more productive research.

LABORATORY SAFETY EQUIPMENT

Research laboratories should be equipped with the following safety equipment: telephone, eye wash station, shower, fire extinguisher, fume hood, first aid kit, spill kit, laboratory coats, and goggles. Be sure to identify each of these and their location before beginning laboratory work. The **telephone** is an essential piece of safety equipment. Numbers should be posted near the phone for the local police and fire departments, building services, radiation safety specialists, and senior investigator's home.

The **eye wash station** is used to rinse the eyes. A squirt bottle containing water is no substitute, because if you are alone and cannot see, it will be difficult to distinguish from bottles containing other substances. The **shower** is designed for emergencies such as when a bad chemical spills on a person or when someone is on fire. Usually, showers are located in a hallway near the laboratory. Pulling on the ring activates the shower, causing delivery of a large quantity of water. Therefore, activate the shower only in a real emergency. The **fire extinguisher** is only intended to be used for small laboratory fires that are easily extinguished. For more serious fires, pull the **fire alarm** immediately and call the Fire Department. A **fume hood** is used for performing reactions that are potentially explosive or that may result in harmful or unpleasant vapors. In the experiments in this book, you will not perform any reactions that have potential for fire or require a fume hood.

The **first aid kit** contains bandages, gauze, antiseptic cream, etc., and is used for treating small cuts and burns. If a more serious injury occurs, call for medical attention immediately. A **spill kit** is used to clean up dangerous spills, such as concentrated acids or bases. The spill kit contains one set of reagents for acid spills and another for alkaline spills. The directions for using the kits are on the kits themselves.

A **laboratory coat** is used to protect the clothing and body from direct contact with aerosols, small spills, etc. The experiments that we describe have been designed to minimize the use of dangerous materials, and do not require the use of laboratory coats. Nevertheless, several materials, such as stains, dyes, and acids, are capable of ruining clothing. Feel free to wear a laboratory coat.

There are several types of **goggles.** Standard goggles, as well as larger face shields, protect the eyes from chemical splashes; these types of goggles typically do not protect the eyes from UV damage. Special goggles are used to protect the eyes from UV damage, but these usually do not provide good protection from chemical spills. The instructors will distribute the appropriate goggles as needed.

A check list of safety items can be found in Appendix 1 at the end of this book This checklist may help familiarize you with the laboratory and methods for carrying out safe research.

SAFE USE OF LABORATORY REAGENTS

No one can know all of the harmful effects of the myriad of chemicals used in laboratories. Nevertheless, it is essential that we use and dispose of chemicals in a way that is not harmful to ourselves, our fellow researchers, or society. The safety of laboratory personnel is regulated by the Occupational Safety and Health Administration (OSHA, http://www.osha.gov/). OSHA has codified safety procedures for laboratory research; for example, regulations for the use of chemicals are contained in a document called the **Federal Hazard Communication Standard**. This standard requires that all laboratories follow four basic procedures: (1) Write a hazard communication plan to deal with potential hazards. (2) Maintain files of **Material Safety Data Sheets (MSDS)** for each laboratory chemical. An MSDS is normally included with purchased chemicals, or can be obtained from the vendor or from the Internet. An example of an MSDS is provided in Appendix 2 at the end of this book. The file of MSDS should be available at all times, so that any researcher in the laboratory can quickly consult this information before performing experiments. (3) Label all chemicals with information indicating the reactivity, stability, and safe use of the chemical. (4) Train all personnel in the proper use of any chemicals they will use.

All laboratory workplaces must have a **Chemical Hygiene Plan**. This plan, which is specific for each individual laboratory, must contain the following: (1) Detailed information on how to respond to emergencies (fire, explosion, radiation spill, chemical spill, biological spill, medical emergency, bomb threat, theft and security, power failure, and other building emergencies) and a list of all laboratory personnel and emergency numbers. (2) Basic safety rules of the institution and the specific laboratory. (3) Standard operating procedures for chemical procurement, chemical distribution and storage, environmental monitoring, housekeeping, maintenance and inspections, and protective apparel and equipment. (4) Detailed work rules for dealing with hazardous chemicals, radioisotopes, or special equipment, and written procedures for laboratory experiments. (5) A training program on safe practices, with records of training. (6) Chemical inventories for all chemicals present in the laboratory. (7) A waste disposal plan. (8) A list of responsible persons and a chemical hygiene officer who is responsible for adherence to the rules. Most institutions have an occupational safety and environmental health unit that can provide help in developing individualized chemical hygiene plans and assist in improving safety procedures.

Safe disposal of waste is critical to laboratory research. Dangerous materials must be disposed of safely so that they do not contaminate the environment or endanger our communities. While it is usually simple to establish the hazardous characteristics of clearly identified waste, unidentified waste materials present serious problems. Therefore, waste must be clearly identified and separated in marked containers. The information required by disposal facilities includes: physical description, water reactivity and solubility, pH, and the presence of oxidizers, sulfides, cyanides, radioactivity, biohazardous material, and other known toxic constituents. Scientists should minimize waste by designing experiments on as small a scale as possible, yet consistent with yielding correct answers. In general, try to separate waste products. Use separate containers for sharp objects, broken glass, plastic ware, organic solvents (be sure to label them clearly), acids (neutralize, if possible), bases (neutralize, if possible), hazardous biological materials, radioactive materials, etc. Most of the waste from the experiments described in this book is not particularly hazardous. However, some wastes are hazardous, and these must be disposed of properly. Use the principles outlined above and consult your instructors to better learn how to deal with these issues.

PHILOSOPHY AND DESIGN OF EXPERIMENTS

As a prelude to discussing the design of experiments, we offer a few general comments on the philosophy of conducting research. The ultimate goal of research is to find the truth. In our studies using the techniques of biochemistry and biotechnology, this means elucidating the laws of nature. Scientific investigations are designed to "find out what is going on" in nature, not to sustain preconceived notions. Understanding this distinction is essential for success in research. To find out what is going on implies that two or more outcomes are possible, and that a result can be obtained that eliminates some of the possibilities. Before carrying out the necessary experiments, the scientist does not know what the results will be. In contrast, with preconceived notions where the outcome is fixed (at least in the experimenter's mind), nothing new can be learned from experiments. Such activities are not really experiments; they are demonstrations.

Hypotheses are the tools of the scientific method. We use them to focus our thinking about a subject and to help us narrow down the possibilities of how to investigate the subject. The value of hypotheses is that they make predictions that can be experimentally tested. If any of these predictions are observed to conflict with experimental data, then the hypothesis must be modified or discarded. When considering a new hypothesis, it is useful to think through all of the predictions derived from the hypothesis and design experiments to test those predictions. Hypotheses that make no predictions that are subject to experimental refutation are not hypotheses at all; they are religions.

If an experimental result appears to invalidate a hypothesis, this may be due to faulty experimental design, or to an artifact or inaccurate measurement. But, if these conditions can be eliminated, then the hypothesis must be modified. Conversely, if all of the predictions (that one can think of) are met, this does not necessarily mean that the hypothesis is correct. There is always the possibility that another hypothesis would also make the same predictions and better describe the phenomenon under study. Thus, it is a lot easier to show that an hypothesis is wrong than it is to show that an hypothesis is correct. Indeed, some philosophers have argued that the latter is not within human capabilities. All we can really say is whether our hypotheses are consistent with the experimental data.

The best way to maintain objectivity in experimental work is to dissociate yourself from the hypotheses under investigation. Human nature leads those who propose hypotheses to become attached to their own ideas. When formulating hypotheses, it is useful to list as many alternative hypotheses as possible, and then work through the predictions of each hypothesis, noting where each differs. Ideally, experiments will become obvious that may result in the elimination of one or more of the hypotheses because the relative strengths of one over the other will become obvious. When multiple hypotheses are formed, the human attachment to them becomes diluted and a more objective approach may result.

There is a place, however, for human ambition in science. If you conduct research as a career, you will conceive many hypotheses, and some (most) of these will be incorrect. An appropriate ego-satisfying activity is to be the first to find out what is going on by designing and conducting the most incisive experiments that refute the hypothesis, even if it is one's own.

Predictions derived from hypotheses are tested through experimentation. It is important to consider the expected results in a qualitative and quantitative sense for every experiment *before* beginning any laboratory work. How will you know if something unexpected happens if you have not thought out the expectations beforehand? For example, the suitability of the proposed measurement method should be considered with regard to the expected results. If the expected results are such that a quick and simple measurement may be unambiguous, then more rigorous methods of conducting the measurement may not be necessary. However, it is pointless to begin an experiment if the proposed method of measurement cannot unambiguously distinguish between the predictions made by opposing hypotheses.

Having expectations of the results allows the researcher to realize quickly when an interesting and unexpected result occurs. The successful researcher uses keen observational skills. Since research does not always turn out as planned (otherwise, it would not need to be done), observational skills, coupled with creativity and further analysis, can lead to redesign of experiments (even midstream) so that more productive results can be obtained. Many very fruitful lines of research have been derived from careful observations made during an experiment.

At several points in this book, you will be called upon to design experiments. Below, we briefly introduce strategies for designing informative experiments.

An important consideration is that the method of measurement is suitable for the study you wish to conduct. Choose a protocol that provides unambiguous data. The ease of the measurement and the number of data points resulting should also be considered. In general, the easier the measurement, the better, because this permits acquiring more data points, simple repetition of the experiment, and good statistical analysis of results. The possibility of false signals must also be considered. For example, if you are measuring a change of some physical property of a solution that is caused by a chemical reaction, it is necessary to show that the measured change actually results from this particular reaction and not some unexpected reaction. To substantiate the result, an independent method for conducting the measurement is highly desirable, even if it is more cumbersome.

For experiments to be most useful, the variables must be isolated; that is, the experiment must be controlled. When we are measuring some change in the physical properties of a solution due to a chemical reaction, a **negative control** might be an analogous experiment in which the reaction does not occur because one or more components necessary for the reaction is intentionally left out. For example, if we were assaying an enzyme that uses two substrates, we might leave out one of the substrates, measure the reaction, and then add that substrate and remeasure. Another common negative control for enzyme assays is to boil the enzyme before adding it to the assay mixture. This presumably inactivates the enzyme so that any activity observed would be due to some heat-stable contaminant. It is useful to consider a number of different ways to provide negative controls for each experiment.

Positive controls are experiments designed to show that the method of measurement is suitable. Positive controls for the example above would show that the conditions were suitable for the reaction to occur and that if the reaction occurred, you would be able to observe it. For example, if you are measuring a change in the physical properties of a solution that is caused by a reaction, then adding a known amount of the reactant(s) to this reaction mixture should result in a predictable response. Our ability to observe the reaction can be checked by testing whether adding a known amount of product gives the appropriate result. Comparison of the results of these two types of positive controls may reveal deficiencies in the assay or the method of measurement.

In many experiments in biochemistry and biotechnology, it is not possible to isolate completely the variables or even know whether you have done so. This is one of the most significant problems to be dealt with in the laboratory. For example, purified enzymes may not be completely pure, and the contaminating proteins may have enzymatic activities that confound the analysis of the purified protein. Also, various solutions may contain unknown activators or inhibitors of the process under investigation. We will see this problem in Chapter 3, where we consider the methods used to quantify protein concentration. Activators and inhibitors may affect the reaction that is being measured, or they may affect the measurement itself. In the latter case, an alternative method of measurement may reveal the problem. The careful scientist remains alert to possible hidden variables and their consequences.

We hope that you use this text as a starting point to develop skill in designing experiments and optimizing laboratory techniques. Ideally, laboratory sessions should include time for this activity. Since the key to conducting useful research and designing good experiments is asking good questions, we have provided below a small sample of the kinds of questions and procedures that could be investigated as extensions of the Chapters in this text. You might reread these questions after you finish each of the following laboratory sessions. Instructors can also use these as starting points for more detailed studies.

Colorimetry. Can the measurements of *p*-nitrophenol (PNP) ionization be made in the presence of a contaminating substance that has a partially overlapping absorbance spectrum? How do you know that the dissociation of PNP involves a single proton?

Protein quantification. Which of the methods in this chapter works best in the presence of various contaminating substances?

Chromatography. Which resin give the best resolution of various substances? How tall must a column be to resolve various substances? How could you test whether your determination of the fully included volume of a gel-filtration column is accurate?

Electrophoresis of proteins. Does increasing or decreasing the percentage of acrylamide in SDS-PAGE give better resolution of particular proteins? What is the effect of salt in a protein sample? For non-denaturing gel electrophoresis, what is the best pH and concentration of acrylamide to use for resolution of alkaline phosphatase? Can hetero-oligomers such as aspartate transcarbamylase from *E. coli* be purified from nondenaturing gels, and then the number of different types of subunits be examined by SDS-PAGE?

Ligand binding. Can the binding of HABA to avidin be used as an assay for the partial purification of avidin from egg white? Are there inhibitors in egg white that prevent detection of complex formation? What is the optimal pH for the binding of HABA to avidin? How would you construct an affinity column for the purification of avidin?

Purification and characterization of alkaline phosphatase and kinetics. Can osmotic shock be used to release the enzyme? Are the yields and purity different if a whole-cell sonicate is used? What is the optimal heat for the heat treatment step? At what NaCl concentration does alkaline phosphatase (AP) elute from the DEAE column? Is AP really a dimer? How could you purify AP further (e.g., gel-filtration, hydrophobic-interaction, hydroxylapatite)? Can AP bind to (and cleave) nucleotides, oligonucleotides, phospho-sugars, or compounds with a C-P bond (phosphonate)?

Enzymatic methods of analysis and coupled assay systems. Choose another enzyme that uses ATP, ADP, NAD, NADH, NADP, etc. from a commercial catalogue, and see if you can develop a reliable coupled assay. Volumes 1-5 of *Methods in Enzymology* are replete with good examples of enzymes that could be used.

Recombinant DNA. Clone the *Eco*RI or *Hin*dIII fragments of bacteriophage lambda into pUC19. Clone fragments from a low-melting-point gel. Is the published restriction map of pBR322 correct for the enzymes *Pst*I, *Bam*HI, *Hin*dIII, *Pvu*II, and *Eco*RI? That is, construct your own restriction map of pBR322 for these enzymes. What is the effect of temperature or other factors on the rate of ligation? At what pH and at what concentrations of salt, buffer, ATP, etc. is the DNA ligase reaction optimal? What is the optimal DNA concentration for an intramolecular ligation (circularization) reaction? Can you develop a method for quantifying the initial rate of the ligation reaction, without using radioactivity? How would you conduct a kinetic characterization of the DNA circularization reaction?

PCR. Is your instructor your parent? Can you design PCR primers for the amplification of genes encoding a class of proteins such as ATP-binding proteins? Using *Drosophila* DNA sequences, can you clone homologous genes from the common housefly (which you must catch yourself and then isolate its DNA)? How about other insects you may find around your home or in the laboratory? Hint: Try genes that are likely to be highly conserved. How could you use PCR to amplify the gene for kanamycin resistance from pACYC177 and clone it into pBR322 or pUC19? Could you use PCR to amplify and clone interesting signal transduction genes (send clones to aninfa@umich.edu) or flavin enzymes (send clones to dballou@umich.edu)?

Using the desktop computer and Internet tools for biochemical research. Now that you have some familiarity with using the Internet, try using it for a variety of professional activities. Look up recent papers of your instructors in PubMed, and read the abstracts. Get the latest paper, read it, and involve your instructor in a discussion about it. Get a list of graduate schools with programs in Biochemistry or Biotechnology (or whatever your interests) and retrieve a list of faculty at the places to which you might want to apply. Look up the faculty research interests and see if you can picture yourself working in those research areas. Choose an interesting protein whose structure is unknown and try to model it by using a known protein with a similar sequence and various tools from the Internet. E-mail government officials at all levels and explain why we need a greater investment in science and education. Create your own home page!

ETHICS IN SCIENCE

The progress of science mandates that each member of the scientific community applies the highest ethics in three distinct general areas. First, science progresses by the accumulation of knowledge over time, and to be most effective, this progress requires that the knowledge be valid. Second, for the scientific community to function coherently, the proper credit (or blame) for accomplishments should be ascribed to the responsible scientist. Third, and most importantly, the activities of scientists must be compatible with the needs and values of the greater society. This point can be partially summarized in the Golden Rule: "Do unto others as you would have them do unto you." Each of these areas will be briefly discussed in this section. A comprehensive discussion of ethics in science is beyond the scope of this work, and the following discussion is intended to stimulate your thinking and interest in this area. Hopefully, your instructors will also embellish and expand this important area of science.

In science, as in life, one should "play fair". However, in science, as in life, there can be subtleties to "playing fair" that require careful consideration of the issues. The root of most ethical problems in science, as in life, is selfishness.

VALIDITY OF SCIENTIFIC DISCOVERIES

The progress of science requires that scientific discoveries be valid. Scientific discoveries describe the laws of nature in an objective and quantitative sense. Because future work will build upon current studies, it is essential that our results be correct.

At the level of the individual, accuracy of conclusions requires that the experiments be properly performed, that the data be properly recorded and treated, that the precision of the data be acceptable and explicitly stated, that the results be reproducible from day to day, and that the materials and methods used in the experiments be understood sufficiently well to permit reasonable conclusions about the results of the experiments. *Scientific work needs to be rigorously documented.* Certainly, no one can know all of the possible artifacts that may affect scientific conclusions, and thus, it is inevitable that some uncertainty will occur in scientific studies. However, it is important that the work be described with sufficient clarity to allow it to be understood at a later time when the presence of an artifact may become known. Furthermore, any uncertainty concerning the materials or methods must be explicitly stated as part of the results and conclusions. Proper documentation is a major deterrent of fraud in science.

At the level of the scientific community, accuracy is insured by the repetition of experiments by other scientists working in different laboratories. For this reason, scientific work must be presented in sufficient detail to permit its repetition. Furthermore, scientific experiments cannot proceed with "secret reagents"; thus, it is often necessary for the scientist to share experimental materials with the community of scientists. Any material essential for conducting a repetition of the experiments must be provided upon request to qualified members of the scientific community. Remember that your goal as a scientist is to find the truth, not to prove that you are right or avoid being wrong.

ETHICAL INTERACTIONS WITH THE SCIENTIFIC COMMUNITY

Scientists are a curious lot. They believe that they should receive credit for their accomplishments, but should not receive (nor desire) credit for the work of others. As strange as it may sound to nonscientists, many scientists believe acknowledgment for their accomplishments is more important than receiving their paycheck. Our paychecks will soon be spent, but our work will hopefully endure for all time. Proper acknowledgment of the accomplishments of other scientists facilitates cooperation towards scientific progress by the community of scientists.

The ideas and accomplishments of scientists are **intellectual property**. The failure to acknowledge the intellectual property of others is essentially theft. Examples of abuse of intellectual property could be the deliberate omission of citations to relevant earlier studies or the minimization of the significance of earlier work to exaggerate your own accomplishments. A good rule of thumb in this regard is to treat the work of others as you would have your own work treated. This does not mean to treat the work of others as they treat your work, for the unethical actions of others may never be used to justify unethical actions by ourselves.

Many scientific discoveries ensue from the collaborative efforts of several individuals. For example, the head of a laboratory may have a number of students working on various aspects of the same project, which in the aggregate lead to a result worthy of publication. In recognition of this, scientific journals permit the inclusion of several coauthors on scientific papers. Frequently, there is some conflict as to whether the contributions of an individual are sufficient to merit inclusion in the list of coauthors. As a guideline, *co-authorship requires that a person has contributed to the work in a direct and essential way*. This does not mean that all of the coauthors participated equally in performing the experiments. It may be ethical to include among the coauthors workers who formulated the hypothesis, designed the experiments, or analyzed the data. However, coauthorship must be reserved for those workers who made essential *scientific* contributions to the work. Coauthorship implies that each of the coauthors is responsible for all of the results and conclusions presented in the paper.

It is important for the progress of science that data be properly communicated to the scientific community. Withholding scientific data for the purpose of providing a personal advantage for future studies slows the progress of science and is therefore unethical.

SCIENTISTS AS MEMBERS OF SOCIETY

The actions of scientists must be compatible with the needs and values of society. Within a laboratory, this requires that each member be a good laboratory citizen. For example, it is highly unethical to permit a dangerous condition to persist that could harm other workers. Also, it is unethical to fail to mention to other workers conditions that may adversely affect their experiments, such as contaminated materials or malfunctioning instruments.

With regard to society at large, it is important that any dangerous materials associated with the experiments be openly acknowledged and disposed of in a responsible way with the knowledge and approval of the appropriate societal institutions. Most importantly, it is necessary that investigations are conducted in harmony with the values of society. For example, it is not permissible in our society to cause humans to suffer from an untreated disease in order to obtain knowledge of the pathology of that disease. (The infamous Tuskegee case dealt with this issue.) In such a case, the value our society places on human life and well-being takes precedence over the need to obtain knowledge.

Furthermore, our society places a value on animal life. In the experiments in this book, we will not be using animals. It is not necessary to sacrifice animal lives to learn the fundamentals of biochemistry and biotechnology laboratory practices. While the use of animal studies and attendant sacrifices of animals are essential to many important areas of research, such as drug toxicology studies, experiments with animals should only be undertaken after careful consideration of their necessity. When deemed necessary, design these experiments to minimize both the number of animals involved and the suffering of the animals.

Society needs to know that our scientific opinions are based on our scientific judgments. Thus, any conflict of interest should be openly stated in scientific publications or presentations. Conditions for conflict of interest are present any time the results of a scientific study promote the material interests of the investigator.

PRESENTATION AND ANALYSIS OF DATA

HANDLING DATA

If you have an electronic calculator, resist the temptation to record all digits that you see. The final calculated values in your experiment should reflect the precision of the measurements; in general, you should report only two or three digits. For example, if the observed absorbance of a solution is 0.623 and the calculated concentration of material in the sample is 3.4567879 mg/mL. Just record 3.46 mg/mL. (Note the round-off rule: if the digit after the last significant digit is 5 or more, round up the last significant digit by 1.) Suppose you want to convert the above concentration into micrograms per milliliter; the reported value would then be 3.46, not 3.457. A zero at the right end of a number is not necessarily a "significant" digit. In this case, where the observed reading contained only three digits, you cannot end up with four significant digits. The above comments refer to calculated values, not observed values.

Graphs most often used in biochemistry are either line or bar types. The purpose of a line graph is to show visually what is going on in the experiment. A table can often relate the same information more quantitatively, but a line graph can show trends more clearly and emphasize possible correlations with known mathematical functions. Line graphs are customarily presented with the independent variable on the horizontal axis (x axis or abscissa) and the dependent variable on the vertical axis (y axis or ordinate). The dependent variable y is the value you measure or some function of what you measure; you choose the independent variable x. Thus, line graphs are useful for expressing $y = f(x)$. For example, if you measure the absorbance at 530 nm as a function of protein concentration, then the absorbance values are the dependent variable set and the protein concentrations that you chose are the independent variable set. Some experiments are plotted on logarithmic scales either because such a plot may linearize the presentation or because the data may cover an extended numerical range. Semilog graphs have a logarithmic scale on one axis and a linear scale on the other. Note that log scales do not pass through zero because $\log(0)$ is undefined.

Bar graphs are frequently used to express visually data that are normally in a table. For example, you might measure an enzymatic activity in different tissues before and after a particular treatment. The results could be demonstrated either in a bar graph or in a table. A line graph would not be appropriate for these data because there is no mathematical relationship ($y = f(x)$) between the dependent and independent variables. Bar graphs can have high visual impact, but the significance of small differences can often be overemphasized. Three-dimensional bar graphs exacerbate this problem, and color further emphasizes small differences. For this reason, colored, three-dimensional bar graphs are frequently used in business presentations.

TABLE 1-1
Reproducibility of Determination of K_d for Binding of Ligand *L* by Various Preparations of Receptor *R*

RECEPTOR PREPARATION	K_d VALUES (µM) AND MEAN VARIATION	DEVIATION FROM MEAN	DEVIATION SQUARED
R1	10.1 ± 0.4	−2.23	4.97
R2	11.2 ± 0.4	−1.13	1.28
R3	12.5 ± 0.4	0.17	0.03
R4	13.2 ± 0.3	0.87	0.76
R5	12.9 ± 0.5	0.57	0.32
R6	11.7 ± 0.4	0.63	0.40
Mean = 12.33	$n = 6$ Average deviation = 0.93		Total = 7.76

STATISTICAL TERMS

Accuracy, precision, reproducibility, and error are statistical terms that are relevant to your biochemical techniques. **Accuracy** refers to the **correctness** of your data or final answer. **Precision** refers to the **variability** of your data (the range of values obtained if you tried to make the measurement repetitively). **Error** refers to accuracy, not to mistakes. It is possible to have good precision in an experiment yet have poor accuracy (for example, your pipette could be out of calibration and consistently give values 10% too high.).

Reproducibility is similar to precision, but it refers to the variability observed when you try to do the entire experiment over again. For example, suppose that you decide to determine the dissociation constant K_d for the interaction of the receptor R and the ligand L. You run the experiment in triplicate with the first sample of receptor R and obtain a value of 10.1 +/− 0.4 µM. This variation denotes the precision of determining K_d for this receptor sample. The value 10.1 is the **mean** or **average** for this determination. When you measure K_d values for five additional preparations of R, you obtain the set of mean values shown in Table 1-1, all with similar precision to the first R sample. These variations between samples are larger than the variation determined on a single sample. This gives you a measure of the reproducibility of preparing R and of determining its K_d for L.

Precision and reproducibility are often described quantitatively with the term **standard deviation** (SD). Consider the set of K_d values in Table 1–1 and their deviations from the mean *d*. The mean "deviation from the mean" or standard deviation is calculated with the following formula:

$$SD = \sqrt{\frac{\sum d^2}{n-1}} \qquad (1\text{-}1)$$

$$SD = \sqrt{\frac{7.76}{5}} = 1.25$$

Thus, the mean is 12.33 ± 1.25. Statistically speaking, you should expect that in this case two thirds of observed values will fall within that range (11.08–13.58). In this example, the values falling within the range of the standard deviation are the following: 11.2, 12.5, 13.2, 12.9, and 11.7 (5 out of 6, or >2/3). Note that one is not always so lucky that fact and probability coincide. When you use a simple statistical analysis like this, it is assumed that the observations follow a bell-shaped distribution curve, with the negative deviations from the mean distributed in the same manner as the positive deviations. Note that the SD calculation is done with one or two extra significant digits, but the final answer is rounded off to match the digits in the mean.

If the data show high variability, the SD will be large. The relative SD provides a measure of the variability that can be compared with other experiments. This is calculated by dividing the SD by the mean and multiplying by 100 to give the answer in percent. In the above example, the relative SD = 1.25(100)/12.33 = 10.1%, which is a quite high value. However, few biochemical data show <1–5% relative SD.

Each measurement you make in the course of an experiment will have an intrinsic error that depends on the quality of your instruments, the variability of room temperature, the consistency with which you carry out the experiments, etc. It can be pointless to work hard to obtain excellent precision in one step and then neglect the precision of another step. To minimize error it is necessary to investigate how each step is affected both by your actions and by the reliability of the instruments that you use. In your experimental design, the use of various instruments will lead to uncertainties in your measurements. Try to understand how the specifications for a given instrument will actually affect your measurement. A simple example is the following: The manufacturer of your 1-mL pipette may guarantee the accuracy to be correct within $\pm$ 0.012 mL. Thus, a 1-mL sample volume may lie anywhere within the range 1.012 and 0.988 mL. Moreover, the relative SD might be 0.7% for a batch of 1-mL pipettes. This introduces a relative error in your data of 0.7 % from this source alone. Consider the case when you use this same pipette to measure 0.1 mL. Now what would your relative uncertainty be?

BASIC PROCEDURES: MEASUREMENT OF WEIGHTS, VOLUMES, AND pH

Most of the experimental materials that you will use will be provided in the form of solutions of known molarity. Nevertheless, it is important for your own experiments that you understand how to prepare properly these materials. Some principles of preparing reagents will be covered in this section.

Concentrations of solutes in solution may be expressed in several ways: for example, as % (wt/vol), % (wt/wt), % (vol/vol), g/mL, or molarity. The term % (wt/vol) refers to the solute in grams/100 mL of solvent, whereas % (wt/wt) refers to grams of solute/100 g of solvent. The clinical term mg% refers to milligrams of solute/100 mL of solvent. The term % (vol/vol) refers to a case in which both substances are liquid measured as milliliters of solute solution/100 mL of solvent solution. In some chromatography applications, mixtures of solvents are listed as vol/vol/vol ratios such as 24:1:1 (methanol:acetic acid:water), where the numbers indicate the relative volumes of each solvent. Of all these expressions of concentration, molarity, signifying moles per liter, is the most useful for biochemical calculations. The abbreviation for molarity is M. Please remember that molarity is a concentration term, whereas mol (moles, molecular weight in grams) is an amount. Thus, 1.0 L of a 1 M solution contains 1 mol of the solute, while 0.5 liters of a 1 M solution contains 0.5 mol of the solute.

To make a solution of defined molarity, the desired amount of the solute is weighed by using a balance and is then dissolved in the solvent (usually water or buffer). After the solute is completely dissolved and the pH is adjusted (if necessary), the final volume of the solution is adjusted to make the desired concentration. Thus, if one wanted to make 1 liter of a solution of 1 M KCl, one would weigh out 1 mol of KCl (74.55 g), dissolve it in ~700 mL of water, and then add water to bring the final volume to 1.0 L. Usually, the best way to do this is to weigh a beaker (this is often called "taring" the beaker), add the desired amount of the solute directly into this beaker on the balance, and then add the water to the same beaker. This avoids transfer of the weighed sample and any attendant losses. It is important to remember that the formula weight of the solute compound should be used in these calculations. For example, the formula weight of the tetrasodium salt of ATP is different from the formula weight for the tetraammonium salt of ATP; consequently, a different amount of these two forms of ATP would be needed to make a 1 M solution of ATP. Be sure to label the solution by including its chemical identity of contents, your name, and date of preparation.

Volume is measured in several ways. Large volumes are measured in containers known as **volumetric flasks** (Fig. 1-1). Various sizes of volumetric flasks are available for measuring volumes ranging from 1 mL to several liters. The flask consists of a large, round-bottom portion and a thin neck with a mark signifying the level of the liquid when the volume equals

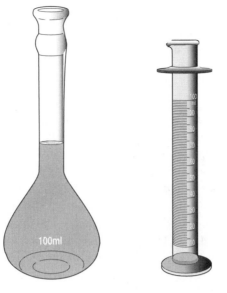

FIGURE 1-1 Volume measurement containers. A volumetric flask is shown on the left, and a graduated cylinder is shown on the right. The plastic ring near the top of the graduated cylinder helps prevent breakage in the event that the cylinder falls over.

the flask capacity. When filling such a flask, be sure that the bottom of the meniscus of the liquid lines up with the mark. For precise measurement, observe the level by placing the flask on a stable level surface, such as the bench, and line up your eyes with the level of the mark. It is advisable to bring the level of the solvent up slowly by adding it from a pipette or squirt bottle so that you do not overshoot the desired volume.

Volumes may also be measured with **graduated cylinders**, but this approach is less accurate than using a volumetric flask. For many applications, we will recommend using graduated cylinders in place of volumetric flasks. You can use graduated cylinders more flexibly than volumetric flasks because they permit the measurement of nonintegral volumes such as 23.5, 75 mL, etc.; however, the accuracy is likely to be only 3–10%.

Small volumes of liquids are difficult to measure precisely, and the best method is usually to weigh the liquid sample while using the solvent density to calculate the volume. Another less accurate way to measure small volumes is to pick up the liquid entirely in a pipette, and read the volume from the position of the lower end of the meniscus, as was done for volumetric flasks. Be sure to hold the pipette vertically and read the meniscus at eye level. One difficulty with this technique is that it is difficult to avoid drawing up air bubbles into the pipette; bubbles will take up some volume, thus making the measurement inaccurate.

Very small volumes (0.001–1 mL) may be estimated by similarly picking up the liquid in a **micropipette** instrument, such as a "Pipetman" or "Finnpipette." These instruments have an adjustable volume setting such that a predetermined volume of liquid can be picked up into a disposable tip. To measure volume with such an instrument, you will have to try several settings until one is found where the instrument picks up all of the liquid but no air. If you know the concentration of a solute in a solution in a small volume, you can determine the small volume by diluting the sample to a defined measurable volume, and then determine the concentration of the solute in that volume, although this method is not always practical. Calculate the unknown volume v_1 as

$$(v_1)(\text{concentration}_1) = (v_2)(\text{concentration}_2)$$

Transferring liquids is one of the most common steps for biochemical experiments. To obtain accurate results, transfers must be correspondingly accurate. To transfer small volumes of liquids, we use glass **pipettes** (for volumes of 1–25 mL) (Fig. 1-2) and **micropipettors** (for volumes of 0.001–1 mL) (Fig. 1-3). It is necessary that you master the use of these instruments to obtain accurate results in this laboratory. There are several different kinds of glass pipettes, and these tend to get mixed together in the laboratory. One type of pipette, called a **serological pipette**, is graduated all the way to the tip. If you fill a serological 1-mL pipette to the 1-mL mark and deliver all of the solution to a tube, you will have delivered 1 mL. A second type of a pipette, called a **Mohr pipette**, has a dead space below the graduation. Thus, if you fill a 1-mL Mohr pipette to the 1-mL mark and deliver all of the liquid to a tube, you will have delivered more than 1 mL. When using a Mohr pipette, the liquid that is delivered to the destination tube should only be from within the top and bottom graduations of the pipette. Of the two types of pipettes, the Mohr pipettes are often easier to use accurately, because it can be difficult to deliver all of the volume contained in a serological pipette to the recipient vessel. This is especially true

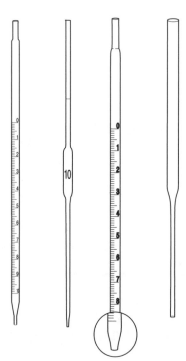

FIGURE 1-2 Various types of pipettes used in laboratories. *(Left to right)* Mohr, volumetric, serological, and Pasteur pipettes.

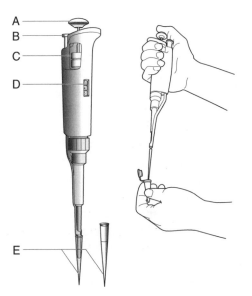

FIGURE 1-3 A typical micropipettor: *(A)* the plunger, *(B)* the tip ejector, *(C)* the volume adjustment knob, and *(D)* the volume indicator. *(E)* The disposable plastic tip is shown both on the micropipettor and separated for visualization. The drawing on the right shows the proper method of holding the micropipettor and of delivering sample to a microfuge tube (see text).

when the solution to be transferred is viscous. A third type of pipette is the **volumetric** pipette, which delivers a defined volume quite accurately, but is useful only for that one volume.

Most liquids that we use in this book are harmless. Nevertheless, to develop proper laboratory habits, we discourage the use of mouth pipetting. Instead, pipetting aids should be provided to permit the picking up and delivery of liquids by pipette (Fig. 1-4). These are fairly simple devices, and an instructor can show you how to use them in the laboratory. The rubber bulb filler has three valves marked A, S, and E (Aspirate, Suck, and Evict,

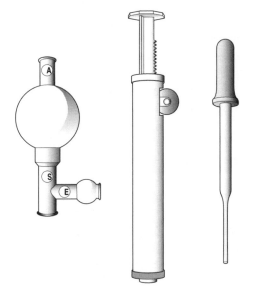

FIGURE 1-4 Various aids for filling and delivering liquids from pipettes. Examples include *(left)* a rubber bulb filler, *(center)* a pipette pump, and *(right)* a rubber pipette bulb.

respectively). Connect the bulb to the pipette through the hole at the bottom of the device (under the S). Simultaneously squeeze the A valve and the bulb to aspirate the air from the bulb. Next, place the tip of the pipette into the solution and squeeze the S valve so that the bulb sucks the solution into the pipette just past the appropriate mark on the stem. Squeeze the E valve to permit air to enter and evict the sample from the pipette until the meniscus falls to the desired graduation. Wipe off the excess liquid on the pipette tip with a clean tissue, which should be discarded, and then deliver the required volume to your receptacle by again squeezing the E valve. This general procedure is used for any of the pipetting aids, but the individual operations may differ.

It is important not to contaminate your pipetting aid with any of the liquids. Contaminating your pipette aid is easier to do than you might expect, so be careful. If you should contaminate your pipetting device, obtain another one and clean out the one you contaminated. Never put a contaminated instrument back into general circulation or you may ruin someone else's (or even your own) future experiment.

Glass pipettes are expensive, but with proper care, they can last for many years. Immediately after use, place the pipettes (with the tips facing up) into the designated container, which should contain a sufficient volume of water to cover the pipette completely. Try to avoid letting solutions dry out in the pipettes. When the container for used pipettes becomes too full, advise an instructor. Do not jam glass pipettes into a full container, since you may break the pipette and hurt yourself.

Several different brands of **micropipettors** (Fig. 1-3) will be used in the laboratory, and it is important to familiarize yourself with the instrument you are using. Typically, these devices use disposable tips to hold the liquid. The tips should snugly fit over the ends of the pipettor. Different instruments may require different sizes of disposable tips. If the tip does not fit snugly to the instrument, then precise measurements cannot be made, because air will leak past the poor seal allowing solution to leak from the pipette. To pick up a desired volume of liquid into the tip, use the volume adjustment knob to set the desired volume. Then press down on the plunger until the first "stop" is encountered, and place the tip into the liquid to be picked up. Slowly release the plunger to pick up the liquid. Place the tip into the receiving vessel, and depress the plunger all the way down, past the first stop. This should result in the complete ejection of the liquid from the tip. Obviously, you should watch carefully to see that no air bubbles have been picked up into the tip while filling, and that all of the liquid has been ejected during delivery.

Several sources of error are encountered when using micropipettors. First, the instruments are often inaccurate because they have been dropped on the floor or otherwise abused. In a research laboratory, these instruments should be inspected and recalibrated frequently, but this is expensive. If the instrument you are using does not hold the liquid (i.e., it drips out when the plunger is not being depressed), tape a label on it, marking it as leaky, set it aside, and obtain another instrument. To lessen the inaccuracy from use of these instruments, it is advisable to conduct all the measurements in a single experiment with a single instrument, if possible. Then, the volumes may all be inaccurate, but hopefully they will be reproducible and proportionately inaccurate. It may be possible to calibrate a set for yourself and keep them at your laboratory bench. Try to devise one or two reasonable methods for calibrating your pipettors.

Another common source of error encountered when using micropipettors is that liquid is picked up on the outside of the tip and transferred along with the measured volume of liquid inside the tip. The same phenomenon can occur with glass pipettes. Thus, it is advisable to stick only the tip of the pipette or pipettor into the liquid to be picked up, so that very little gets on the outside of the pipette or pipettor tip. Touching the outside surface of the pipette or pipettor tip to the edge of a test tube or to a piece of tissue paper can also be used to reduce this source of inaccuracy.

In addition to the methods noted above, we will often use uncalibrated disposable glass pipettes for transfer of liquids when the volume transferred is not important (Figs. 1-2 and 1-4). For example, when layering a liquid onto the surface of an acrylamide gel, the precise amount of liquid is not important. These disposable, uncalibrated pipettes are referred to as **Pasteur pipettes**. They frequently have a long, drawn-out tip to enable the manipulation of small volumes. To use a Pasteur pipette, slip a rubber squeeze bulb over the wide end of the pipette and pick up and eject liquid by squeezing or releasing the bulb as you would with an eye dropper. It is very easy to break off the drawn out tip of the pipette, so be careful. Dispose of the used Pasteur pipettes in the designated broken-glass box.

Complete mixing of the solutions that you are adding to one another is important for most of your experiments. There are several ways this can be done. For very small volumes, picking up and ejecting the solution from the tip of a micropipettor several times is usually sufficient. Another good way to mix solutions is to hold the test tube in one hand and repetitively tap on the bottom with the other hand or a finger. This imparts a rapid spinning to the solution and causes mixing. For larger volumes it may be useful to cover the vessel with parafilm and invert the vessel several times. This method is unsuitable if the solutions are precious, warm, caustic, or otherwise dangerous, since Parafilm can leak. Another way to mix adequately the larger volumes is to place the liquid into a flask and swirl the flask or stir with a magnetic stirring apparatus.

Accurate **timing** will be important in many of your experiments, such as when studying the rate of an enzyme-catalyzed reaction. In the research laboratory setting, such experiments would typically be performed using a stopwatch or an electronic timer. Simpler devices such as a wristwatch or a wall clock with a second hand will also work. Plan the experiment ahead and have all the necessary materials within easy reach. Try to be as consistent as possible with your manipulations.

Measurements of **pH** are made with a potentiometer known as a **pH meter** (Fig. 1-5) or, less accurately, with pH paper. The pH sensor is usually made of glass and is therefore fragile. Take care that you do not bump

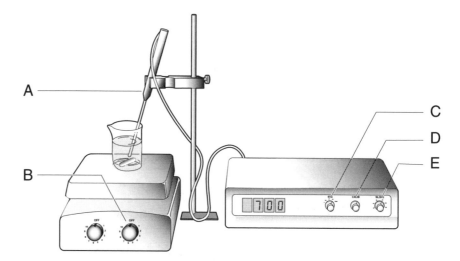

FIGURE 1-5 A typical modern pH meter. *(A)* The fragile electrode is immersed in the sample (or, when not in use, in the storage solution). *(B)* The stirring control rheostat controls the speed of the stirring. *(C)* The standby switch is used whenever the electrode must be moved to a new solution. *(D)* The calibration knob is used to calibrate the instrument with solutions of known pH. *(E)* The temperature calibration knob is used to correct for the temperature of the sample.

the electrode or it may break. Some pH systems have a direct sensor electrode and a separate calomel reference electrode. Others have both electrodes built into a single electrode stem. To use the pH meter, first, remove the electrode(s) from the storage solution and rinse them off thoroughly with a squirt bottle. Next, immerse the electrode(s) into a reference solution of known pH that is near the desired pH. For example, if you want to make a solution of pH 7.2, use the reference solution marked pH 7.0 to calibrate the instrument. If you want to make a solution of pH 4.5, then use the reference solution marked pH 4.0 to calibrate the instrument. With the electrode(s) immersed in the reference solution, switch the pH meter from standby mode to pH mode. At this time, be sure the temperature dial on the instrument is set to the temperature at which you are making the measurement (usually 25 °C). Correct the pH reading so that it matches the pH of the reference solution by adjusting the calibration dial. Next, put the instrument on standby and remove the electrode(s) from the reference solution. Rinse the electrode(s) thoroughly with distilled water and place it (them) in the solution whose pH you want to measure. Then switch to the pH mode and record the pH. When you are adjusting the pH of a solution, you should add small amounts of acid or base, as appropriate, and mix the solution until a stable reading is obtained. Continue this operation until you achieve the desired pH. If only slight adjustments are needed, use dilute acid or base so that you do not overshoot the mark. Then place the instrument on standby, rinse the electrode(s) again, and store the electrode(s) in the storage solution. Do not let the electrode(s) dry out. A quick but inaccurate way to check the pH of a solution is to use **pH paper**. These papers contain different strips that change color at various pH values. A reference set of the colors obtained at various pH values is shown on the container. Immerse a strip in the solution to be measured and allow ~1 min for the development of color. Then determine the pH by comparing the color with the reference illustrations.

VARIOUS INSTRUMENTS USED

SPECTROPHOTOMETERS

Basic instruments such as the Bausch & Lomb **Spectronic 20®** or "**Spec 20®**" spectrophotometer are sufficient for the experiments in this book. The following describes rudimentary operations for the Spectronic 20® (Fig. 1-6), but similar instructions also apply to other basic spectrophotometers (see Chapter 2). Warm up the instrument at least 15 min prior to taking any measurements. Set the wavelength by adjusting the dial at the top right (C). Next, you must calibrate the instrument. With no sample tube in the chamber, close the shutter so that no light reaches the detector. Set the meter to read infinite absorbance (0% T) by adjusting the dial (B) on the left front

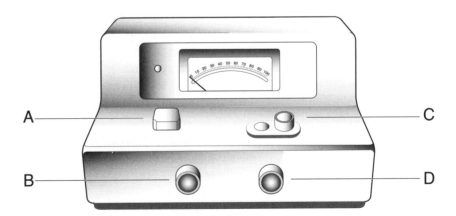

FIGURE 1-6 A Spectronic 20® spectrophotometer: *(A)* Cuvette compartment (lid is closed in the diagram) and *(B)* on/off switch and zero light calibration knob. When there is no cuvette in the instrument, a shutter prevents light from reaching the sample, and this knob is used to set the instrument to infinite absorbance. The Spectronic 20® spectrophotometer has *(C)* a wavelength adjustment knob and *(D)* a transmittance adjustment knob, that is used to calibrate the absorbance of control solutions to zero.

of the instrument. Then open the chamber lid (A), and insert a test tube (which is the cuvette) containing either water or a reagent blank into the tube holder. Insertion of the test tube opens the shutter. Close the lid of the chamber and adjust the reading to zero absorbance (100% T) by turning the dial (D) on the right front of the instrument. This adjusts the intensity of the light on the sample. Now the instrument is set for use. It is advisable to check the calibration periodically by ensuring that the instrument reads infinite absorbance when no tube is inserted and zero absorbance when a tube with water or reagent blank is inserted.

It is important that there be no stray light impinging on the tube when readings are taken. The top of the tube holder has a light shield that must be closed when readings are taken. Sometimes this part becomes cracked or broken off completely. If this is the case, the chamber can be made dark by using an improvised lid constructed of cardboard or black cloth. If your readings change when you hold your hand over the closed light shield, then your light shield is leaking light.

There should be at least 3 mL of sample in the tube to obtain a proper reading. This is necessary so that the light beam will pass through the portion of the tube that contains the solution. (The light beam does not pass through the lowest part of the tube.) In an ideal world, all the readings would be recorded with the same test tube or with matched tubes. However, for practical reasons, we will simply use ordinary 13 by 100 mm test tubes. Nevertheless, it is better to avoid using scratched or blemished tubes.

Avoid breaking tubes in the tube holder. If this should happen, the tube holder can be removed and cleaned out. Also avoid spilling liquid onto the outside of tubes to be used for measurements, since this will compromise the measurement; indeed, if liquid gets into the tube holder, it may ruin the readings for an entire experiment. Even a very small amount of spilled liquid can produce a condensate that fogs up the detector and blemishes future measurements. You may wish to gently wipe off the outside of the tubes with a tissue such as a Kimwipe. However, if you do this, be aware that it is quite easy to break thin-walled glass tubes in your hand.

Be aware of the limitations of the instrument. The elementary circuitry in these instruments and the imprecision of the sample tubes do not permit reliable readings much below $A = 0.05$ or above $A = 0.9$. Below $A = 0.05$, there is little difference in the amount of light passing through the reference tubes and the sample tube so that irregularities in the tubes become significant. If $A = 0.05 = \log(1/T)$, we can calculate $T = 0.89$. A change of 0.01 A will result in only a 1.1% change in the intensity, which is likely to be within the uniformity of the test tubes. Above $A = 0.9$, there is < 13% of the incident light passing through the tube ($A = 0.9 = \log(1/T)$; $T = 0.126$. Therefore, only 12.6% of the incident light is transmitted) so that small light leaks and other factors affect the measurements. For example, if a change of 0.1 A is observed during a reaction when A initially is 0.9, we can calculate that the new $A = 1.0 = \log(1/T)$; $T = 0.10$. Therefore, a 0.1 A change (which would be a 20% change in T at $A = 0$) will be only a 2.6% change in T at $A = 0.9$, so that readings from the meter would be imprecise and subject to outside influences.

As noted above, the instrument can be blanked with a tube containing water or with a reagent blank. The reagent blank should contain all substances present in the test samples except the substance being measured. For example, if the unknown substance is measured in dilute acid, then the reagent blank should contain the same dilute acid at the same concentration. If the substance to be measured is converted to a colored compound upon heating, the reagent blank should be heated in exactly the same way. Alternatively, you could just calibrate the instrument with water and read the reagent blank as a sample. The absorbance of this reagent blank must then be subtracted from the values obtained for the experimental samples to permit determination of sample concentration. This is often a better way to proceed, since you obtain a value for the reagent blank itself. Comparison of these values from day to day may alert the researcher that a given batch of reagents is not behaving properly.

CENTRIFUGES

We describe three types of centrifuges. Very small volumes will be centrifuged in plastic 1.5-mL "**microtubes**" (Fig. 1-3) in a "**microfuge**" (Fig.

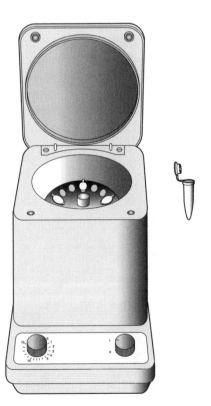

FIGURE 1-7 A typical microfuge. The microfuge tubes (Fig. 1-3) are placed in the rotor after balancing as described in the text.

1-7). These instruments are robust, and because the samples have low mass, precise balancing of the opposing tubes is not required. It is sufficient if the opposing tubes have approximately the same volume of liquid, within ~0.1 mL. The time and speed of the centrifugation can be set using the dials on the front of the instrument. Do not cover the ventilation port on the top of the lid while the instrument is running. These microfuges are capable of high *g*-forces and are very convenient to use.

Volumes in the range 1–15 mL that do not require high-speed centrifugation can be spun in a **clinical table-top centrifuge**. These instruments are so named because they are designed to separate blood cells from serum in clinical laboratories. Usually, we use disposable plastic conical-bottom centrifuge tubes for this instrument. The tubes should always have a counterbalancing tube, and both tubes should weigh the same. A pan balance and a squirt bottle of water will be used to weigh the proper amount of water into the counterbalance tube in order to balance the instrument. After the centrifugation run is complete, be sure that the rotor has come to a complete stop before opening the chamber.

High-speed centrifugation that requires refrigeration can be performed with Sorvall Superspeed® centrifuges (Fig. 1-8). These instruments have the potential for being dangerous, and they cost >$15,000 each; therefore, it is important that you use them properly. If you need to adjust the temperature setting, seek the help of an instructor. Samples must be accurately counterbalanced with a pan balance. Also, it is important that the tubes be filled to the proper level; overfilling a tube will result in spillage during the run and a consequent imbalance, while underfilling tubes permits the tubes to collapse inward, leading to breakage during the run, especially at high speeds. Let your instructor help you for the first few times you use the instrument, until you have mastered the technique.

The rotors that we will use in the high-speed centrifuges (Fig. 1-8) are designed to hold tubes of various sizes with the aid of plastic or rubber adapters. In all cases, the tube or tube plus adapter must fit snugly into the rotor so that they do not deform during centrifugation runs. When weighing

FIXED ANGLE

FIGURE 1-8 Typical high-speed centrifuge. The Sorvall SS-34® rotor is shown sitting on top of the centrifuge. A cross section of this rotor is depicted to the right. This is a fixed angle rotor that holds 40-mL tubes or with appropriate adapters, smaller tubes. The relevant dimensions are given in the text.

tubes to balance them, include the adapters with the tubes, since they are not all the same weight. If you should break a tube in a rubber adapter, be sure that you remove all traces of broken glass from the adapter. Otherwise, the small bits of broken glass on the bottom of the adapter will ensure that the next tube used in that adapter will also break.

Time and speed are set using the dials on the front top of the instrument.

Never spin at a higher speed than the speed for which the rotor is rated; e.g., do not spin a rotor rated at 10,000 rpm at 20,000 rpm.

Otherwise, a rotor failure that could result in severe damage to the instrument and/or serious injury to yourself and your colleagues may occur. Charts are available for determining the safe range of speeds and consequent *g*-forces for given rotors and tubes. If there is even a fragment of doubt, consult an instructor.

The speed required to achieve a given *g*-force can be calculated by the following formula:

$$\text{rpm} = 299\sqrt{\frac{g}{\text{radius}}} \qquad (1\text{-}2)$$

The relative centrifugal force is *g*. Suppose that we are aiming for 12,000*g*. The radius refers to the distance in centimeters between the center of the rotor and some point in your sample. Assume that the rotor in the centrifuge (Sorvall SS-34®) is a "fixed angle" rotor, in which the centrifuge tubes are held at a fixed angle to the ground, not vertical or horizontal (see Fig. 1-8). Therefore, the value of the radius will vary in different parts of your liquid. For Sorvall centrifuges, the estimation of the *g*-force is obtained on a chart furnished by the Sorvall Company, which is a subsidiary of DuPont, Newton, Connecticut. This is calculated on the basis of the maximum radius (10.8 cm) of the rotor, i.e., the bottom of the centrifuge tube. The formula reduces to

$$\text{rpm} = 91\sqrt{g}$$

The rpm required for 12,000*g* = 9969 rpm. Therefore, choose about 10,000 rpm.

The rotors for superspeed centrifuges are quite massive, so that it takes a long time for them to cool down in the instrument. Thus, when you need to centrifuge your sample at 4 °C, it is a good idea to have the rotor precooled by keeping it in the cold room when not in use. When you are done using the rotor, put it back in the cold room so that it remains cold for your colleagues. Be sure that the rotor is clean before putting it back. Usually, the rotor can be cleaned by rinsing it with distilled water followed by drying. Because of their mass these rotors are cumbersome to manipulate.

Take special precautions and seek help if a rotor seems too heavy for you to move. If you drop the rotor (which costs several thousand dollars), it will be ruined.

The rotors sit in the high-speed centrifuges on a spindle that has two drive pins. On the bottom of the rotor, there is a seat that also contains two drive pins. You should inspect the spindle to be sure the drive pins are intact, and the rotor to be sure that its drive pins are intact. When you set the rotor onto the spindle, it should sit level. The lid to the rotor attaches it securely to the spindle. The bottom screw on the lid attaches the lid to the bottom portion of the rotor. The top screw on the lid will attach the rotor to the spindle. Note that both top and bottom screws have left-handed threads so that you tighten by turning counterclockwise. It is important that the rotor be attached to the spindle, especially for high-speed runs. Never use the rotor without a lid. Before use, examine the rotor assembly carefully so that you understand these instructions. Check with one of your instructors before using the centrifuge.

Occasionally, there will be a spill of liquid in the centrifuge, for example, when a tube breaks. In that case, use a paper towel to wipe up the spill as much as possible. Then clean the rotor and the inside of the centrifuge.

Newer centrifuges have a number of safety features that are lacking in the older models. For example, on the newer centrifuges, it is not possible to open the chamber door while the rotor is in motion. **Do not attempt to thwart this safety feature. Never open the chamber while the rotor is in motion.**

On the older centrifuges, it is possible to open the chamber door while the rotor is in motion; therefore, be sure to check that the meter reads zero revolutions per minute before opening the chamber.

Centrifuge rotors and tubes. There are many different types and sizes of rotors and tubes used in centrifugation, each with specific characteristics tuned to particular functions. The fixed-angle rotor shown in Fig. 1-8 is a very functional "workhorse" rotor found most commonly in laboratories. The Sorvall SS-34® rotor holds eight tubes at an angle, each containing ~40 mL. There are no moving parts, so the rotor is robust and easy to clean and maintain. Larger versions of fixed-angle rotors hold four or six larger tubes (bottles) and have total capacities of as much as 4 liters. These cannot be spun at velocities to give as high a g-force as the smaller rotors. These larger rotors are very useful during early stages of protein purification when large volumes must be handled. Swinging bucket rotors hold the tubes in a jig that swings to the horizontal position when spinning. They have the advantage that the meniscus and the precipitate are perpendicular to the length of the tube; thus the precipitate is not forced to the side of the tube as in a fixed-angle rotor. The disadvantage is that they have moving parts and are therefore more fragile and require more maintenance. Continuous flow rotors enable application of material while the rotor is spinning. Thus, as precipitates are centrifuged to the bottom of the rotor, the supernate is replaced with new material. This results in a large increase in effective capacity. Such rotors are often used to spin down cells, or at early stages in preparations, to spin down cell walls and other larger particles. These types of rotors usually do not have tubes; the precipitate is collected in the rotor itself.

Rotor tubes can be constructed of a variety of materials, depending on the requirements for size and g-forces. Regular glass tubes can be used in low-speed (<3000 rpm) centrifuges such as the clinical centrifuges. Heavy-walled Corex® glass tubes can be used to handle speeds as high as 15,000 rpm. For most higher speed ranges, tubes made of polypropylene, polycarbonate, and other synthetic materials are used. Ultracentrifuges, which can achieve g-forces >100,000, routinely use nitrocellulose or polyallomer tubes. Polycarbonate tubes can also be used if they are appropriately specified.

ELECTROPHORESIS

We use several different types of electrophoresis for experiments described in this book, including polyacrylamide gel electrophoresis and agarose gel electrophoresis. For this purpose, gel casting chambers and electrophoresis tanks will be used, and small power supplies will provide the necessary direct current. Descriptions of the instruments and their use are found in Chapters 5 and 11.

Even though we will not conduct any experiments at very high voltages, it is advisable to use responsible laboratory techniques with the power supplies. First, never turn on the power without first checking to ensure that the leads are properly connected to the electrophoresis tank. Second, be sure that the current dial is turned down to zero current before turning on the power supply. Then, slowly raise the current to the desired setting. To stop the run, first, turn the current dial to zero, then turn off the instrument. This protocol will help the next user avoid an accident. Last, disconnect the leads to the electrophoresis tank.

OTHER GENERAL TECHNIQUES

DIALYSIS

We will frequently use dialysis to alter the composition of small molecules present in samples, e.g., to add or to remove salt or to change the buffer. Dialysis is used to separate molecules on the basis of size. A semiperme-

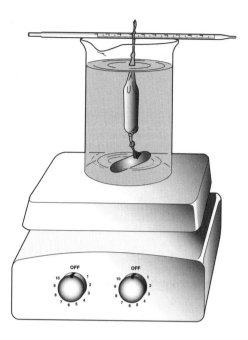

FIGURE 1-9 A typical dialysis setup. Note the double knots to ensure that the sample does not leak.

able membrane material of cellophane (cellulose acetate) or nitrocellulose that is formed as tubing and tied off at the ends (to form a bag) usually encloses the sample (Fig. 1-9). The pores of the membrane allow small molecules such as solvents, salts, products, and substrates to diffuse freely through the membrane, whereas proteins and other large molecules are retained. Therefore, if you place your sample in a dialysis bag in a volume of 10 mL and then place the bag in a beaker containing 1 liter (called the dialysate), the solvent solution in your dialysis bag will equilibrate with the 1-liter dialysate and will become diluted 100-fold. Equilibration with gentle stirring usually requires 2–3 h. Refreshing the dialysate and reequilibrating will produce an overall dilution of 10,000-fold.

Dialysis tubing is supplied on a roll. If you are going to dialyze sensitive samples, it is advisable to unroll the dry dialysis tubing and boil it in a solution of EDTA and $NaHCO_3$ for a few minutes to remove metals and other contaminants. Then store it in a fresh solution of EDTA and $NaHCO_3$ to keep it hydrated and ready for use. For the samples we will use in these laboratories, it is sufficient to use the dialysis tubing straight from the roll as noted below. While wearing disposable gloves, peel off the desired length of tubing and cut it with a pair of scissors. Next, rinse the tubing thoroughly, inside and out, with distilled water at the sink. To close the bottom of the tube, make two overhand knots. The procedure for tightening the knots is to pull the portion of the tube that will be outside the knot with one hand while holding the knot in the other hand. This avoids stretching and weakening the bag portion of the tubing. Next, place the sample into the tubing, by using a pipette. To be sure that the tubing is not accidentally pierced during this step, use a blunt pipette, not a Pasteur pipette. Tie two knots in the top of the bag to seal it shut. If you are dialyzing a sample with high salt against a buffer with low salt, the sample will swell during dialysis as the solvent fills the bag to compensate for the osmotic pressure. In that case, it is important to leave room inside the tubing (i.e., use a bag larger than the sample volume but not containing air) so that it does not burst during this swelling. Very precious samples, such as the final step in an enzyme preparation that took days to carry out, are best dialyzed in two separate dialysis bags so that all is not lost if one bag leaks.

Usually, you will put your dialysis samples into a large flask filled with the desired final buffer and containing all of the dialysis samples from the laboratory. Be sure that your sample is adequately labeled. You can either tie a string to one end of the bag and put a label on the end of the string or tie your bag loosely to a pipette and put your label on the end of the pipette that sticks out of the liquid.

COLUMN CHROMATOGRAPHY

You will conduct many column chromatography experiments in which you prepare your own columns. Figure 1-10 shows a typical chromatography setup used in our laboratory. The column is mounted vertically and has a stopcock at the bottom to regulate the flow. Inside the bottom of the column is a plastic frit that helps to collect the eluant from the bottom of the column matrix and to direct it to the stopcock in a minimum volume. Thus, eluant traveling down the edges of the column reaches the stopcock at very nearly the same time as does eluant from the middle of the column. This avoids formation of funnel-shaped bands. The chromatography matrix to be used will be supplied as a slurry in buffer.

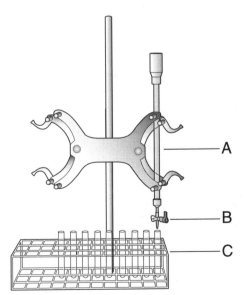

FIGURE 1-10 A typical manual chromatography setup suitable for experiments in this laboratory: *(A)* column, *(B)* stopcock, and *(C)* test tube rack with test tubes to collect effluent.

To pour an effective column, which has a uniform dispersion of the column matrix, first close the stopcock at the bottom of the column and add a small volume (~1/10 of the column volume) of the starting buffer to the empty column. Swirl the flask to resuspend the settled slurry, and with a pipette, add the appropriate amount to the column. Then open the stopcock and allow the slurry to settle into a bed while the column is running. Do not let the mobile phase (the buffer) meniscus fall below the column bed (this is called letting the column run dry). If the column does run dry, wash it out and start over. When the slurry has settled to form a bed, you will usually wash the column with many volumes of the starting buffer to equilibrate the column matrix with the appropriate buffer. For our experiments, this will be done by manually adding buffer to the top of the column. Do not add buffer too vigorously, or you will disturb the bed, which can cause formation of channels and pockets in the column matrix. This will lead to uneven flow and, consequently, poorer chromatographic separation. This is especially important for gel filtration experiments (see Chapter 4).

In the research laboratory setting, fractions from chromatography columns are frequently collected by a device called an automatic fraction collector (Fig. 1-11). These instruments hold a series of test tubes and collect either a fixed volume of eluate into each tube, a fixed number of drops per tube, or eluate for a fixed time into each tube. The most advanced of these units are computer controlled and, when interfaced with a system containing a flow-through spectrophotometer cell, may be programmed for different purposes. Thus, they can collect peak fractions into separate tubes or collect only a single desired peak, discarding the others.

In the experiments in this book, we will collect column fractions by hand (Fig. 1-10). To collect fractions of ~1 mL, a good way is to take a

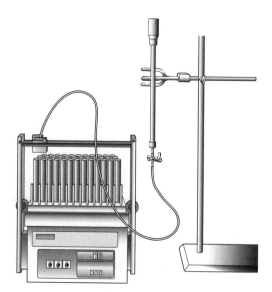

FIGURE 1-11 Setup used for chromatography in the research laboratory. An automatic fraction collector is shown connected to the column.

small test tube and add 1 mL of water. Then collect drops from the column into tubes until the height of the liquid matches the height of the liquid in the tube containing 1 mL of water. Another way is to count drops as they fall into collection tubes and to collect the same number of drops into each collection tube. Then, you can measure the volume in one of the tubes to determine the volume collected per tube. Surprisingly, equal numbers of drops may not give equal volumes as the column is eluted, because as the protein concentration changes, the surface tension changes, causing the drop size to vary.

STORAGE AND STABILITY OF SAMPLES

Some of the experiments that we describe will extend over several laboratory sessions, so you will store your samples. Usually, samples will be stored at 4 °C. Be sure to label properly each sample with your name and the description of what it is. Often, you will have different fractions that you are storing. Instead of writing the history of each sample on a tiny label, devise some type of nomenclature or code, and record careful notes in your laboratory notebook about the nomenclature that you are using. When removing samples from storage, be sure not to disturb anyone else's samples.

Samples in tubes or in small flasks can be sealed with Parafilm, a flexible waxy material that is furnished in a roll, "co-rolled" with a sheet of clean paper. Use scissors to cut a piece ~2.5 inches square. Peel off the backing and press one corner against the upper end of the tube with your left thumb. With your other hand, pull the opposite corner of the square across the top and then down on the opposite side. Now wrap the Parafilm 180°+ around the tube and repeat the process to produce a second layer over the top. The second layer is initiated from the original starting point, held down by the same thumb. Use a little tension while doing this so that the Parafilm stretches a bit. Smooth down the edges with your palm. If the tube is cold and wet on the outside, the Parafilm will not stick well. Drying and warming the tube with your hand will help the Parafilm to seal. Parafilm cannot be used with tubes that are to be heated or that contain organic solvents.

WASHING GLASSWARE

At some point, you may have to wash glassware. For most purposes, it will suffice to rinse out the glassware with hot water ~10 times and then with distilled water 10 times. If you use detergent to clean the glassware, be sure to rinse the glassware very well, since even traces of detergent can affect many of the samples that you will be using.

TABLE 1-2 Common Prefixes		
TERM	**DEFINITION**	**ABBREVIATION**
kilo	10^3	k
milli	10^{-3}	m
micro	10^{-6}	μ
nano	10^{-9}	n
pico	10^{-12}	p
femto	10^{-15}	f

SAMPLE CALCULATIONS

SOLUTIONS AND DILUTIONS

For most of our experiments, small amounts of materials and dilute solutions will be used. Thus, you should be familiar with the terms defined in Table 1-2 below.

From Table 1-2, you can conclude that $1\ \mu g = 10^{-6}$ g and $1\ ng = 10^{-9}$ g. One microgram can also be expressed as 10^3 ng or 1000 ng. Recall the relationship (concentration (C))(volume) = amount. For example, 5 mL (5×10^{-3} L) of a 10 μM (10×10^{-6} M) solution contains

$$(5 \times 10^{-3}\ \text{L})(10 \times 10^{-6}\ \text{mol/L}) = 50 \times 10^{-9}\ \text{mol or 50 nmol.}$$

This relationship is also useful for calculating the dilutions required for making up solutions (volume = V_1) of required strength (C_1) from stock solutions of higher concentrations (C_2).

$$(C_1)(V_1) = (C_2)(V_2) = \text{amount}$$

These principles are illustrated in the following examples. (Note that M_r is the abbreviation for molecular weight.)

EXAMPLE 1-1

Ninety milligrams of glucose ($M_r = 180$) is dissolved in 100 mL of H_2O. What is the molarity of glucose in this solution?

$$90 \times 10^{-3}\ \text{g/180 g/mol} = 0.0005\ \text{mol} = 0.5\ \text{mmol}$$

$$0.5\ \text{mmol/100 mL} = 5\ \text{mmol/1000 mL} = 0.005\ \text{mol/L} = 5\ \text{mM}$$

EXAMPLE 1-2   How would you make 20 mL of 25 mM glucose?

$$(20 \text{ mL})(25 \text{ mM}) = (20 \times 10^{-3} \text{ L})(25 \times 10^{-3} \text{ M}) = 500 \times 10^{-6} \text{ mol} = 0.5 \text{ mmol}$$

$$(0.5 \times 10^{-3} \text{ mol})(180 \text{ g/mol}) = 90 \times 10^{-3} \text{ g} = 90 \text{ mg}$$

Answer: 90 mg of glucose is dissolved in water, and the final volume is brought to 20 mL. Note that

$$(\text{L})(\text{M}) = \text{mol}$$

$$(\text{mL})(\text{M}) = \text{mmol}$$

$$(\text{mL})(\text{mM}) = \mu\text{mol}$$

Therefore, an alternative calculation for this problem would be

$$(20 \text{ mL})(25 \text{ mM}) = 500 \text{ } \mu\text{mol} = 0.5 \text{ mmol}$$

EXAMPLE 1-3   How would you dilute the above solution (25 mM) to make 2 mL of 5 mM glucose?

$$(C_1)(V_1) = (C_2)(V_2)$$

$$C_1 = 5 \text{ mM}, V_1 = 2 \text{ mL}, C_2 = 25 \text{ mM}, V_2 = ?$$

$$V_2 = (C_1)(V_1)/C_2 = (5 \text{ mM})(2 \text{ mL})/(25 \text{ mM}) = 0.4 \text{ mL}$$

Answer: Take 0.4 mL of 25 mM glucose and dilute to 2 mL with water (i.e., add 1.6 mL of H_2O, assuming no volume change on mixing).

PROBLEMS   **1-1.**   How many mmol and μmol of glucose are present in 9-mg glucose?

1-2.   If you dissolve 9 mg of glucose in 5 mL of H_2O, what is the concentration of glucose in mmol and μmol?

1-3.   If you take 2 mL of the above solution and dilute it to 50 mL, what is the concentration of glucose in the resulting solution in μmol?

1-4.   How many nmol of glucose are present in 2 mL of the above (diluted) solution?

1-5. Fasting normal blood glucose concentration is between 70 and 100 mg % (mg per 100 mL of blood). Express this concentration range of blood glucose in mM.

BUFFERS AND pH

Most organic ionizable compounds are weak acids or bases, and therefore their dissociation from protons is reversible. The extent of dissociation depends on various factors such as the hydrogen ion concentration $[H^+]$, the ionic strength, and the temperature of the medium. Weak acids such as RH reversibly dissociate as

$$RH \rightleftharpoons R^- + H^+ \tag{1-3}$$

Therefore, the equilibrium constant expressed for dissociation K is described as

$$K = \frac{[R^-][H^+]}{[RH]} \tag{1-4}$$

NOTE✔

where RH is the acid (proton donor) and R^- is the conjugate base (proton acceptor). **It would be more accurate to use the chemical activities of the various species in place of the concentrations in equation 1-4. Concentration terms such as shown in equation 1-4 assume that the chemical activity is the same as the concentration of the ion. However, the chemical activity is the product of the activity coefficient and the concentration of the species, and the activity coefficient is dependent on the concentration of the species, temperature, ionic strength, nature of the species, and other parameters. In very dilute solutions, most ions have activity coefficients of nearly 1. At higher ionic strengths, such as found in most biological environments (0.1–0.3 is typical), the activity coefficients decrease to values ranging from 0.2 to 0.8, depending on the particular species involved. This is thought to be largely due to the ionic species associating at higher concentrations with species of the opposite charge, rather than acting as the pure ionic species in solution. Therefore, keep in mind that K in equation 1-4 is a constant only when the concentration of each species is equal to the activity of that species, i.e., for dilute solutions usually of low ionic strength. For higher concentrations, when the chemical activities differ from the actual concentrations, it is not a true constant, which is noted by using the symbol K'. Also note that for dissociation of acids (generating hydrogen ions) the K is conventionally written as K_a (as opposed to K_b, where OH^- ions are formed).**

Taking the logarithm (base 10) and rearranging, we can rewrite equation 1-4 as

$$-\log[H^+] = -\log K_a' + \log \frac{[R^-]}{[RH]} \qquad (1\text{-}5)$$

$$pH = pK_a' + \log \frac{[R^-]}{[RH]} \qquad (1\text{-}6)$$

By definition $-\log_{10} [H^+] = pH$ and $-\log_{10} K' = pK_a'$. Equation 1-6 is known as the **Henderson–Hasselbalch equation**, which describes the relationship between the pH, the pK_a, and the relative composition of the weak acid. For example, from equation 1-6 you can see that when the concentration of the acid RH is equal to the concentration of the conjugate base (salt) R^-, then $[R^-]/[RH] = 1$. Since $\log_{10}(1) = 0$, we see that the pH of the solution will be equal to the pK_a' of the acid. Therefore, the pK_a' of a weak acid can be determined by measuring the pH of a solution containing equal concentrations of the acid (RH) and its salt (R^-), such as when the acid solution is half neutralized with a strong base. From equation 1-6, you can also see qualitatively that when the pH of the medium is below the pK_a', the protonated form (acid) predominates, and when the pH is above the pK_a', the salt form (nonprotonated or the conjugate base) predominates.

Buffers are generally mixtures of weak acids and their salts. Because of the reversible dissociation of weak acids, these solutions resist change in pH when either acid or alkali is added to the solution. For example, when acid is added, the salt form (R^-) takes up a proton to form the acid (RH). When base is added, the acid form (RH) dissociates to produce more conjugate base:

$$RH + OH^- \rightarrow R^- + H_2O \qquad (1\text{-}7)$$

The titration of a weak acid is quite instructive to the understanding of buffer systems. Figure 1-12 shows titrations of acetic acid and the buffer MES. Acetic acid has a pK_a' of 4.76, and MES has a pK_a' of 7.47. Note that in the first part of the titration, the pH is quite low and changes significantly with each addition of strong base (NaOH). When the pH is near the pK_a', the addition of alkali changes the pH of the solution only gradually until the acid is mostly neutralized. The change of pH per addition of base is smallest when the acid is half neutralized; that is, pH = pK_a' of the acid. After the acid has been largely converted to its conjugate base, the pH again changes significantly with each addition of hydroxide. This is because there is very little conjugate acid left to "buffer" the addition of base. Therefore, the pK_a' is at the point that the solution has the highest buffering capacity; that is, the change in the pH of the solution with the addition of OH^- (or H^+) is

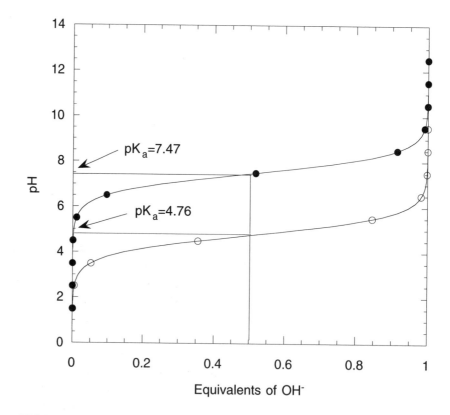

FIGURE 1-12   Titration of acetic acid (O) and the buffer, MES (●). Note the pK_a values and how the (conjugate acid)/(conjugate base) pairs act as buffers at pH values near their pK_a' values. The shapes of the titration curves are the same but are shifted vertically by the value of the pK_a.

least. For this reason, buffer solutions are made up with acids (plus the salt of the acid) whose pK_a' values are close to the desired pH of the buffer. *As a rule of thumb, the pH of buffers should be within ± 1 pH unit of the pK_a' of the acid used to make up the buffer.*

Beyond that range there will be very little buffering capacity. It is important to remember that the buffering capacity of a buffer depends on the concentrations of the buffering ions. If you want to carry out an experiment in which an acid is one of the reactants or products and the concentration of this reactant is 1 mM or greater, then a buffer concentration of 10 mM would probably not be very effective. You would want a buffer concentration of 100 mM or higher. These points are illustrated in some of the practice problems given at the end of this section.

The Henderson–Hasselbalch equation is useful for calculating either the buffer compositions or the pH of solutions. These are illustrated by the following examples:

EXAMPLE 1-4   How would you make up 100 mL of 0.1 M acetate buffer, pH 4.0? (pK of acetic acid is 4.76).

This buffer could be prepared by mixing solutions of 0.1 M acetic acid (RH) and 0.1 M sodium acetate (R$^-$). Substituting the desired pH value (4.0) in equation 1-6,

$$4 = 4.76 + \log \frac{[R^-]}{[RH]}$$

$$\log \frac{[R^-]}{[RH]} = -0.76 \qquad\qquad \frac{[R^-]}{[RH]} = 0.17$$

Therefore, the ratio of [acetate] to [acetic acid] is 0.17, or for each 1 mL of 0.1 M acetic acid, 0.17 mL of 0.1 M sodium acetate should be added (final volume = 1.17 mL). For 100 mL total volume, the volume of acetic acid would be 100/1.17 or 85.5 mL and the volume of 0.1 M sodium acetate would be 100 − 85.5 or 14.5 mL.

Answer: Mix 85.5 mL of 0.1 M acetic acid and 14.5 mL of 0.1 M sodium acetate to make 100 mL of 0.1 M acetate buffer of pH 4.0. **The molarity of a buffer solution is expressed as the combined molarity of the buffering ions (in this case, acetate + acetic acid). In the above buffer, the concentration of acetic acid is 0.0855 M and acetate is 0.0145 M. Hence, the combined concentration is 0.1 M.**

NOTE✔

EXAMPLE 1-5   What is the pH and molarity of the buffer prepared by adding 20 mL of 0.2 M NaOH to 50 mL of 0.1 M acetic acid?

Answer: The initial amount of acetic acid was 50 mL × 0.1 M = 5 mmol. The NaOH (strong base) added (20 mL × 0.2 M = 4 mmol) to the acetic acid will convert an equivalent amount of acetic acid to acetate ions.

$$CH_3COOH + Na^+ + OH^-\ CH_3COO^- + Na^+ + H_2O$$

In the resulting mixture (70 mL), there are 4 mmol of acetate and 1 mmol of acetic acid present. From the Henderson-Hasselbalch equation,

$$pH = 4.76 + \log \frac{(4\,mmol/70\,mL)}{(1\,mmol/70\,mL)}$$

or

$$pH = 4.76 + \log 4 = 5.36$$

The molarity of acetate (combined acetic acid and acetate) will be 5 mmol in 70 mL or (5 mmol/70 mL) × 1000 mL/L = 71.4 mmol/L, i.e., 71.4 mM

TABLE 1-3
The pK_a Values of Common Buffers

COMPOUND	pK$_a$ VALUE	pK
Oxalic	1.27	(pK_1)
Histidine	1.82	(pK_1)
Phosphoric	2.15	(pK_1)
Glycine	2.35	(pK_1)
Citric	3.13	(p$K_{1)}$
Citric	4.76	(pK_2)
Acetic	4.76	
Histidine	6.04	(pK_2)
MES	6.09	
Citric	6.40	(pK_3)
Phosphoric	6.82	(pK_2)
MOPS	7.15	
TES	7.40	
HEPES	7.47	
TRIS	8.08	
BICINE	8.26	
Pyrophosphate	9.95	(pK_4)
CHES	9.50	
Glycine	9.78	(pK_2)

MES, 2-(*N*-morpholino) ethanesulfonic acid; MOPS, 3-(*N*-morpholino) propanesulfonic acid; TES, *N*-((Tris-hydroxymethyl) methyl)-2 aminomethanesulfonic acid; HEPES, *N*-2-hydroxyethyl piperizine-*N*´-2-ethanesulfonic acid; TRIS, Tris(hydroxymethyl) aminomethane; BICINE, *N*,*N*-(Bis-2-hydroxyethyl) glycine; and CHES, 2-(cylohexylamino) aminomethanesulfonic acid.

NOTE✔

(0.0714 M). **Because both the RH and R⁻ are present in the same volume of solution, the volume terms cancel out in the above equation and the amount of each ion (instead of concentration) present in the solution can be used to calculate the pH.**

A number of weak acids and bases have been used to make biochemical buffers. Such compounds, with their pK_a values, are listed in many textbooks and biochemical handbooks; a few, including most of the buffers we will use in the experiments in this book, are also listed in Table 1-3. With the Henderson–Hasselbalch equation 1-6, we can use these pK_a values to calculate how to make up buffers at a desired pH or to calculate pH values of solutions of known composition. It should be noted, however, that pK_a' values are sensitive to the temperature and ionic strength of the solution. Also, as discussed above, solutions at high concentrations do not behave ideally. These properties must be taken into consideration for accurate calculation of the pH of any solution. Consult textbooks of physical chemistry for such calculations.

When referring to tables of pK_a' values for compounds containing more than one ionizable group, the pK_a' values are numbered in increasing order (the higher pK_a' values imply weaker acids). For example, phosphoric acid (H_3PO_4), a polyprotic acid, has three ionizable groups and therefore has three pK_a' values as indicated below:

$$H_3PO_4 \leftrightarrow H_2PO_4^- + H^+ \qquad\qquad pK_1 = 2.15$$

$$H_2PO_4^- \leftrightarrow HPO_4^{2-} + H^+ \qquad\qquad pK_2 = 6.82$$

$$HPO_4^{2-} \leftrightarrow PO_4^{3-} + H^+ \qquad\qquad pK_3 = 12.38$$

The same principles of titrations hold for polyprotic acids such as H_3PO_4. Referring to Fig. 1-13, it can be seen that there are three regions of good buffering capacity coinciding with each of the pK_a' values of phosphoric acid. Also note that there are sharp inflections as each equivalent is completely titrated. It is difficult to titrate completely a strong acid or a very weak acid (strong base) to the end points, because the addition of OH^- (or

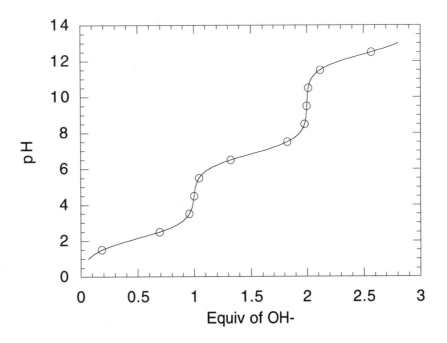

FIGURE 1-13 Titration curve of phosphoric acid.

H$^+$) would have to be at very high concentrations to achieve the necessary pH values. For example, if [OH$^-$] = 0.01 M, the pH would only be 12.

Some compounds, such as amino acids, will have at least two ionizable groups: one acidic and one basic. The basic group (R–NH$_2$) may be treated as a weak acid that will ionize as shown below for glycine:

$$^-OOC–CH_2–NH_3^+ \leftrightarrow {}^-OOC–CH_2–NH_2 + H^+ \qquad pK_2 = 9.78$$

and the carboxylic group will ionize as

$$HOOC–CH_2–NH_3^+ \leftrightarrow {}^-OOC–CH_2–NH_3^+ + H^+ \qquad pK_1 = 2.35$$

Note that in either case the protonated form (R–NH$_3^+$ or R$'$–COOH) is the acid form and the nonprotonated form (R–NH$_2$ or R$'$–COO$^-$) is the conjugate base. Because of the presence of both positive and negative charges on the same molecule, these molecules are called **zwitterions** (hybrid or amphoteric ions). At one particular pH, the positive charge on the amino group will be exactly equal to the negative charge on the carboxylic acid, thus making the net charge of the molecule zero. The pH at which the net charge of an amphoteric molecule is zero is known as the **isoelectric point (pI)** of the molecule. When the pH of the medium is below the pI, there will be more positive charges than negative charges on the molecule, and the molecule will be cationic. Accordingly, if the pH is above the pI, then the net charge of the molecule will be negative. Thus, the pH can determine the charge on such molecules (including proteins), and this feature can be used to separate molecules, especially in chromatography. Table 1-3 provides information for calculating the extent of ionization of different groups as well as the net charges of several molecules at any given pH by using the Henderson–Hasselbalch equation as shown in the example below.

EXAMPLE 1-6   Calculate the net charge of glycine at pH 6.0. (pK$_1$ of glycine is 2.2 and the pK$_2$ of glycine is 9.8.)

Answer: The carboxyl group will ionize as

$$R–COOH \leftrightarrow R–COO^- + H^+$$
$$\text{(acid)} \qquad \text{(conjugate base)}$$

Substituting the values (pH = 6.0, pK = 2.2) into the Henderson-Hasselbalch equation,

$$6.0 = 2.2 + \log \frac{[RCOO^-]}{[RCOOH]}$$

$$\log \frac{[RCOO^-]}{[RCOOH]} = 3.8$$

$$\frac{[RCOO^-]}{[RCOOH]} = 10^{3.8} = 6310$$

Thus, only 1 out of 6311 glycine molecules will be in the RCOOH form, while the fraction in the RCOO⁻ form is 6310/6311 = 0.9998. Therefore, for all practical purposes, it may be assumed that the carboxyl group is completely ionized at pH 6.0; that is, the fractional charge on it is –1.0.

The amino group ($pK = 9.8$) will ionize as

$$R\text{–}NH_3^+ \leftrightarrow R\text{–}NH_2 + H^+$$
$$\text{(acid)} \quad \text{(conjugate base)}$$

and therefore at pH 6.0

$$6.0 = 9.8 + \log \frac{[R\text{-}NH_2]}{[R\text{-}NH_3^+]}$$

$$\log \frac{[R\text{-}NH_2]}{[R\text{-}NH_3^+]} = -3.8$$

$$\frac{[R\text{-}NH_2]}{[R\text{-}NH_3^+]} = 10^{-3.8}$$

Therefore, it is clear that the amino group is almost completely in the R–NH₃⁺ form, i.e., the fractional change on the amino group is 1.0. The net charge of glycine at pH 6.0 = +1.0 + (– 1.0) = 0. From this calculation it is clear that pI of glycine is 6.0. Note that pH 6.0 is exactly at the midpoint between the two pK_a' values of glycine.

The following problems will illustrate calculations involving pH, buffer, and charges on ionic species. Try and work them out to make your concepts about these biochemical parameters more clear.

PROBLEMS

Use the pK_a' values given in Table 1-3.

1-6. How many mL of 0.1 M sodium acetate should be added to 100 mL of 0.1 M acetic acid to make a buffer of pH 5.1? What is the molarity of the resulting buffer with respect to acetate (acetate plus acetic acid)?

1-7. How many mL of 0.1 M HCl should be added to 100 mL of 0.1 M Tris base (Trishydroxymethyl aminomethane) to make a buffer of pH 8.2? What is the molarity of Tris in this buffer?

1-8. Calculate the pH of a mixture containing 50 mL of 0.1 M NaH_2PO_4 and 150 mL of 0.1 M Na_2HPO_4. How many mL of 0.1 M H_3PO_4 should be added to the above buffer to lower the pH by one unit?

1-9. To each of 100 mL of the following three (A, B, and C) 0.1 M acetate buffer solutions (pH of $A = 4.4$, pH of $B = 4.76$, and pH of $C = 5.2$), 10 mL of 0.1 M NaOH solution was added. Calculate the resulting change in the pH and the buffering capacity ($\Delta OH^-/ \Delta pH$) of each solution. Calculate also the change in the pH if 0.1 M HCl was used instead of NaOH.

1-10. Calculate the net charge of glutamic acid ($pK_1 = 2.1$, $pK_2 = 4.07$, and $pK_3 = 9.47$), alanine ($pK_1 = 2.35$ and $pK_2 = 9.87$), and lysine ($pK_1 = 2.16$, $pK_2 = 9.06$, and $pK_3 = 10.54$) at pH 5.0. What are the pI values of these amino acids?

KEEPING A LABORATORY NOTEBOOK

In order to make progress in research, it is essential that you keep a proper laboratory notebook to provide a systematic and accurate record of your observations and data. There are two reasons for this. First, you simply will not be able to remember all the details of experiments performed months or years ago (unless you are remarkably unproductive). Thus, valuable past experience will be lost without accurate records. Second, in science, it is simply not enough to do good experiments and obtain interesting results. The results must be communicated to the scientific community by publishing in journals, books, on the Web, etc., and by giving oral presentations. To publish results, it is mandatory to document rigorously those results. These results, including your design of the experiments, any hypotheses you formulate, your conclusions formed from the data, and your inventions of devices or processes, are all considered your intellectual property. Thus, you obviously will need good records to document these properties. Moreover, in the industrial world, patent rights and liability information are based on the accurate keeping of records. If you are employed in an industrial setting, your job will depend on keeping good records. Complete and accurate documentation of research efforts is one of the key factors in deterring fraud in science.

As noted above, the purpose of a laboratory notebook is to document the efforts and results of the researcher. Laboratory notebooks are not meant to be works of art; therefore, there is no need at all to "copy over neatly" notes that were taken on separate sheets of paper, etc., in an effort to pretty up the notebook. Of course, a neat notebook is certainly an asset. Planning how you are going to record your information, perhaps by leaving room for notes and prerecording procedures to be used, may help you to keep a better notebook (as well as prepare you for the experiment). Be sure to obtain a bound laboratory notebook and clearly write your name, address, and phone number on the front and inside the cover. It is often useful to have cross-hatched pages so that you can easily graph results as your experiment progresses.

Notes should always be taken directly into the notebook. If for some reason notes are taken on a separate piece of paper, then that piece of paper should be stapled into the notebook. Do not copy the information into the notebook from scattered sheets of paper, since information can be lost or altered in the copying. The laboratory notebook should be an up-to-the-minute record of your laboratory activity. The only restriction concerning appearance is that the penmanship should be sufficiently clear that you and other scientists can discern the information. It is a good idea to use a ball point pen for notebooks. Pencil tends to disappear over the years, and many entries have been lost when small spills smear notes taken with water-soluble marker pens.

What should be in the laboratory notebook entry? First, the **date** should be clearly noted. The **purpose** of the exercise should be briefly noted. For example, you might put a heading such as "Purification of alkaline phosphatase from *E. coli*" at the top of that day's work. This will remind you of what you were trying to do if you looked at the book years later. If you are working on more than one project on a given day, this will inform you about what notes go with what experiment. Next, it is wise to record a **brief overview** of what you are going to try to do. For example, you might put something like the following: "Today, I will try to perform an osmotic shock of intact *E. coli* cells and determine if alkaline phosphatase activity is released from the cells." Extensive expositions of the theory or the methods to be tried are usually not necessary if these are already established. However, if you are developing a new method or a new theory, you would want to document your thinking in the notebooks as proof of your knowledge as of that date. Next, you should enter the **details of your activities**. All of the important details must be included, such as the volumes, concentrations, time, temperature, centrifugation time, flow rate, fraction size, wavelength, what samples were loaded into what lanes of the gel, visual appearance of the samples, etc. It is really important to record in the notebook your calculations for making dilutions and other operations. That way, if you make a mistake and obtain a bizarre result, you can retrace your notes and determine the cause. It is especially important to have good notes when the experiment seems not to have worked, since some clues about the problem are probably right there in your notes. Also, put some information about how you carried out the manipulations, especially if these are complicated, so you can understand the experiment years later. The net result is that it should be possible for a competent scientist to work from your notes and obtain the same results (which may or may not be valid).

Sometimes, you will perform a routine operation many times with the identical procedure. In that case, you can skip the details of the operation (if they are identical to those used previously) and just say, for example, "performed Lowry assay for protein using the standard procedure as on 6/1/95." The pages of the notebook should be numbered, so that this kind of cross-referencing is simplified.

All of the experimental **results** must be in the notebook. If your experimental results are a printout from a spectrophotometer or scintillation counter, or a gel photo or autoradiograph, etc., then these should be directly attached into the notebook. If there is a sample such as a gel that is not photographed, then a drawing of the gel result or a photocopy, or a record of the measurements of the positions of the bands, or both can be attached into the notebook. A table is useful to help keep track of the data and prepare it for future plotting. To help the observation of what is happening in an experiment, it is often effective to plot the data as the experiment progresses. In this way, aberrant trends are immediately obvious, and outlying points can be repeated and checked. It also enables one to redirect an experiment so that it will be more valuable. It is not wise to enter points directly on the graph without having a table, because if plotting errors occur, they are difficult to recognize. Graphs in laboratory notebooks are mainly for the purpose of obtaining a quick understanding of what happened. A figure worthy of publication can always be made later. Therefore, do not worry too much about the appearance, but make sure all graphs have the axes labeled correctly, and specify clearly which data are used in the graph.

It is important to record impressions that are obtained during the experiment. For example, "upon addition of solution B the sample became cloudy, but then after ~1 min, it cleared up" is the type of observation that should definitely be noted. Finally, if there are any **conclusions** that might be important later, they should be recorded. For example, you might note that "Alkaline phosphatase activity was clearly present in the sample that was applied to the column, but no activity could be eluted from the column. Thus, the enzyme is either still on the column or was somehow lost or inactivated during the procedure." It is not necessary to include an extensive discussion of the results in the laboratory notebook unless it involves critical thoughts that you might forget. The proper place for discussion of the results is in the laboratory reports (or the discussion section of manuscripts).

Inevitably, there will be occasions when you write something down incorrectly. In that case, do not erase or "white-out" the wrong information; instead, draw a line through it, and add a little note such as "this is wrong, see below for what actually was done."

At the end of a day's work, it is a good idea to make a mark such as a line across the bottom of the page, indicating the end of that session. Sometimes, you will need to or want to go back and add something to previous notes. For example, you might later find out that there was a problem with one of the materials used in an experiment. Then, you should go back and note this, but *be sure that your note is dated and initialed so that it is clear it was added later.* For example, you might want to note the following on an experiment done 6/1/95: "6/20/95-I now realize that this experiment was done at pH 5.0: apparently, I forgot to adjust the pH of the starting buffer because it is pH 5.0 today-A.N."

Finally, it is useful to leave a few pages blank at the beginning of the notebook so that a table of contents can be developed as the notebook is used or an overview of the project can be inserted. Of course, the date that these insertions are added must be noted.

For the experiments in this book, the instructors will check laboratory notebooks at various times, and this will constitute an important part of your grade. Obviously, it could be a disaster if you came to the laboratory without your notebook.

LABORATORY REPORTS

Regardless of how cleverly and carefully it has been carried out, laboratory work that is not shared with the community of scientists is of little scientific value. Laboratory reports are prepared as if they are research communications. We use the format of the *Journal of Biological Chemistry*, although our reports are considerably shorter. If you have never seen this journal, it will be an excellent exercise to go to the library and examine a few papers to get an idea of the format, as well as the types of science that are published there.

Some of your laboratory reports will be detailed and complete as described below. Others will be shorter but will emphasize the collection of accurate data and the proper reporting and calculating of results. The assignment of which type of laboratory report is determined by your instructor.

The **Title** of the laboratory report will simply be the title of the experiment as in this laboratory manual.

The **Abstract** will be a brief description of the project and the results. The abstract should usually be less than 200 words. Despite its brevity, all the important results of the work should be noted. The abstract is sometimes the hardest part to write, so it is often best to leave it until last.

The **Introduction** should be a succinct statement of the state of knowledge of the project at the time of its inception, the purpose of the research, and the approaches taken. Statements of previous knowledge in the introduction should be supported by citations of the appropriate primary research publications. You may also wish to include the basis of the experimentation and the major conclusion of the project. This can give the reader a reason to read on.

The **Materials and Methods** section should succinctly describe how the work was performed, how the measurements were made, and where the materials were obtained. Often, you will use literature citations for standard procedures. For example, if protein concentration was determined using the method of Lowry, you need not write out the method. It would suffice to say "Protein was measured by the method of Lowry *et al.* (2)" with reference 2 being the original Lowry paper. Sometimes you change the original protocol a bit, in which case you would say, for example, "Protein was determined by the method of Lowry *et al.* (2), except that the incubation with solution 1 was for 10 min instead of 30 min."

The **Results** section has the data that were obtained, and these results are often shown as graphs or photographs. Do not put tables of the raw data in the report if these can be better represented in the form of graphs. All figures must be adequately labeled on both axes and have proper figure legends that describe the experimental conditions and the nature of the experiment. Try to avoid extensive discussion in the results section.

The **Discussion** section is where you talk about the significance of the results. For example, if an experiment failed, you might want to discuss what you think went wrong. The discussion is the appropriate place to go over the theory that is supported by or refuted by your data. Limitations in the data should be clearly noted.

For some short reports, you might want to write a combined **Results and Discussion** section, just like is often found in the *Journal of Biological Chemistry*. However, this is not always any easier than writing separate sections and should definitely not be done in those cases where there are a great number of figures.

The **Literature Cited** or reference section should contain the citations to the primary literature that have been referred to in the text. Use the format as found in the *Journal of Biological Chemistry*. Avoid putting personal communications or other inaccessible references in the literature cited section; these should instead be placed in **footnotes**.

Abbreviations used more than five times should be defined in a footnote at the beginning of the paper, where the first abbreviation is used. *Journal of Biological Chemistry* discourages abbreviation when the term is used less than five times; for those cases, they prefer avoiding the use of the abbreviation entirely. Exceptions to this rule are standard abbreviations that are widely used; a list of these may be obtained in the "instructions to authors" section of the *Journal of Biological Chemistry*. This is found in the first issue of the year and can be obtained via the Internet (http://www.jbc.org).

We hope that you go on to write many great *Journal of Biological Chemistry* papers, which is why we will try to coach you on how to go about preparing a manuscript.

The calculation of species curves and ligand titration curves for receptors with more than one binding site can be somewhat complicated. The approach is illustrated below for a tri-protic acid such as phosphoric acid. The approach will apply to any such ligand system and can be contracted or extended to other numbers of sites. Figure 1-13 was calculated by this procedure.

$$H_3A \leftrightarrow H^+ + H_2A^- \qquad K_1 = \frac{[H^+][H_2A^-]}{[H_3A]} \qquad H_3A] = \frac{[H^+]^3[A^{3-}]}{K_1K_2K_3}$$

$$H_2A^- \leftrightarrow H^+ + HA^{-2} \qquad K_2 = \frac{[H^+][HA^{2-}]}{[H_2A^-]} \qquad [H_2A^-] = \frac{[H^+]^2[A^{3-}]}{K_2K_3}$$

$$HA^{-2} \leftrightarrow H^+ + A^{-3} \qquad K_3 = \frac{[H^+][A^{3-}]}{[HA^{2-}]} \qquad [HA^{2-}] = \frac{[H^+][A^{3-}]}{K_3}$$

Now set up the equations to calculate the fractions of each species.

$$\alpha_1 = \frac{[H_3A]}{[H_3A] + [H_2A^-] + [HA^{2-}] + [A^{3-}]}$$

$$\alpha_1 = \frac{\dfrac{[H^+]^3[A^{3-}]}{K_1K_2K_3}}{\dfrac{[H^+]^3[A^{3-}]}{K_1K_2K_3} + \dfrac{[H^+]^2[A^{3-}]}{K_2K_3} + \dfrac{[H^+][A^{3-}]}{K_3} + [A^{3-}]}$$

$$\alpha_1 = \frac{[H^+]^3}{[H^+]^3 + K_1[H^+]^2 + K_1K_2[H^+] + K_1K_2K_3}$$

The denominator is the same for all of the fractional species equations.

$$\alpha_2 = \frac{\dfrac{[H^+]^2[A^{3-}]}{K_2K_3}}{\text{denominator}}$$

$$\alpha_2 = \frac{[H^+]^2}{\dfrac{[H^+]^3}{K_1} + [H^+]^2 + [H^+]K_2 + K_2 K_3}$$

$$\alpha_3 = \frac{\dfrac{[H^+][A^{3-}]}{K_3}}{\text{denominator}}$$

$$\alpha_3 = \frac{[H^+]}{\dfrac{[H^+]^3}{K_1 K_2} + \dfrac{[H^+]^2}{K_2} + [H^+] + K_3}$$

$$\alpha_4 = 1 - \alpha_1 - \alpha_2 - \alpha_3$$

where α_4 is the fraction of free receptor, A^{3-}.

Prescribe a series of $[H^+]$ (pH) values and calculate the various α values. If you want to have a titration of H_3A with OH^-, you will need to account for the number of protons for each species. Thus, the x axis in a typical titration will be OH^- while the y axis will be pH. Calculate the x axis as $3*\alpha_1 + 2*\alpha_2 + \alpha_3$. Because of the way these equations are set up, this will give a titration that is backward. Thus, the appropriate x axis will be $3 - 3*\alpha_1 + 2*\alpha_2 + \alpha_3$, which will give a titration of H_3A versus OH^-.

The Kaleidagraph equations for calculating titration curves with binding constants of $m1$, $m2$, and $m3$ are given below. This is set up for phosphate with $pK_a{'}$ values of 2.15, 6.82, and 12.38. The first equation is for α_1, the second for α_2, the third for α_3, the fourth for α_4, the fifth for titration quantities, and the sixth is for converting to a titration with OH^-.

$m1$=7.08e–3; $m2$=1.514e–7; $m3$=4.17e–13; c2=(c1)^3/((c1^3) + (m1)*(c1^2) + (m1*m2)*c1 + m1*m2*m3)

$m1$=7.08e–3; $m2$=1.514e–7; $m3$=4.17e–13; c3=(c1)^2/((((c1)^3)/m1) + c1^2 + (c1)*(m2) + m2*m3)

$m1$=7.08e–3; $m2$=1.514e–7; $m3$=4.17e–13; c4=c1/((c1^3/m1*m2) + c1^2/m2 + c1+m3)

c5=1–c4–c3–c2

c6=3*c2+2*c3+c4

c7=3–c6

EQUIPMENT USED IN THIS BOOK

✓ Spectronic 20® spectrophotometers (one instrument per two students)

✓ Protein minigel apparatus (one apparatus capable of two simultaneous gels per four students)

✓ Agarose minigel apparatus (one apparatus per two students)

✓ Power supplies for gel electrophoresis

✓ Refrigerator for student samples

✓ Water baths that must be able to hold 80 °C. One instrument per ten students

✓ pH meter

✓ Scale

✓ Microfuge

✓ Refrigerated superspeed centrifuge. (A dedicated instrument is not necessary; the students will use it only for the purification of alkaline phosphatase.)

✓ Camera and UV transilluminator for photography of agarose gels. (A dedicated instrument is not necessary; the students will use it for a few hours at the end of two to three laboratory sessions.)

✓ Autoclave (for the preparation of bacteriological media by the instructors)

2 Spectrophotometry

INTRODUCTION

Spectrophotometry or colorimetry is the process that determines the amount of light absorbed by colored compounds. It is used for quantitative estimation of compounds based on their color intensities and for measuring the spectral properties of molecules and atoms. Many biochemicals are colored; that is, they absorb visible light and thereby can be measured by colorimetric procedures. Even colorless biochemicals can often be converted to colored compounds by **chromogenic** (color-forming) reactions to yield compounds suitable for colorimetric analysis. In the experiments in this book, you will make use of spectrophotometry often in measurements of enzyme activities, determinations of protein concentrations, determinations of enzymatic kinetic constants, and measurements of ligand-binding reactions. The exercises in this chapter are to familiarize you with the use of spectrophotometry by measuring spectra of compounds, quantifying compounds by their absorbance properties, and using the spectral properties of *p*-nitrophenol (PNP) to determine its pK_a.

The absorption of light is due to the interaction of light with the electronic and vibrational modes of molecules. Molecules have discrete energy levels (i.e., they are quantized), and only light of exactly the correct energy for causing transitions from one state to another is absorbed. The energy of light is given as $E = h\nu$, where h is Planck's constant (6.6×10^{-27} erg•s) and ν is the frequency of the light in Hertz (s^{-1}). **Frequency** is related to **wavelength** (λ) by $\lambda = c/n\nu$, where c is the speed of light in a vacuum and n is the refractive index of the medium. Normally, light is retarded in various media; n is usually slightly greater than 1.

Each type of molecule has an individual set of energy levels associated with the makeup of its chemical bonds and atomic masses, and thus will absorb light of specific wavelengths (energies) to give unique spectral properties. The energies of the transitions, as well as the propensities of the molecule to absorb a given wavelength of light, are dependent on the specific nature of the molecule, so that the spectral shapes and absorption intensities are signatures of the molecules. These spectral properties can thereby be used for qualitative analysis to identify and distinguish particular compounds. For such qualitative analysis, spectrophotometers are used to record spectra of compounds by scanning broad wavelength regions to determine the **absorbance** properties (the intensity of the color) of the compound at each wavelength. Table 2-1 shows the approximate wavelengths and frequencies of various regions and the most frequent kinds of interactions light of such energy has with molecules; it also shows some of the types of spectroscopy used to measure such energies. **Ultraviolet-visible (UV-vis)** spectroscopy uses the range 200–800 nm and involves energy levels that excite **electronic transitions**. The energy of light at 400 nm is ~250 kJ/mol (~60 kcal/mol), which is of the order of the energy of chem-

TABLE 2-1
Spectral Regions

WAVELENGTH (nm)	FREQUENCY (Hz)	SPECTRAL REGION	INTERACTIONS
10^6–10^{10}	3×10^{11}–3×10^7	radio waves	spin orientations[a] NMR and EPR
10^3–10^5	3×10^{14}–3×10^{12}	infrared	vibrations, rotations, bending
4×10^2–8×10^2	7.5×10^{14}–3×10^{14}	visible	electronic transitions
2×10^2–3×10^2	1.5×10^{15}–1×10^{15}	ultraviolet (UV)	electronic transitions
10^{-3}–10^{-0}	3×10^{20}–3×10^{17}	x-rays	inner shells

[a]NMR, nuclear magnetic resonance; EPR, electronic paramagnetic resonance.

ical bonds. Absorption of UV-visible light excites electrons that are in **ground-state orbitals** to **excited-state molecular orbitals**.

Quantification of colored compounds is accomplished with a spectrophotometer, which compares the amount of light transmitted through a blank sample (lacking the colored compound) to that transmitted through a colored sample. For quantification, this is generally accomplished by passing through a solution of the compound light of the particular wavelength that is absorbed maximally by the compound to be measured. This usually gives the best sensitivity for the analysis. You will carry out such experiments in the exercises in this chapter.

Most common laboratory spectrophotometers can measure color in the visible region of light (e.g., between wavelengths of 350 and 800 nm). Others can extend this range into the UV region (200–350 nm), where nearly all compounds absorb light. The Spectronic 20® spectrophotometer limits us to the range 340–650 nm for experimentation.

When light of a particular wavelength λ is passed through a path length l of a colored solution of a defined concentration, a certain proportion of the light is absorbed by the solution. If the incident light energy is I_0 and the transmitted light (after passing through the solution) is I, then the fraction of light transmitted is I/I_0, which is known as the **transmittance T**. If we were to double the light path to be $2(l)$, the newly added fraction of colored solution would receive only I/I_0 light with which it could interact, and it would absorb a similar fraction of light, or a total of $I/I_0 \times I/I_0$. This argument can be extended to further fractions. For homogeneous samples, each successive layer will absorb the same fraction of the incident light as the previous layer. This is **Lambert's law**. Similarly, it can be reasoned that if the concentration of the absorbing species were twice that of the original solution, each successive layer of the solution would receive fractionally less light than did the original solution. This reasoning gives rise to **Beer's law**, which states that light absorption is proportional to the number of molecules in the path. Such fractional loss of light by successive layers of sample is described mathematically by an exponential decay of the transmitted light versus path length and versus concentration. Beer, Lambert, and others have shown that T decreases exponentially with respect to the concentration c of the colored compound and to the length of the light path, l, in the following manner:

$$T = 10^{-\varepsilon cl} \tag{2-1}$$

$$-\log_{10} T = \log_{10}(1/T) = \varepsilon cl = A \tag{2-2}$$

Equation 2-2 is the **Beer–Lambert equation** in which ε is the **absorption coefficient (absorptivity)** that is characteristic of the compound at the particular wavelength of light under a defined set of conditions. Another term for absorptivity is **extinction coefficient**.

The absorbance A is defined as $-\log_{10} T$. Absorbance is also referred to as **optical density (OD)**. From equation 2-2, it can be seen that the

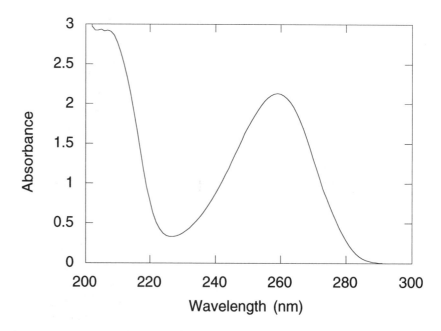

FIGURE 2-1 Absorbance spectrum of 0.138 mM of ATP (calculated from the absorbance) at pH 7.0 in 100 mM phosphate buffer.

absorbance of a solution is directly proportional to the concentration of the absorbing material when the light path is kept constant. A is dimensionless; thus, ε has units of per mole per centimeter.

When the concentration is expressed in molarity M and the light path in centimeters, ε is known as the **molar absorptivity** (units are $M^{-1}cm^{-1}$) or molar **extinction coefficient** for that compound. Therefore, ε is the absorbance of a molar solution when the light path is 1 cm. For example, ATP has a molar absorbance of 15,400 M^{-1} cm^{-1} at 259 nm (pH 7.0); therefore, a 1 M solution of ATP (at pH 7.0) in a cuvette of 1-cm path length would give an A value of 15,400 at 259 nm. This solution, of course, would be too opaque to read with any practical instrument. Even a solution 1/1000 as concentrated, 1 mM, would give a value of 15, which is still out of the measuring range of spectrophotometers. (An absorbance of 15 implies that the transmittance is 10^{-15}, which cannot be distinguished from zero by practical instruments.) A solution that is 0.1 mM would still give a strong absorbance, 1.5 A (3.2% transmission), but this low a value can nevertheless be reliably measured with a good instrument (Fig. 2-1).

Molar absorptivity is not the only term used to describe the absorptivity of a compound; you may frequently see terms written as $E^{1\%}_{280nm}$ or $\varepsilon^{1\%}_{280nm}$. These terms indicate that at 280 nm a 1% solution of the substance (for example, a protein) will have an absorbance of E or ε. Such terms are often used in clinical work and for the characterization of protein concentration with solutions of impure proteins.

The experiments described in this book illustrate some of the principles of spectrophotometry discussed above, such as the absorption spectrum, molar absorbance, estimation of the concentration of an unknown by colorimetry, effects of pH on the color of a compound, the calculation of the dissociation constant, and also some of the problems encountered in a chromogenic analytical procedure.

DESIGN AND PROPERTIES OF SPECTROPHOTOMETERS

Spectrophotometers have a light source, a means of dispersing the various wavelengths of light so that only one wavelength reaches the sample at one time, a sample compartment, and a detector. Figure 2-2 shows a schematic of the optics of the Spectronic 20® spectrophotometer. The **light sources** most commonly used for spectrophotometry are tungsten (for the visible region, 340–1000 nm) and deuterium lamps (for the UV region, 180–350 nm). These lamps each have characteristic spectral outputs and do not provide the same I_0 for each wavelength. Therefore, the variable I_0 must be accounted for to determine spectral properties of the sample.

A **monochromator** is used to select particular wavelengths of light for measurement in the spectrophotometer. Either a prism or a grating disperses light, and a narrow exit slit (Fig. 2-2) selects only a small range of the dispersed spectrum. The grating or prism is rotated so that the slit selects the desired wavelength and all other wavelengths are blocked from the sample and the detector. The Spectronic 20® spectrophotometer used in our laboratory utilizes a grating to disperse the light. In some very simple instru-

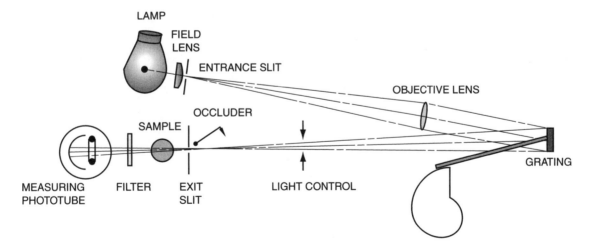

FIGURE 2-2 Optical diagram of a Spectronic 20® spectrophotometer from Bausch & Lomb. This is a very basic single-beam instrument. See Chapter 1 and Fig. 1-6 for operating instructions.

ments, an optical filter that permits light of only a small range of wavelengths to pass through is used. Although simple and inexpensive, the filter does not offer much flexibility.

The sample compartment is designed to hold reproducibly cuvettes for the light to pass through to the detector. **Detectors** are usually phototubes, photomultipliers, or photodiodes. When phototubes or photomultipliers receive light of sufficient energy, they emit electrons, which can be measured in the form of current by sensitive electronic devices. The emission of electrons upon irradiation with light is called the **photovoltaic effect**, which was first expounded by Albert Einstein in 1905. The photomultiplier is a phototube with an ingenious cascade amplifier that can achieve amplification of photocurrents of more than 10^9-fold. The Spectronic 20® uses a phototube for its detector. A photodiode is a solid-state detector that changes its current-carrying capacity upon reception of light. One of the desirable features of photodiodes is that they can be aligned into linear arrays of 500–1000 minuscule diodes. Thus, each diode can serve as a slit as well as a detector. The entire spectrum can be projected on the diode array and thereby detected all at once. This permits very rapid acquisition of spectra in so-called diode array spectrophotometers, and it also permits measurements without any mechanical movements of the grating.

Each of the parts of the spectrophotometer has optical properties that vary with wavelength. The light source has its emission spectrum, the

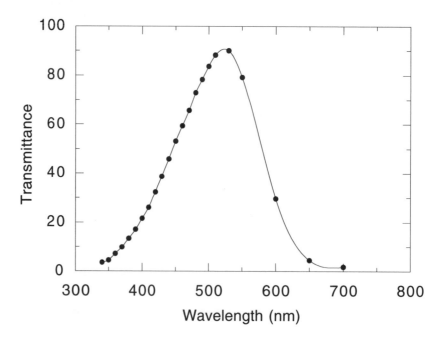

FIGURE 2-3 Output of a Spectronic 20® with respect to wavelength. This is an I_0 curve obtained with a single setting of the light adjustment knob.

detector has variable sensitivity based on wavelength, the focusing optics have different efficiencies at each wavelength, and the dispersing device (e.g., monochromator) has regions of the spectrum in which it is more or less efficient. These properties lead to I_0 being different at each wavelength. One way to demonstrate this for the Spectronic 20® is to read the transmittance of a blank solution at various wavelengths without adjusting the optical attenuator (the right knob (D) on the instrument). This is shown in Fig. 2-3. To compensate for this variability, we adjust the light intensity so that a blank signal of 100% T is achieved at each wavelength. Thus, after placing a cuvette containing the blank or background materials (usually everything except the sample) into the sample compartment, we adjust the light control (the right knob shown in Fig. 1-6) until the detector gives a signal of 100% T.

Figure 2-4 shows an uncompensated I_0 for a buffer blank and I for a solution of flavin mononucleotide (FMN) as a function of wavelength. Figure 2-4 also shows the absorbance ($A = \log_{10}(I_0/I)$) for each wavelength in the form of an absorbance spectrum. You can see how the variable optical behavior of the instrument must be compensated for to be able to determine a true absorbance spectrum. It also indicates how the instrument is likely to be more reliable at wavelengths between 450 and 600 nm; there is more signal available.

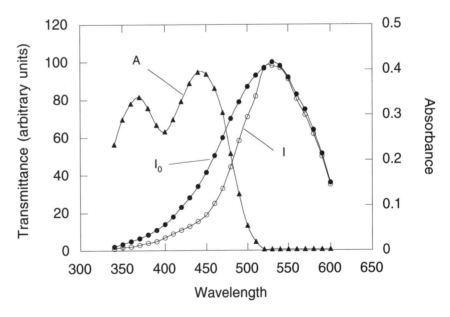

FIGURE 2-4 I_0 of buffer blank, and I and A for FMN (0.036 mM) in phosphate buffer (50 mM) at pH 7.0, measured on a Spectronic 20® instrument. Note that absorbance is the log(I_0/I) and that at shorter wavelengths this is a ratio of small numbers owing to the optical properties of the instrument.

EFFECTS OF SPECTRAL BANDPASS AND STRAY LIGHT

When recording spectrophotometric data, it is assumed that only light of the particular chosen wavelength reaches the sample and is measured. However, monochromators are not perfect devices, and, in fact, a range of wavelengths always reaches the detector with the maximum intensity of light at the selected value and progressively lesser amounts of light on either side of the maximum in a shape approximated as a triangle of intensity (Fig. 2-5). The characteristic bandwidth of a monochromator is usually given as the width of the band at half height of light emerging from the monochromator. This is the width of the "triangle" of intensities at its half height (Fig. 2-5), which is centered on λ_0. Ideally, three fourths of the total light coming from the monochromator is within the area of the triangle as shown in Fig. 2-5, i.e., within the half bandwidth of the monochromator.

The Spectronic 20® has a bandwidth of 20 nm. If a compound has a narrow absorption band and the bandwidth of the "monochromatic" light is not significantly less than that of the absorption band, in addition to light of λ_0, light of wavelengths that are not absorbed will impinge on the sample and reach the detector. Thus, measurements of transmission will be inflated and will not determine the true absorbance, but will yield an apparently diminished absorbance. Figure 2-6 shows the effects of bandpass on samples of cytochrome *c* and FMN.

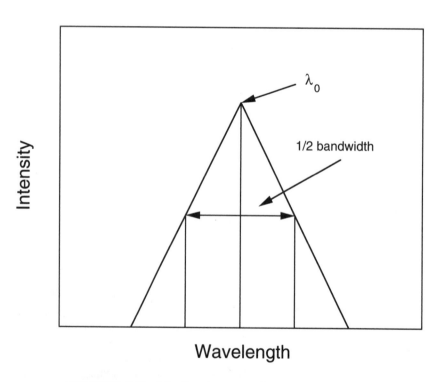

FIGURE 2-5   Idealized output of a monochromator.

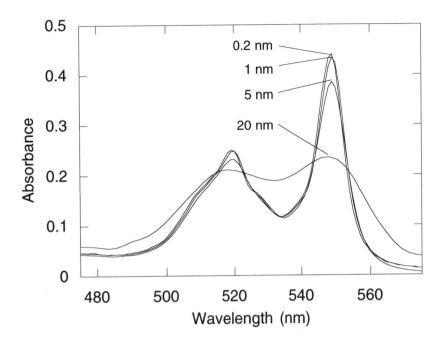

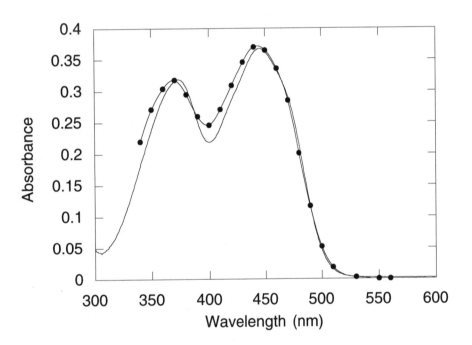

FIGURE 2-6 *(A)* Spectra of reduced cytochrome *c* (13.4 μM) at various spectral resolutions. The first three scans were recorded with a Shimadzu UV-2501 PC spectrophotometer, while the last scan was performed on a Spectronic 20®. *(B)* Spectra of FMN recorded with the Shimadzu spectrophotometer with a spectral resolution of 1 nm (no symbols) and with the Spectronic 20® (●).

The triangle function shown in Fig. 2-5 is only an approximation of the output of a monochromator. Indeed, the shape is better approximated by a Gaussian curve, and small amounts of light of wavelengths from across the entire spectrum will reach the detector. Light leaks from the room will also have a similar effect. This extra light that falls upon the sample is referred to as **stray light** and compromises the ability to accurately measure samples with high absorbance. A sample with an absorbance of 2.0 allows only 1% of the light at that wavelength to reach the detector. Stray light of only 1% will change the apparent absorbance to ~1.7. Very high quality spectrophotometers have very low stray light specifications (e.g., 0.00001%) and can accurately measure samples with high absorbances.

SPLIT BEAM RECORDING SPECTROPHOTOMETERS

The measurements of spectra you make using the Spectronic 20® instruments require you to change manually the wavelength and adjust the I_0 for each wavelength selected. Recording spectrophotometers frequently split the optical beam, so that the instrument alternately measures automatically the I and I_0 and computes the absorbance from $\log_{10}(I_0/I)$ as it scans through the wavelengths. See Figs. 2-1 and 2-6 for examples of spectra.

CHROMOGENIC REACTIONS USED FOR ANALYSIS

Chromogenic reactions are used to convert a colorless chemical to one that has color so that it can be measured by spectrophotometry. Chromogenic reactions are frequently used in biochemistry laboratories. Several of the tests described in Chapter 3 for the determination of protein concentrations use chromogenic reactions. A useful chromogenic reaction should have the following properties:

1. The color yield should be proportional to the weight or concentration of the unknown in the sample over a reasonable range of values. Some reactions produce more than one colored product, or side reactions destroy some of the reagent (or unknown) in an unproductive way. At higher concentrations, these competing reactions may become more significant, and the direct proportionality of the measurement to the unknown will be lost (i.e., a graph showing the relationship between absorbance and concentration will not be a straight line and may not be reproducible from sample to sample). In such a case, one cannot use a single concentration of the pure standard as a reference sample; instead, one must use a set of standards covering a range of concentrations to

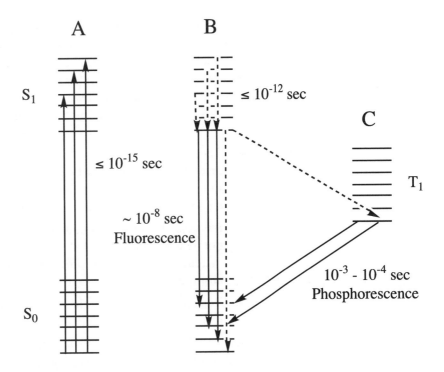

FIGURE 2-7 Jablonski diagrams of absorption and emission processes. The dotted arrows represent radiation-less energy transfers. *(A)* The absorption of radiation ($\leq 10^{-15}$ s) to promote the molecule from ground electronic *(S_0)* and vibrational levels to higher vibrational levels within the excited electronic state S_1. *(B)* Very fast relaxation (internal conversion) of the excited vibrational levels ($\leq 10^{-12}$ s) to the ground vibrational levels of the excited electronic state. This can be followed by rapid radiation-less transfers to the environment, photoexcited chemistry (not shown), or fluorescence ($\sim 10^{-8}$ s) to various vibrational levels within the ground electronic state. *(C)* Occasionally, the molecule converts to a triplet state (via a process called intersystem crossing) from which it gives off light (phosphorescence) after comparatively long times (10^{-4}–10^{-3} s).

produce a standard curve with a range that includes the concentration of the unknown. You will frequently be generating standard curves in your analyses as described in this book.

2. The color yield should be high, so that very small amounts of unknown can be determined (measured). Restated, this means that the molar absorptivity of the product should be high. Many useful chromogenic reagents have molar absorptivities between 10,000 and 20,000 $M^{-1}cm^{-1}$; a few compounds (including hemoproteins) have molar absorptivities >70,000 $M^{-1}cm^{-1}$.

3. The colored product should be reasonably stable, so that precise timing of measurements is unnecessary. This is important when using simple instruments, such as the Spectronic 20®. However, there are instru-

ments such as stopped-flow spectrophotometers that are designed to perform measurements quickly and can therefore reliably detect compounds that are unstable, e.g., intermediates in a reaction sequence. Stopped-flow instruments efficiently mix small volumes of reactants into an observation cell where the reaction can be spectrophotometrically monitored. Such instruments can make observations in times as short as 1–2 ms after mixing.

4. There should not be any other substances present in the sample being analyzed that interfere with the reaction. In many cases, it is necessary to carry out one or more separation steps to remove interfering substances before the chromogenic reaction can be carried out.

5. It is desirable that the conditions required for the chromogenic reaction not be too stringent so that small variations in temperature or volume of reagents or reaction time do not produce a significant change in absorbance.

6. Reagents required for the reaction should be stable, so that fresh reagents need not be prepared for every analysis.

7. The absorbance of the blank solution should be low. If the blank solution has a high absorbance, the absorbance obtained with a low concentration of unknown will be only slightly greater and the inherent instrumental error will be significant. This sort of error is a problem in all quantitative methods when one is subtracting a relatively large number from a number only slightly larger.

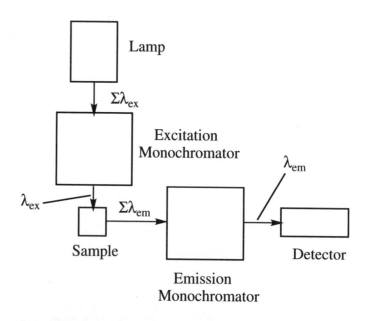

FIGURE 2-8   Diagram of a spectrofluorimeter.

OTHER FORMS OF SPECTROSCOPY

There are many forms of spectroscopy used in the elucidation and quantification of biological compounds. Described below is a small fraction of the more commonly used spectroscopic techniques.

FLUORESCENCE SPECTROSCOPY

Fluorescence spectroscopy is frequently used for studying biological systems because it is very sensitive, has high specificity, and is usually not destructive to the sample. When a sample absorbs UV or visible light, electrons are promoted to a higher-energy electronic orbital. The absorption process usually initiates from a ground electronic and vibrational state and leads to an excited electronic and vibrational state. This is the basic process involved in absorption spectroscopy (photometry). Figure 2-7 shows a Jablonski diagram representing the absorption process for an electronic transition. This diagram shows the ground and the excited electronic states with discrete vibrational energy levels for each of the electronic states. Upward arrows represent the transitions from the ground to the excited states. Many transitions of similar energy (three are shown) give rise to an envelope of absorption energies (the absorbance spectrum) for the compound.

Absorption occurs in a very short time ($\sim 10^{-15}$ s), such that the atomic positions remain essentially unchanged during the process. This is known as the **Franck–Condon principle**. After absorption, the vibrational modes relax very quickly ($\leq 10^{-12}$ s) by dissipating heat to the solvent (this radiation-less transfer is referred to as **internal conversion**) (Fig. 2-7B, top). The molecule resides briefly in the lowest vibrational mode of the excited electronic state and returns to the ground electronic state by one of several competing mechanisms. Most molecules are not fluorescent because they interact efficiently with their environment to return to the ground state rapidly ($<10^{-10}$ s) by radiation-less energy transfer (in the form of heat) to the solvent. Some molecules can undergo chemical transformations from the excited state. This is called **photochemistry** (not shown).

However, fluorescent molecules have a finite lifetime in the excited state S_1 ($\sim 10^{-8}$ s), and give off energy in the form of light in returning to the ground state, S_0. The **lifetime** of the excited state is dependent on the nature of the fluorophore, the effects of solvent, the effects of interaction with nearby molecules (quenching), and other competing processes. Because the vibrational energy has already been dissipated by internal conversion, the emitted fluorescent light is of lower energy and therefore of longer wavelength than the absorbed light. Transitions to many different vibrational levels of the S_0 state (three are shown) produce an emission spectrum. On rare occasions molecules pass into a triplet state (T_1) where the lifetime is much longer (e.g., 10^{-4}–10^{-3} s). This process is called intersystem crossing. The triplet state has unpaired electrons, and the emission wavelength is longer than that for fluorescence. Emission from the triplet state is called **phosphorescence**.

Figure 2-8 shows a schematic of a typical spectrofluorimeter. The source of excitation is usually a xenon or mercury lamp (or a laser), which provides high-intensity light in the UV-vis range. The high intensity source is chosen because the intensity of the fluorescence is directly related to the intensity of the exciting radiation. A **monochromator** selects the wavelength of the excitation light and delivers it to the **sample**. The sample fluoresces, and an **emission monochromator** selects the emission wavelength (which is of longer wavelength than that of the excitation) and passes the light to a sensitive **detector**. The direction of emission is usually chosen to be at right angles to that for excitation so that the very bright excitation light does not pass directly to the detector and swamp out the weak fluorescence emission. If one excites the fluorescent molecule with light of a wavelength within the envelope of the absorption spectrum and then scans the emission monochromator, the **emission spectrum** of the fluorescing molecule is obtained. Similarly, if one sets the emission monochromator to a wavelength within the envelope of the emission spectrum and then scans the excitation monochromator, the **excitation spectrum** is obtained. The excitation spectrum should correspond to the absorbance spectrum of the molecule that fluoresces.

The emitted light in fluorescence is usually not very intense. However, instruments have been developed with highly sensitive photomultiplier (and other) detectors to detect such low levels of light. The high sensitivity of fluorescence derives from the fact that all of the emitted (detected) light is due to the sample. In contrast, in absorbance spectroscopy (which is actually the measurement of transmission), the detector must distinguish between I_0 and I, two numbers that may be large and nearly the same. Thus, in transmission, the detector "measures" the light that does not get to it $(I_0 - I)$.

The **specificity** of fluorescence measurements derives from the fact that only certain types of molecules fluoresce. Thus, for example, in a mixture of several types of molecules that might be colored, only one or two are likely to fluoresce. The fluorescent compound can often be recognized in this mixture because the excitation spectrum looks like its absorbance spectrum.

Proteins are often fluorescent because tryptophan is a good fluorophore. Cofactors such as flavins, pyridine nucleotides, and pyridoxal, as well as some substrates or products of reactions, are fluorescent. In addition, it is often possible to label proteins, membranes, or nucleic acids with fluorescent dyes and measure various properties. One practice labels macromolecular structures with two fluorescent dyes and measures **fluorescence energy transfer** from one to the other. Such measurements give information about the distance between the two labels, contributing to knowledge of the structure.

Spectrofluorimetry can be used very similarly to spectrophotometry. The disappearance of fluorescent substrates or the appearance of fluorescent products can be monitored. Fluorogenic reactions can often be devised for compounds that are not fluorescent. The same principles apply to such reactions as for chromogenic reactions (see above). In using spectrofluorimetry, it should be noted that the numbers measured for fluorescence are dependent on the particular instruments used and the particular settings of those instruments. The optics, detector sensitivities, and light source are unique for each instrument. Therefore, it is incumbent on the spectroscopist to standardize the instrument with known samples in order to compare results with those obtained elsewhere.

There are many variations and uses of fluorescence techniques. Lackowicz (1983, 1991a, 1991b, 1992) provides an excellent survey of approaches and gives both theoretical and practical information.

INFRARED SPECTROSCOPY

The principles of infrared (IR) spectrophotometry are the same as those inherent in UV-vis spectrophotometry. However, the wavelength region is from ~1000 to 200,000 nm, and the energy transitions are due to changes in the vibrational, rotational, and kinetic energy of molecules. The light sources, detectors, and monochromators are chosen to optimize this wavelength range. The energies involved are from 0.5 to 50 kJ/mol. IR spectroscopy is useful for qualitative analysis, because substituent groups such as carbonyls, alcohols, aromatic rings, carbon-halogen bonds, etc., each have group frequencies that serve as signatures for their identification. Although IR spectroscopy can be used for quantifying materials, the precision and accuracy are not usually as good as UV-vis spectrophotometry. IR has the added disadvantage in that water is not usually a useful solvent because it absorbs too much of the incident energy.

NUCLEAR MAGNETIC RESONANCE SPECTROSCOPY

Nuclear magnetic resonance (**NMR**) is one of the most useful forms of spectroscopy for organic and biological chemistry. It measures very low energy transitions of the spin states of protons (or other nuclei) caused by placing molecules in a magnetic field. Atomic nuclei such as ^1H, ^{13}C, ^{19}F, or ^{31}P have nuclear spin quantum numbers of $1/2$ and have associated quantized spin angular velocities. When one of these nuclei is placed in a magnetic field, spins aligned with the field have a slightly lower energy than those aligned opposite the field. This energy, ΔE, is directly related to the strength of the field, H. The resonance or absorption is described by

$$\Delta E = \gamma_i hH = h\nu$$

(2-3)

where γ_i is the gyromagnetic ratio, which is unique for each type of nucleus, h is Planck's constant, and ν is the frequency of the light (electromagnetic energy). At higher magnetic fields, it can be seen from equation 2-3 that the separation of spin energies is greater than at lower fields. The exact position (energy or field, H) at which resonance occurs will also be affected by the chemical environment. This **chemical shift** is defined as the position where the absorption appears in the spectrum relative to some standard signal. Protons attached to an aromatic ring will have different chemical shifts than those attached to an aliphatic carbon, and even more subtle differences can be detected that are due to smaller differences in the chemical composition. In addition to the chemical shift, **spin-spin coupling** arises from the interaction of the measured spin system with local magnetic fields from nearby spins of protons and other nuclei. Spin-spin coupling is dependent on both distance and angle; its measurement gives further information about the molecule. **Nuclear Overhauser effects** derive from the spin of one nucleus sensing a change in the spin of another. These effects are related to r^{-6}, where r is the distance between the two nuclei. Nuclear Overhauser effects are thereby observed only when r is small.

These three types of magnetic resonance interactions, chemical shift, spin-spin coupling, and Nuclear Overhauser effects, are exploited in NMR spectroscopy to derive detailed structural information about molecules. Small molecules can often be unambiguously identified and quantified. Large molecules, such as proteins, can also be studied and their detailed three-dimensional conformations determined, but, currently, the upper limit for detailed structural analysis is on proteins of ~25,000 Da. NMR methods have developed significantly in the past 20 years and are used widely in biology and chemistry. Currently, **Fourier transform NMR** is almost exclusively used. A significant disadvantage of the method is that the instrumentation is very expensive. Nevertheless, there are numerous specialized centers all over the world where researchers can carry out experiments under the tutelage of experts.

ELECTRON PARAMAGNETIC RESONANCE SPECTROSCOPY

Electron paramagnetic resonance (**EPR**) spectroscopy is widely used for the study of systems that have unpaired electrons. Such unpaired electrons have associated quantized spins that can align either with or against a magnetic field. This is analogous to NMR, but the energy differences are considerably larger. EPR is frequently used to investigate metals in biology, because metals such as iron, cobalt, copper, manganese, and molybdenum in one or more redox states are paramagnetic and have unpaired electrons. The environment of the metal can be monitored by EPR. **Hyperfine interactions** due to nearby nuclei that have nuclear spin, such as ^{14}N, also give information about the ligands to the metal. Therefore, EPR is extensively used for the study of metalloproteins. **Free radicals** occurring in reactions can also be monitored by EPR. One application is the use of **spin traps** in which molecules that efficiently form stable free radicals can trap nascent radicals in reactions. The EPR spectra of such free radical traps can be diagnostic of the type of radical that transferred the electron to the trap. In addition, **spin labels** (which are molecules bearing a stable free radical) can be linked to biological membranes and macromolecules and measured by EPR. Such measurements give information about the conformation and mobility of these biological structures.

REFERENCES

Cantor, C. R., and P. R. Schimmel. 1980. Biophysical chemistry, part 2. Techniques for the study of biological structure and function. W. H. Freeman & Co., New York.

Lackowicz, J. R. 1983. Principles of fluorescence spectroscopy. Plenum Press, New York.

Lackowicz, J. R. (ed.). 1991a. Topics in fluorescence spectroscopy, vol. 1. Techniques. Plenum Press, New York.

Lackowicz, J. R (ed.). 1991b. Topics in fluorescence spectroscopy, vol. 2. Principles. Plenum Press, New York.

Lackowicz, J. R. (ed.). 1992. Topics in fluorescence spectroscopy, vol. 3. Biochemical applications. Plenum Press, New York.

Skoog, D. A., D. M. West, and F. J. Holler. 1996. Fundamentals of analytical chemistry, 7th ed. Saunders College Publishing, New York.

EXPERIMENT 2-1

MEASURING THE ABSORBANCE SPECTRA OF THE BASIC AND ACIDIC FORMS OF PNP

MATERIALS

1. Buffers at 0.1 M at pH 5 and 10. (Your instructor will inform you of the identity and exact pH of each of these buffers.)
2. *p*-Nitrophenol (PNP) solution, 0.1 mM

PROCEDURE

The relative amount of light absorbed by a colored material depends on the wavelength of the impinging light. Changing the position of the instrument grating with the wavelength control knob can set this. Many chromophoric groups, such as pH indicators and dyes, are sensitive to pH. We will measure the spectrum of 0.1 mM PNP (O_2N–C_6H_4–OH). The ionized form of PNP is colored. Place 1 mL of this solution into each of two test tubes and add 3 mL of one of the above buffers to one tube and 3 mL of the other to the second tube.

Using the blank tube and the most highly colored of the two tubes above, measure the absorbance from 360 to 500 nm in 10-nm increments. Do not forget to adjust the instrument to zero absorbance (100% *T*) at each wavelength with the water blank tube. This step is necessary because the amount of electromagnetic energy (light) passing through the sample depends on the wavelength of the light that reaches it; moreover, the detector sensitivity is also dependent on the wavelength as discussed above. Now repeat the task using the less colored tube.

Instructions for operating the Spectronic 20® spectrophotometer are given in Chapter 1.

PLOT ABSORBANCE OF PNP VERSUS WAVELENGTH

Plot the absorbance values for each sample versus wavelength to produce the visible spectra of PNP at pH 5 and 10. These pH values yield nearly minimum and maximum fractions of the anionic form. Is there a wavelength at which no changes in absorbance occur as the pH is varied? (Such a wavelength would be called an isosbestic point.) Why would the isosbestic point be useful to measure for each of the samples?

DETERMINING THE pK_a OF PNP

MATERIALS

1. Buffers (0.1 M) at nominal pH values of 5, 6, 7, 7.5, 8, 8.5, 9, and 10. (Your instructors will inform you of the identity and exact pH of these buffers.)
2. *p*-Nitrophenol (PNP) solution, 0.1mM

PROCEDURE

In Experiment 2-1, you measured the spectra of the ionized and the protonated forms of PNP. Using those results, in this experiment, you will choose a wavelength that is useful to distinguish the protonated and unprotonated forms of PNP. You will use 0.1 mM PNP ($O_2N–C_6H_4–OH$) as in Experiment 2-1. Place 1 mL of this solution into each of eight test tubes and add 3 mL of the above buffers, one to each tube, using the stock buffers above. Two of these tubes should already be made up from Experiment 1. You might make them up again to determine with what precision you can make up such solutions and measure their absorbance. Read the absorbance *A* at the most reasonable wavelength determined from Experiment 1, adjusting the Spectronic 20® to zero *A* with a water blank. Plot the absorbance of PNP versus the pH.

QUESTIONS

1. How many micromoles of PNP were placed into each tube?

2. Draw the structure of the *p*-nitrophenolate ion formed in the alkaline buffer. From the plot of *A* versus pH, calculate the approximate pK_a of the compound (if necessary, reread the section on pH and buffers in Chapter 1).

3. What was the concentration of the PNP and the PNP anion in each of the tubes after dilution with buffer? Use the units micromolar (μM) and millimolar (mM).

4. What is the molar absorptivity of PNP at pH 8.5? Use the Beer–Lambert equation and assume that the path length $l = 1$ cm.

If you use the calculated concentration from above in millimolar, multiply your answer by 1000 to get the molar value. If you use the micromolar value in the equation, multiply your value by 1,000,000. Remember that the units in the final answer are $L \cdot cm^{-1} \cdot mol^{-1}$.

5. Suppose that a test tube containing 7 mL of PNP in alkali gave an absorbance of 0.282 at 400 nm. If the molar absorptivity under these conditions is 15,000 $L \cdot cm^{-1} \cdot mol^{-1}$, calculate the concentration of PNP. Give the concentration in molar, millimolar, and micromolar. Assuming the path length $l = 1$ cm, calculate the amount of PNP in the tube.

6. PNP (2.8 mg) (MW = 140) was dissolved in 20 mL of 0.1 M NaOH. A small aliquot of this solution (0.1 mL) was diluted with 4.9 mL of pH 8 buffer (0.1 M), and the absorbance at 415 nm was determined. If the molar absorptivity is 15,000 $L \cdot cm^{-1} \cdot mol^{-1}$, at this wavelength and pH and the path length $l = 1$ cm, what reading do you expect?

7. The buffer used in this experiment is commonly called Tris, but its true name is tris (hydroxymethyl) aminomethane and its structure is shown below.

The word "Tris" means three, referring to the three hydroxymethyl groups. Draw the other structure, the one that forms in acid (the protonated form). The pK of Tris is ~8.0; in other words, it is a weak base, and it buffers usefully over the range 7.0–9.0, depending on the temperature. Note that the amino group is a primary amine, as in most amino acids.

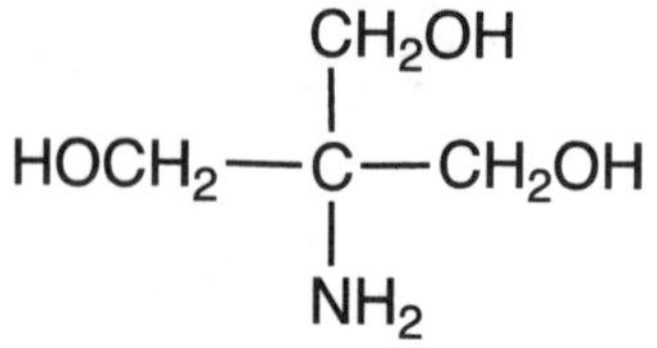

$$\underset{\displaystyle NH_2}{\overset{\displaystyle CH_2OH}{HOCH_2 - C - CH_2OH}}$$

OPTICAL ASSAY OF PNP

In this experiment, you will determine the concentration and the amount of PNP in an unknown sample by comparing the absorbance of the sample with the absorbance of standards containing known amounts of PNP. This will be an opportunity to develop a standard curve and to use it for determination of an unknown concentration.

MATERIALS

1. Standard PNP solution (0.1 mM)
2. pH 10.0 buffer
3. Unknown PNP solution (A, B, C, or D)

METHOD

Dilute the standard PNP solution appropriately with pH 10 buffer (4 mL final volume) to make five known standards containing 10, 20, 30, 50, and 75 µM PNP. Measure the absorbance at the wavelength where PNP absorbs maximally (A_{max}, which you determined in Experiment 2-1). Dilute 1 mL of one of the unknowns with 3 mL of buffer, and measure A at the same wavelength.

RESULTS AND CALCULATIONS

Plot the known concentrations of PNP versus A_{max} for the data obtained with the standards. This plot is known as a standard curve. Read the concentration of the unknown from the standard curve.

QUESTIONS

1. Why would you want to use the wavelength of maximum absorption for this determination?

2. Calculate the amount of PNP in nmol present in the 4 mL of the diluted PNP solution.

3. Calculate the concentration of PNP in the original unknown that was used.

EXPERIMENT 2-4

COLORIMETRIC DETERMINATION OF pH

In this experiment, you will use the absorbance properties of PNP as a function of pH to determine the pH of a solution.

MATERIALS

1. 0.1 M Tris base
2. 0.1 M HCl
3. 0.1 mM PNP

METHOD

Using 0.1 M Tris base, 0.1 M HCl, 0.1 M KH_2PO_4, and 0.1 M K_2HPO_4, make up 10-mL solutions that will have final pH values of 7.4, 7.8, and 8.2. Using data from Table 1-3 in Chapter 1 and the Henderson–Hasselbalch equation, you can calculate how much of each reagent to use. Take 3 mL of the mixture, and add 1.0 mL of 0.1 mM PNP. Measure A for each solution. (What wavelength should you use?) Use your curve from Experiment 2-2 to determine the pH of each of your solutions. Also, measure the pH of the remainder of the solutions using a pH meter to check your determinations. The latter will be done with the help of an instructor. How do these values compare with those calculated with the Henderson–Hasselbalch equation? Are your variances within the expected uncertainties of these measurements?

REAGENTS NEEDED FOR CHAPTER 2

✓ *p*-Nitrophenol (PNP): Sigma 104-8, 1 g

✓ Glycine (used to make pH 9.0 and 10.0 buffer): Sigma G8898, 1 kg

✓ Potassium phosphate (monobasic) and potassium phosphate (dibasic) (used to make pH 7.0 and 7.5 buffers. Make a 0.5 M solution of each and mix in different proportions to get buffers of various pH). Monobasic, Sigma P5379, 1 kg; Dibasic, Sigma P5504, 1 kg

✓ Tris base: Sigma, T8404, 1 kg

✓ Sodium acetate: Sigma, S7545, 250 g

✓ Sodium citrate: Sigma, S4641, 500 g

✓ HCl

3 Quantification of Protein Concentration

PURPOSES OF PROTEIN QUANTIFICATION

It is often important to know the concentration of protein in a sample. For example, during the purification of an enzyme, the progress of the purification may be tracked by comparing the total amount of the desired enzymatic activity with the total amount of protein after each fractionation step. If the purification is proceeding properly, as the enzyme becomes more pure, the amount of enzyme activity relative to the total protein concentration will increase. The ratio of the enzymatic activity to the total amount of protein in a sample is known as the **specific activity** and is measured in units per milligram of protein. The amount of enzyme activity in a sample is usually designated in "units," with one unit being the quantity of enzyme necessary to bring about the formation of 1 μM of product/min under some defined set of assay conditions. In another case, suppose you treat cells with a reagent, such as a hormone or drug, and want to determine if the specific activity of a particular enzyme was affected. This will again require you to determine the protein concentration. You will be determining the specific activity of proteins in some of the experiments in this book.

COMMON PROCEDURES FOR PROTEIN QUANTIFICATION

In this chapter, we examine several methods used to quantify total protein concentration. For each of the methods, we use chromogenic assays to construct a standard curve from samples containing known amounts of a purified protein, bovine serum albumin (BSA). We then determine the amount of protein in unknown samples with the same assays by comparing the results to those on standard curves obtained with BSA. Thus, the experiments are analogous to those performed in Chapter 2 to determine the concentration of PNP in unknown samples. However, there is an important difference in these groups of experiments. The measurement of PNP involved reading the unknown concentration of pure PNP from a standard curve that was constructed on the basis of known quantities of pure PNP. In contrast, when determining protein concentration, one rarely has a pure protein of unknown concentration. Instead, mixtures containing different proteins and other substances (carbohydrates, lipids, detergents, nucleic acids, etc.) are usually examined. Therefore, if the chromogenic reactions are affected by the composition of the proteins or by the other substances present, the resulting answer may not be an accurate determination of the total protein concentration. The presence of many compounds, such as sulfhydryl derivatives, detergents, carbohydrates, amine derivatives, nucleic acids, lipids, salts, etc., is known to affect the chromogenic reactions that we describe in this chapter. Thus, the results obtained with these procedures will be only an estimation of the true protein concentration. Nevertheless, these methods are commonly used because they are reproducible, simple to perform, and inexpensive, and the answers they give are useful. One exercise in this chapter is to compare the results obtained when several different chromogenic methods are used to determine the concentration of protein in unknown samples.

A more reliable way to determine the concentration of a pure protein is to conduct an amino acid analysis of the protein. This is typically performed by acid hydrolysis of the protein followed by separation of the amino acids and determination of their quantity. Certain amino acids such as tryptophan and cysteine are partially destroyed by acid hydrolysis, and acid hydrolysis converts asparagine and glutamine to aspartate and glutamate. Therefore, the concentrations of the other amino acids will provide the basis for the assessment of the protein concentration. It is not necessary to know the amino acid sequence of the protein for this method to be useful, but it is important that the protein be pure. Since accurate amino acid analysis requires special equipment, it is usually done at a specialized facility. Receipt of data takes at least 2 days, and the analysis costs about $40 per sample. Therefore, it is not realistic to use this method for routine determination of protein concentration. However, if you have a small amount of pure protein, determining its concentration by amino acid analysis may be useful. Then you can use known concentrations of the protein in the various chromogenic reactions and compare them with results obtained using BSA. From these results, you can calculate the appropriate factor for determining the true concentration of your protein with BSA as a standard in future experiments.

Another way to determine accurately the concentration of protein is to determine the dry matter weight after drying the sample at constant temperature to constant weight. This method is not rapid and is sensitive to the presence of impurities. For these reasons, this method is not frequently used.

COLORIMETRIC PROCEDURES FOR QUANTIFICATION OF PROTEINS

ABSORBANCE AT 280 AND 260 nm

The concentration of protein can be estimated by measuring the absorbance of protein-containing solutions at 280 nm (UV). This method is commonly used because it does not destroy the sample and it is very rapid. Indeed, in many cases, the absorbance at 280 nm of the eluate from chromatography columns is continuously monitored to determine when the proteins are eluting from the column. The minimum instrumentation required for this procedure is a spectrophotometer capable of measurements in the UV region of the spectrum and quartz cuvettes (glass cuvettes will not transmit 280-nm light). Most proteins have an absorption maximum at 280 nm, owing to the presence of tryptophan (W), tyrosine (Y), and phenylalanine (F). Since the amino acid composition of proteins varies greatly, the molar absorptivity will also vary greatly, depending on the content of these amino acids. Proteins that do not contain W, Y, and F will have no absorption maximum at 280 nm, while proteins that contain many W, Y, and F residues will have high molar absorptivities, with absorption maximum 280 nm. Thus, the method is not very accurate unless the protein is pure and the molar absorptivity is known, for example, by calibration with a sample of known concentration (e.g., an aliquot of the same material that had separately been analyzed by amino acid analysis). The absorptivity at 280 nm (extinction coefficient) of proteins is usually expressed as $E^{1\%}$ (10 mg/mL) or $E^{1mg/mL}$, where the latter may range from 0.0 to 10 cm^{-1}. Unless the protein is free of other compounds that absorb light at 280 nm, the results will be inaccurate. Nucleic acids are particularly troublesome because the purine and pyrimidine rings have absorption maxima near 260 nm with considerable absorption extending as high as 280 nm. If nucleic acids are the only contaminant, the protein concentration may be estimated by using the formula below (Groves et al., 1968), which corrects reasonably well for the nucleic acid content:

$$\text{protein (mg/mL)} = 1.55\,A_{280} - 0.76\,A_{260}$$

NOTE✔ **Methods that use UV light require spectrophotometers capable of operating in the UV range.**

BIURET METHOD OF PROTEIN DETERMINATION

In alkaline conditions, copper(II) is thought to bind to the peptide nitrogen of proteins (Fig. 3-1), and this complex absorbs light maximally at 550 nm. In addition, it is thought that in the Biuret reaction some of the Cu^{2+} is reduced to Cu^{1+}; this property is used in the bicinchoninic acid (BCA) protein determination. Since the copper reacts with the peptide bond, there is little interference by free amino acids, and the amino acid composition of the proteins is not very important. Certain buffers, such as Tris and ammonia, react with copper(II), so that they interfere with the assay. The interference by ammonia makes this assay impractical to use on protein samples obtained from ammonium sulfate precipitation because these samples contain high concentrations of ammonia. In addition, the assay is fairly insensitive, limiting its usefulness.

LOWRY METHOD OF PROTEIN DETERMINATION

The Lowry method (Lowry et al., 1951; Peterson, 1979) is one of the most commonly used methods of protein determination; it is inexpensive, easy to perform, very sensitive, and highly reproducible. Nevertheless, the assay has serious limitations: it is sensitive to a variety of contaminants, the standard curves are linear only at low protein concentrations, and the timing and mixing of reagents with the samples must be precise. It is essential to create a standard curve each time protein determinations are performed.

The Lowry reaction consists of the Biuret reaction (described above), followed by the reduction under alkaline conditions of the Folin–Ciocalteu reagent (phosphomolybdotungstate mixed acids). Copper ions facilitate the reduction process. The principal chromogenic groups are the peptide linkages in complex with copper (Biuret) and the blue-reduced molybdotungstates, which are largely reduced by Tyr, Trp, and polar amino acids. Therefore, the sensitivity of the test depends on the composition of the pro-

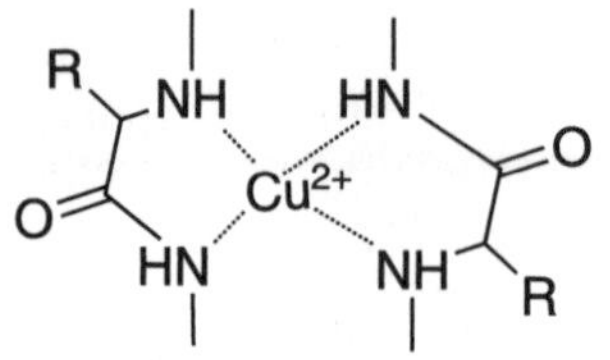

FIGURE 3-1 Putative cupric complex with peptide bond.

tein. The product of the reaction, heteropolymolybdenum blue, is strongly blue with an absorption maximum of ~750 nm. This wavelength is usually out of the range of interfering colors.

BCA METHOD OF PROTEIN DETERMINATION

The BCA reaction (Smith et al., 1985) is similar to the Lowry reaction, except that BCA is used in place of the Folin–Ciocalteu reagent (Fig. 3-2). Cu^{2+} is reduced to Cu^{1+} by protein in alkaline solution (the Biuret reaction). Two molecules of BCA chelate to a cuprous ion, resulting in an intense purple color with an absorbance maximum at 560 nm. The sensitivity of the method is similar to that of the Lowry method (20–2000 $\mu g\ mL^{-1}$ of the unknown sample), but it is not as susceptible to certain contaminants such as detergents. However, the reaction is more sensitive to interference from carbohydrates than the Lowry reaction is. The BCA method is not a true end point method, because the color continues to develop slowly over time. However, after an incubation of 30 min at 37 °C the color is sufficiently stable for reliable measurements (2.5% drift per 10 min). Some of the substances reported to interfere with the BCA method are the following: catecholamines, tryptophan, lipids, phenol red, cysteine, tyrosine, impure sucrose or glycerol, H_2O_2, uric acid, and iron.

DYE-BINDING METHOD (BRADFORD METHOD) OF PROTEIN DETERMINATION

The binding of the dye Coomassie® Brilliant Blue G-250 (Fig. 3-3) to proteins causes a shift in the absorption maximum of the dye from 465 nm

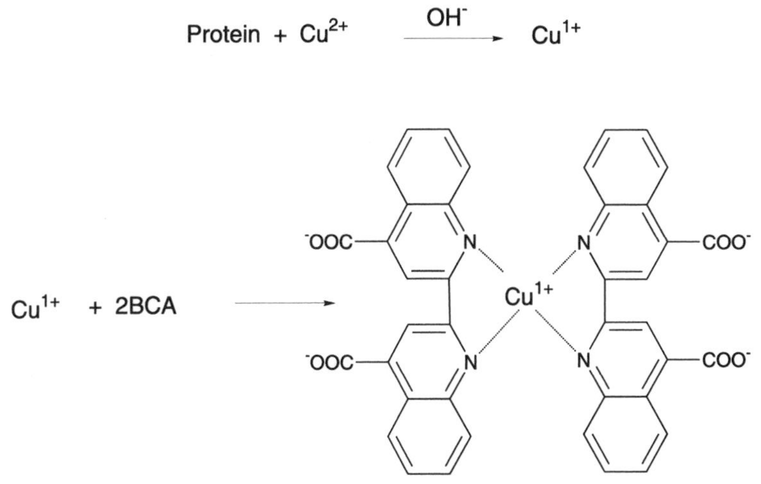

FIGURE 3-2 BCA reaction.

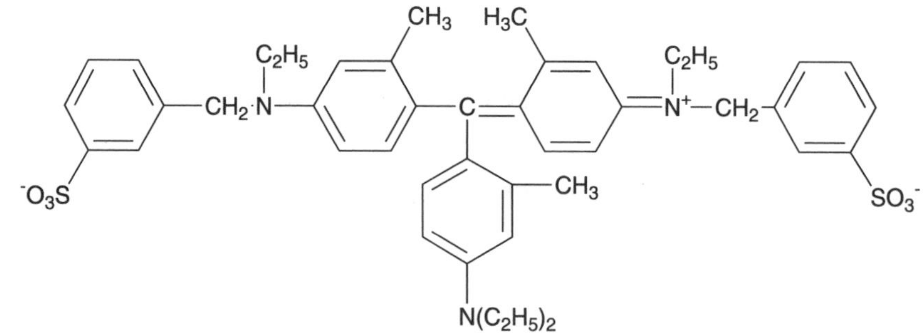

FIGURE 3-3   Coomassie® Brilliant Blue G-250.

(red) to 595 nm (blue) in acidic solutions (Bradford, 1976; Sadmark and Grossberg, 1977). This dye forms strong, noncovalent complexes with proteins via electrostatic interactions with amino and carboxyl groups and via van der Waals forces. Since the color response is nonlinear over a wide range of protein concentration, it is strongly recommended that a standard curve be run with each assay. The dye is prepared as a stock solution in phosphoric acid. The method is a simple one-step procedure in which the dye reagent is added to the samples and the absorbance is determined at 595 nm. The method is quite sensitive and accurate, but the reagent stains cuvettes and is somewhat difficult to remove. Dye-binding protein assays are compatible with most common buffers, chaotropic reagents such as 6 M guanidine·HCl, and 8 M urea, and preservatives such as sodium azide. However, high concentrations of detergents can interfere with this assay.

REFERENCES

Bradford, M. 1976. A rapid and sensitive method for the quantitation of microgram quantities of protein utilizing the principle of protein dye binding. Anal. Biochem. **72:**248.

Groves, W. E., F. C. Davis, Jr., and B. H. Sells. 1968. Spectrophotometric determination of microgram quantities of protein without nucleic acid interference. Anal. Biochem. **22:**195.

Lowry O. H., N. J. Rosebrough, A. L. Farr, and R. J. Randall. 1951. Protein measurement with the Folin phenol reagent. J. Biol. Chem. **193:**265.

Peterson, G. L. 1979. Review of the Folin phenol quantitation method of Lowry, Rosebrough, Farr and Randall. Anal. Biochem. **100:**201.

Pierce Chemical Co. 1997. Instructions for Coomassie® Protein Assay Reagent 23200, Rockford, Ill.

Sedmark, J. J., and S. E. Grossberg. 1977. A rapid, sensitive and versatile assay for protein using Coomassie brilliant blue G250. Anal. Biochem. **79**:544.

Smith, P. K., R. I. Krohn, G. T. Hermanson, A. K. Mallia, F. H. Gartner, M. D. Provenzano, E. K. Fujimoto, N. M. Goeke, B. J. Olson, and D. C. Klenk. 1985. Measurement of protein using bicinchoninic acid. Anal. Biochem. **150**:76.

EXPERIMENT 3-1

DETERMINATION OF PROTEIN CONCENTRATION

In this experiment, you will be provided with a protein standard (BSA) at a known concentration (10 mg/mL) and several samples containing protein at unknown concentrations. Some of these samples may contain pure BSA, while others may contain different proteins or mixtures of different proteins. In addition, some of the samples may be spiked with sulfhydryl-containing compounds, detergents, carbohydrates, etc., which will interfere with one or more of the assays. Your objective will be to measure protein concentration by the Biuret, Lowry, BCA, and Bradford methods for each of these samples. By comparing the results from these experiments, you may be able to deduce which samples are protein mixtures and which contain interfering substances.

The first step in the analysis is to prepare dilutions of the BSA standard and to construct a BSA standard curve for each assay method. Be sure to have several points within the sensitivity range of each assay method (see below). To determine the apparent protein concentrations in the unknowns, you should examine several dilutions of each. This step is necessary to ensure that at least one of the dilutions used results in a value falling within the linear portion of the standard curve. It is good experimental design to perform the assays on the standards and unknowns side by side and at the same time. This ensures that variables such as temperature, time, etc., are internally controlled. In addition, it is useful to perform the assays in duplicate, so that large pipetting errors may be readily recognized and experimental error can be ascertained. We will discuss the results obtained by the different methods and compare the results that were obtained from different groups using the different techniques. This will give you some real results to better understand the differences between precision, accuracy, and reproducibility.

For additional experiments, you may decide to test directly how substances such as detergents affect the validity of the tests. These experiments should be designed using the principles learned in the colorimetry experiment, as well as the principles explained in Chapter 1.

BIURET METHOD

MATERIALS

Biuret reagent

Dissolve 1.5 g of copper sulfate ($CuSO_4 \cdot 5H_2O$) and 6 g of sodium potassium tartrate in 500 mL of H_2O. Add 300-mL 10 % (wt/vol) NaOH and add water to bring the final volume to 1 L. Store the reagent in a plastic container in the dark. It will keep indefinitely if 1 g of potassium iodide is added to inhibit the reduction of copper.

PROCEDURE

1. Add 2.5 mL of Biuret reagent to 0.5-mL samples that contain ≤3 mg of protein.

2. Let stand at room temperature for 30 min. The unknowns and the standards should be treated side by side.

3. Measure the absorbance at 550 nm. Calibrate the instrument by using a blank that contains 0.5 mL of buffer or H_2O (instead of protein solution) plus 2.5 mL of Biuret reagent.

LOWRY METHOD

MATERIALS

1. Stock reagents

    Solution A: 1% wt/vol copper sulfate

    ($CuSO_4 \cdot 5H_2O$)

    Solution B: 2% wt/vol sodium potassium tartrate

EXPERIMENT 3-1

Solution C: 0.2 M NaOH.

Solution D: 4% sodium carbonate.

2. Lowry reagent I

Add 49 mL of solution D to 49 mL of solution C. Then add 1 mL of solution A and 1 mL of solution B. This is the copper-alkali reagent (**Lowry reagent I**), which must be prepared freshly for each lab session.

3. Lowry reagent II

Dilute the Folin–Ciocalteu reagent twofold (i.e., dilute with an equal volume of H_2O). This is **Lowry reagent II**.

PROCEDURE

1. Add 2.5 mL of Lowry reagent I to 0.5-mL samples containing ≤ 0.5 mg of protein.

2. Mix well and let stand at room temperature for 10 min.

3. Add 0.25 mL of Lowry reagent II.

4. Mix well *immediately* and let stand at room temperature for 30 min.

5. Measure absorbance at 750 nm (or with the Spectronic® 20, at 650 nm) against a blank consisting of 0.5 mL of sample buffer (or H_2O) processed as the samples.

BCA METHOD

MATERIALS

Stock solutions

BCA Solution A: 1% sodium BCA, 2% $Na_2CO_3 \cdot H_2O$, 0.16% sodium tartrate, 0.4% NaOH, and 0.95% $NaHCO_3$, and sufficient solid sodium bicarbonate to adjust the pH to 11.25 (stable at room temperature)

BCA solution B: 4% copper sulfate·$5H_2O$

Working reagent: Mix 50 volumes of BCA solution A with 1 volume of BCA solution B.

PROCEDURE

1. Make up a set of standards with BSA in the range of 20–2000 µg mL^{-1}.

2. Into labeled test tubes, add 0.1 mL of sample (or standard).

3. Add to these tubes 3 mL of BCA working reagent.

4. Mix well.

5. Incubate tubes at 37°C for 30 min.

6. Cool to room temperature and read the absorbance at 560 nm against a blank of sample buffer processed as above. Set up a standard curve.

BRADFORD METHOD

MATERIALS

Dye reagent

Use vigorous homogenization or agitation to dissolve 100 mg of Coomassie® Brilliant Blue G-250 in 50 mL of 95% ethanol. This solution is mixed with 100 mL of 85% phosphoric acid, diluted with H$_2$O to 1 L, and filtered. The reagent is stable for 2 weeks at room temperature. (The mixed reagent can be purchased from Pierce as Coomassie® Protein Reagent or from BioRad as BioRad Protein Assay mixture.)

PROCEDURE

1. Add 3 mL of dye reagent to 60 µL of protein sample containing 100–1500 micrograms of protein per milliliter.

2. Mix and let stand at room temperature for 5 min.

3. Read the absorbance at 595 nm against a blank consisting of 60 µL of sample buffer or H$_2$O and 3 mL of the dye reagent.

REAGENTS NEEDED FOR CHAPTER 3

✓ Copper sulfate: Sigma C6283, 250 g

✓ Sodium potassium tartrate: Sigma S6170, 500 g

✓ Potassium iodide: Sigma P2963, 100 g

✓ Sodium carbonate: Sigma S7795, 1 kg

✓ Sodium hydroxide: Sigma S0899, 500 g

✓ 2N Folin–Ciocalteu reagent: Sigma F9252, 500 mL

✓ BCA (4, 4′-dicarboxy-2, 2′-biquinoline): Sigma D8284, 10 g

✓ Sodium bicarbonate: Sigma S6297, 1 kg

Instead of preparing the BCA solution, you may purchase this reagent from Sigma or Pierce (see below). In that case, BCA and sodium bicarbonate are not required.

✓ BCA pre-mixed solution: Sigma B6943 or Pierce 23225

✓ Coomassie® Brilliant Blue G-250: Sigma B5113, 100 mg·L^{-1} of working solution

✓ 95% Ethanol

✓ Phosphoric acid

Instead of preparing the dye reagent, this reagent may be purchased from Sigma, Pierce, or BioRad (see below). In that case, Coomassie® Brilliant Blue G-250, 95% ethanol, and phosphoric acid are not required.

✓ Bradford reagent (pre-mixed): Sigma B6916, Pierce 23200, and BioRad 500-0006

4 Chromatography

INTRODUCTION

PRINCIPLES OF CHROMATOGRAPHY

Chromatography is a process used to separate molecules on the basis of a chemical property, such as molecular mass, charge, or solubility. The process uses a **stationary phase** and a **mobile phase**, which may be either liquid or gas. The sample, which is often a complex mixture of molecules, is passed over or through the stationary phase by the flow of the mobile phase, which is called the eluting solvent or eluant. Molecules with different physical properties partition differently between the stationary and mobile phases, resulting in a separation. Molecules that are strongly attracted to the stationary phase will be retarded or retained, relative to molecules that are not attracted strongly to the stationary phase. By choosing the appropriate stationary and mobile phases, it is possible to obtain effective separations of molecules that are only slightly different from each other.

TYPES OF CHROMATOGRAPHY

The same principles of separation, based on differential attraction to the stationary and mobile phases, apply to many different forms of chromatography. These forms differ in the chemical composition of the stationary and mobile phases and in the geometry of the chromatographic setup. In **paper chromatography**, the stationary phase consists simply of a piece of filter paper and the mobile phase travels through the dry filter paper by capillary action. In **thin-layer chromatography** (TLC), the stationary phase is a thin layer of silica, cellulose, or other inert material that coats the surface of a glass or plastic plate. The sample is spotted near the bottom of the plate and allowed to dry. The plate is then placed vertically into a closed tank with enough eluting solvent (the mobile phase) to wet the bottom of the plate. Capillary action draws the solvent upward to separate the various components in the mixture (Fig. 4-1).

In **column chromatography**, the stationary phase (also called the matrix or column packing) consists of a fine granular or bead-like matrix that is packed into a uniform tube referred to as a column. In the experiments described in this book, we will use column chromatography to purify enzymes from complex mixtures of proteins. Figure 4-2 illustrates some principles of column chromatography. The column matrix is packed in the chromatographic tube and is equilibrated with the desired solvent. The

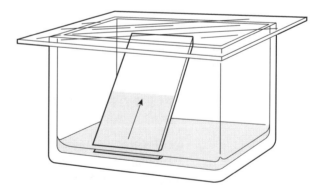

FIGURE 4-1   Thin-layer chromatography. The sample is applied to the plate at a position just above where the solvent pool reaches. Then the plate is placed into the closed tank and the sample is separated in the direction of the arrow. TLC is particularly useful for resolving small molecules; the solvent frequently consists of acetic acid/methanol/water in various proportions.

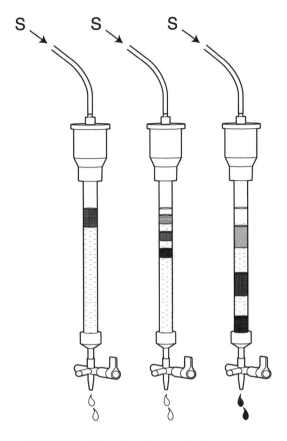

FIGURE 4-2 Principles of column chromatography. *(Left)* A sample containing three compounds is layered at the top of the column (see Fig. 4-6). Solvent (S) is fed to the top of the column to bring about elution of the sample. *(Middle)* As eluant is passed down through the column, the sample begins to separate. *(Right)* The first component begins to elute. The eluant and sample are then collected in a fraction collector (Figs. 1-10, 1-11, and 4-3).

sample, containing a mixture of substances, is added to the top of the packing in a small volume of solvent. The sample solution is allowed to percolate into the packed column. Additional solvent is added to the top of the column, and it also percolates into the column packing, displacing the sample molecules down the column. As the sample comes into contact with the packing particles, the differences in the attraction of the various molecules to the column matrix and to the solvent lead to fractionation of the various kinds of molecules in the sample.

The chemical and physical properties of the matrix are critical for the separation. The matrix must be robust, so that it does not change during the process, and uniform, so that the chromatography proceeds consistently.

The matrix must also provide a large surface area that is accessible to both the mobile phase and the sample molecules. Collecting small fractions of the effluent from the column permits different molecules in the starting sample to be separated from each other. In some applications, mobile phases of increasingly harsh character (**gradients**) are used to release selectively (elute) molecules from the stationary phase. A typical setup for gradient elution column chromatography in a research laboratory is shown in Fig. 4-3. The principles remain the same, but the solvent is pumped through the column, the eluant is monitored with a sensitive detector connected to a recording device, and the fractions are collected automatically with a fraction collector (also see Fig. 1-11).

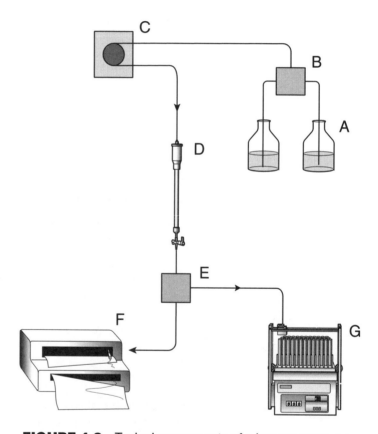

FIGURE 4-3 Typical components of a low-pressure gradient elution column chromatography system. The solvent bottles *(A)* contain the components of the solvent, which are combined in the mixer *(B)* to give a prescribed varying ratio of the two solvents. A pump *(C)* drives the eluant to the column *(D)* where separation occurs. A detector *(E)* monitors the eluant and the recorder *(F)* gives a record of the samples that pass through to the fraction collector *(G)*. The detector can be a photometer, a refractive index monitor, a fluorometer, or another suitable device.

Various chemical properties of biomolecules have been exploited for the purpose of effecting chromatographic separations. In this chapter, we discuss column methods commonly used with a liquid mobile phase: gel-filtration (size exclusion) chromatography, ion-exchange chromatography, hydrophobic-interaction chromatography, and affinity chromatography. We will use all of these methods, except hydrophobic-interaction chromatography, in experiments described in this book.

GEL-FILTRATION (SIZE EXCLUSION OR GEL PERMEATION) CHROMATOGRAPHY

Gel-filtration chromatography is a form of column chromatography in which molecules are separated on the basis of their molecular mass or, more properly, their Stokes radius. The **Stokes radius** is the effective radius a molecule has as it tumbles rapidly in solution. A long, extended molecule has a larger Stokes radius than a compact molecule of the same molecular mass. The stationary phase in gel-filtration chromatography consists of fine beads that contain pores of controlled size. The space between the beads is referred to as the void space or "outer voids," V_o. The space within the beads is referred to as V_i. The inert volume of the stationary phase is referred to as V_g. The mobile phase fills all of the space between the beads (V_o) and within the beads (V_i). The chemical and physical properties of both the stationary and mobile phases in gel-filtration chromatography are chosen to prevent, as nearly as possible, interactions of the proteins with the beads that are due to properties other than molecular size (e.g., to suppress hydrophobic or electrostatic interactions). This is in direct contrast to other forms of chromatography in which interactions with the stationary phase provide the basis of separation.

The sample is applied in a narrow band at the top of the column and then washed through the column by the mobile phase. Large molecules in the sample that cannot pass through the pores of the beads are excluded from the beads and are restricted to the outer voids; they elute from the column after an amount of the mobile phase equal to V_o has passed through the column. Molecules that are much smaller than the pores equilibrate with the entire liquid volume and elute at a volume equal to V_t, which is the sum of V_o and V_i. Molecules that are small enough to pass through some of the pores of the beads, however, elute at various volumes (V_e), depending on how small they are and what fraction of the pores of the beads are accessible to them. Such molecules are referred to as partially included. Molecules in the sample can be separated in order of their size by collecting fractions as the mobile phase is eluted through the column, with the largest molecules eluting first and the smallest last. Figure 4-4A shows the separation of excluded, partially included, and fully included molecules in a gel-filtration column.

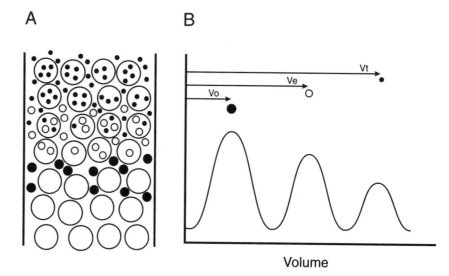

FIGURE 4-4 Principle of gel-filtration chromatography. *(A)* The large filled spheres represent molecules that are completely excluded from the pores in the beads. The small dots represent molecules that have access to all of the volume in the pores of the beads. Note that for completely included molecules (small dots), $V_e = V_t = V_0 + V_i$. Molecules that are partially included are indicated as open spheres. *(B)* shows an elution profile of the column. The various *V*-values are determined as shown.

A partition coefficient, designated K_{av}, can be calculated from these values with equation 4-1.

$$K_{av} = \frac{V_e - V_0}{V_t - V_0} \tag{4-1}$$

Semilogarithmic plots of the relationship of K_{av} to molecular weight are given in Fig. 4-5. Assuming that all molecules are of similar shape, the separation of molecules on the basis of molecular mass will be greatest in the molecular weight range for which the curve is linear, where K_{av} is ~0.2–0.8. The steeper the slope of the sigmoidal relationship in the fractionation range, the greater the resolving power of the stationary phase. Steepest slopes are found with beads of highly uniform pore size. However, gels with the greatest resolving power will have a more restricted range of molecular sizes that can be resolved.

STATIONARY-PHASE MATERIALS

A variety of gel-filtration stationary-phase materials are available for the separation of biomolecules in various size ranges. Some of the most com-

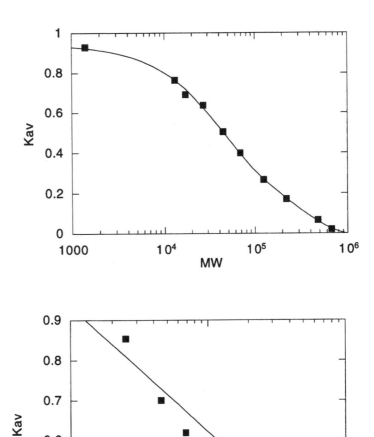

FIGURE 4-5 The sigmoidal dependence of K_{av} on the logarithm of the molecular mass. *(Top)* Separation of a series of proteins of known molecular mass on a Pharmacia Sephadex G-200 column is shown. *(Bottom)* Separation of a series of proteins on a BioRad BioGel A 1.5 M column is shown. All of the proteins used were partially included in the bottom panel and are in the resolving range of the resin.

monly used gels are listed in Table 4-1, along with the size range of molecules that are best separated by each material. Stationary phases consisting of cross-linked oligosaccharides, such as agaroses and dextrans, or cross-linked acrylamides are most common. As shown in Table 4-1, gel-filtration chromatography can be exploited for separations of molecules with molecular masses ranging from $\sim 10^2$ to $\sim 10^6$ Da.

TABLE 4-1
Low-Pressure Gel-filtration Chromatography Media

GEL PRODUCT	EXCLUSION SIZE	MW RANGE	MATERIAL	SOURCE
A15M	15×10^6	8×10^4–12×10^5	Agarose	Biorad
A5M	5×10^6	4×10^4–4×10^5	Agarose	Biorad
A0.5M	5×10^5	1×10^4–5×10^4	Agarose	Biorad
Sepharose 4B	2×10^7	8×10^3–12×10^6	Dextran	LKB
Sepharose 6B	1×10^7	4×10^4–4×10^6	Dextran	LKB
Sephadex G200	5×10^5	4×10^4–2×10^5	Dextran	LKB
Sephadex G150	2×10^5	2×10^4–12×10^4	Dextran	LKB
Sephadex G100	1×10^5	8×10^3–8×10^4	Dextran	LKB
Sephadex G75	6×10^4	4×10^3–4×10^4	Dextran	LKB
Sephadex G50	4×10^4	2×10^3–3×10^4	Dextran	LKB
Sephadex G25	3×10^4	0.8×10^3–2×10^4	Dextran	LKB
Sephadex G10	8×10^3	0.4×10^3–6×10^3	Dextran	LKB
P100	1×10^5	8×10^3–8×10^4	Acrylamide	BioRad
P60	6×10^4	3×10^3–4×10^4	Acrylamide	BioRad
P30	4×10^4	2×10^3–2×10^4	Acrylamide	BioRad
P10	2×10^4	2×10^3–1×10^4	Acrylamide	BioRad
P6	6×10^3	0.8×10^3–4×10^3	Acrylamide	BioRad
P2	8×10^3	1×10^2–2×10^3	Acrylamide	BioRad

LIMITATIONS OF GEL-FILTRATION CHROMATOGRAPHY

Certain problems inherent with the process make gel filtration less than ideal for purifying biomolecules. First, even if the sample is applied as a narrow band to the column, the turbulence associated with passing the mobile phase through the column results in a broadening of the bands of the applied molecules as they travel through the column. Thermal diffusion of the molecules and frictional effects from the glass walls also contribute to the broadening of bands of the eluting molecules, such that even under optimal conditions, dilution of the desired species is about threefold. Often, the investigator must be willing to accept dilution of the sample of as much as 10-fold. The broadening of the bands works against the purification of the individual components of the sample, since the broad peaks can partially overlap each other. Thus, it is advisable to apply the sample (usually fairly concentrated) as a narrow band on the top of the column. This can limit the usefulness of the method, because the sample may in some cases be difficult to concentrate sufficiently to make the method optimal. The technique is most useful when the desired component in a mixture is much larger or much smaller than the majority of the contaminants. Gel-filtration chromatography is frequently used as a late step in a purification scheme that may consist of several different fractionation methods that exploit other chemical properties of the desired molecule. It is also useful for combining the separation on the basis of size with an equilibration with a new buffer, such as described below.

COMMON USES OF GEL-FILTRATION CHROMATOGRAPHY

Gel-filtration chromatography can be used to rapidly exchange the solvent in which large molecules are dissolved. For this purpose, you should choose a stationary phase that completely excludes the large molecules (i.e., the large molecules all elute in V_0), and equilibrate the column in the desired new solvent. The large molecules in the undesired solvent are then loaded onto the column and eluted with the new solvent. The large molecules elute ahead of the original solvent and come off the column in the void volume (V_0) in the desired new solvent. This procedure is also useful for desalting a sample after an ammonium sulfate precipitation step (Chapter 6).

Another common use of gel-filtration chromatography is to determine the approximate molecular mass of a compound. Since the behavior of a molecule in gel filtration is a function of the Stokes radius of the molecule and therefore is dependent on the shape of the molecule, the determination of mass by this method will be only approximate. With proper care, a gel-filtration column will yield reproducible results, so that the column can be calibrated by determining elution volumes of proteins with known molecular masses. A plot of K_{av} versus log(MW) for the known proteins will yield a calibration curve from which the molecular mass of molecules of interest can be estimated (Fig. 4-5). Alternatively, a plot of normalized elution volume (either V_e/V_t or V_e/V_0) versus log(MW) can be used for the calibration curve. Both methods are frequently mentioned in published research. The column is usually run separately with each of the proteins, and a curve such as shown in Fig. 4-5 is constructed from the composite data. Another way to perform this type of application is to mix the molecule of interest and several other molecules of known size. The mixture is then applied to a column, and the volume at which each of the molecules elutes from the column is determined. If these proteins are enzymes, it may be possible to assay the activity of each enzyme to positively identify each elution peak.

GUIDELINES FOR GEL-FILTRATION CHROMATOGRAPHY

The following guidelines are offered to help obtain the best separations in gel-filtration chromatography.

1. The column should be vertical. After setting up the column, stand back and look at it from two directions, at right angles to one another, to see whether it is vertical. In more exacting research applications, you would use a carpenter level to ensure that the setup is vertical. Alternatively, you could drop a string with a small weight on it, which acts as a plumb, near the column to help you judge whether it is vertical.

2. Prevent air bubbles from forming in the column bed, and pack the matrix material uniformly with no troughs or channels. The bed material should never be allowed to go dry (i.e., the elution buffer should not be allowed to drop below the top of the column bed). If the bed does go dry or bubbles are obvious in the column, the column will have to be emptied and repacked.

3. Choose a column of appropriate size for the application. The volume of the column bed should be 20- to 50-fold that of the sample to be applied. Usually, tall thin columns with lengths 20–40 times their diameter are used. Although in theory the shape of the column should not matter, in practice, tall thin columns are used so that small deviations in sample application and flow rate have little effect on the results (see Scopes, 1987)

4. Try to maintain a constant temperature with no drafts on the column so that turbulence and thermal gradients do not disturb the flow.

5. An effective way to pack a small column, as you will do in the laboratory exercise, is as follows (also see Chapter 1 and Fig. 1-11). Fill the column about one third of the way with the elution buffer. Open the valve at the bottom of the column, so that the liquid starts to drip out. Then, with a Pasteur pipette, add a dilute slurry of the column matrix material to the top of the column while stirring gently. (Gel-filtration columns perform best if packed while running, because this prevents the separation of beads of different sizes during settling of the bed.) Keep adding dilute slurry to the top of the column and occasionally stir gently the unsettled portion of the bed with a Pasteur pipette or stir rod. The objective is to produce a uniform bed of the appropriate height, with as few additions of slurry as possible. In a research application, the desired amount of slurry for the whole bed would be added at one time by placing a large funnel on the top of the column. If you think that you have added too much slurry to the column, you can remove some that has not yet settled without harming the settled bed. The upper surface of the slurry should be flat. After the desired bed has been obtained, equilibrate it with several column volumes of elution buffer.

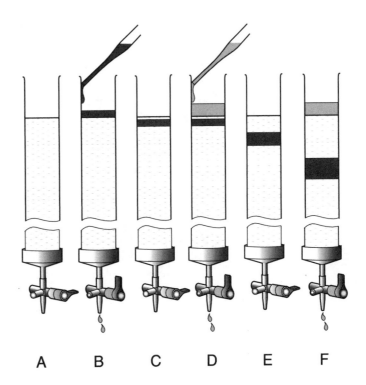

FIGURE 4-6 Procedure for manually loading gel-filtration columns. See text for description.

6. The sample should be applied in a narrow band at the top of the column as shown in Fig. 4-6. This is performed as follows: Stop adding buffer to the top of the column, and allow the buffer to flow until the level of the elution buffer reaches the top of the settled bed without allowing any of the bed to go dry. Stop the flow by closing the valve at the bottom of the column. Next, add the sample gently to the wall of the column above the bed. Touch the pipette tip to the glass surface, and make a circular motion with the tip as you slowly eject the sample, so that it is evenly loaded across the top of the column (Fig. 4-6, column B). Try to avoid disturbing the settled bed. After the sample has been loaded onto the top of the column, open the valve and let the sample flow into the column, again closing the valve as the last of the sample enters the bed (Fig. 4-6, column C). Next, gently add a small volume of the elution buffer, about equal to the volume of the sample, to the top of the column with the same technique used to load the sample (Fig. 4-6, column D). Open the valve at the bottom of the column, and allow this buffer to enter the column as you did with the sample, again closing the valve before the column runs dry (Fig. 4-6, column E). Finally, add a larger quantity of buffer over the bed (doing so gently at first so that you do not disturb the bed), and open the valve at the bottom to begin the separation (column F). As the buffer is running, keep adding buffer to the top of the column to prevent it from going dry.

In order to isolate the various molecules as they come off of the column, you will collect fractions. Working in pairs is the best way to do this. Set up a rack of clean empty tubes under the column (Fig. 1-10). Start collecting fractions as soon as the sample is applied to the column. There are two ways that fractions can be collected. You can put a desired volume, for example, 1 mL, into a test tube. Then you can hold this tube behind the tube into which you are collecting sample, and switch to a fresh tube when the volume collected is equal (by eye) to the volume in the tube containing 1 mL. Another way is to collect a fixed number of drops, for example, 25 drops, into each collection tube. Then you can measure the volume corresponding to 25 drops. The problem with this method is that the size of the drops is influenced by several factors, such as the protein concentration of the liquid, and thus can change during the chromatographic run. Nevertheless, this method will work well for the applications described in this book. In a research application, you might collect fractions more precisely by using a device designed to collect a fixed volume into each tube or by pumping the eluting solvent at a constant rate with a metering pump and collecting fractions at constant timed intervals.

AFFINITY CHROMATOGRAPHY

PURPOSE AND PRINCIPLES

Affinity chromatography is frequently used to purify specific biological macromolecules such as proteins and nucleic acids. In nature these biological macromolecules are often involved in highly specific interactions with other proteins or nucleic acids, or with small molecules, such as substrates. The specificity of these physiologically relevant interactions derives from complementary surfaces on the interacting molecules and often involves several different types of interactions, such as hydrogen bonding and hydrophobic interactions. The precise spacing of the reactive groups on the interacting molecules results in great specificity of binding. When these biologically relevant interactions between molecules are exploited, the selectivity of affinity chromatography is potentially the highest of any form of chromatography. Indeed, the biologically relevant interactions exploited by affinity chromatography have been optimized by millions of years of evolution. In affinity chromatography, the stationary phase contains an immobilized molecule that binds with high specificity to a single protein or a small number of similar proteins, nucleic acids, or other molecules. When a crude mixture of macromolecules, such as a whole cell extract, is brought into contact with the stationary phase, most of them are not attracted to the stationary phase and are easily washed away. A few macromolecules interact with high specificity with the immobilized molecule and are retained on the stationary phase. After washing the noninteracting molecules away, the desired molecules are eluted with specific substances, often resulting in highly purified material. Highly specific elution of the desired macromolecule from the stationary phase is usually effected by adding to the eluting buffer a gradient of the same molecule that was immobilized; thus, it competes for the binding sites on the macromolecule and displaces it. Affinity chromatography is frequently an excellent choice for a first or an early step in purifying a protein or nucleic acid from crude cell extracts (Fig. 4-7).

Affinity chromatography can sometimes be useful, even if you know only a little about the molecular weight, hydrophobicity, charge, etc., of a protein. For example, if you are searching for an enzyme that has a particular activity, it may be possible to construct an affinity column with an attached ligand that is identical or similar to the substrate. This affinity column may selectively remove the desired enzyme from the mixture on the basis of the strong interaction of enzyme and the immobilized substrate. Then you can elute the enzyme with the appropriate substrate or ligand.

PRACTICAL CONSIDERATIONS

Many different affinity chromatography approaches have been devised. One of the most common approaches is to couple chemically a small mol-

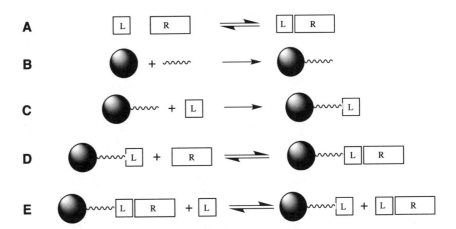

FIGURE 4-7 Principles of affinity chromatography. *(A)* The biological binding reaction. *(B)* An inert spacer arm is chemically attached to the column matrix. *(C)* The ligand is chemically immobilized to the spacer arm to produce an affinity matrix. *(D)* The crude sample is passed through the column, resulting in the adsorption of the desired protein to the immobilized ligand. The column is then washed with elution buffer to remove impurities that do not bind to the immobilized ligand. *(E)* The sample is eluted from the affinity matrix by including sufficient free ligand in the elution buffer to compete for the sample binding sites.

ecule that is the substrate of an enzyme (or, better still, an analog of the substrate that binds the enzyme but is not altered by it) to an inert material such as agarose, dextran, or polyacrylamide. These matrixes can be formed into beads with a large number of reactive groups. The inert matrix material is selected, so that it provides a stable open gel structure, which large proteins can penetrate, and a large surface area to which the affinity ligands can be linked. Linkage is usually to free hydroxyl groups on the matrix material, which confers to an affinity resin both high capacity and good flow characteristics. Often a spacer arm is used to facilitate the interaction of the immobilized ligand with the target protein in the mobile phase (Fig. 4-8).

The enzyme to be purified may bind to the immobilized substrate with high specificity and can often be eluted from the column by adding a high concentration of the substrate or substrate analog to the elution buffer. In other cases, the interaction of the enzyme with the immobilized ligand may be so strong that elution of the enzyme from the column requires partially or completely denaturing conditions. In experiment 4-2, you will purify the enzyme β-galactosidase from crude extracts by affinity chromatography with a stationary phase that contains an immobilized substrate analog. You will elute the β-galactosidase under harsh conditions involving high pH.

A variety of affinity chromatography matrixes containing immobilized small molecules that are substrate analogs for various enzymes are commercially available, e.g., the column matrix described in the laboratory exercise to purify β-galactosidase. Many other matrixes that are coupled to amino acids, nucleotides, nucleotide analogs, coenzymes, etc.,

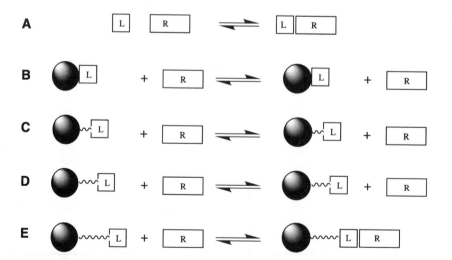

FIGURE 4-8   Use of spacer arms. *(A)* The biological reaction; *(B)* ligand coupled directly to matrix does not bind R; *(C and D)* ligands coupled to the matrix via short spacer arms do not bind R; and *(E)* a ligand coupled to a long spacer arm (10–12 atomic bonds) binds R well.

are also available. If the desired affinity resin is not available, you can make your own affinity chromatography media by using various commercial kits. A large amount of literature is available to help you design the process (Cuatrecasas, 1970; Steers et al., 1971; Steers and Cuatrecasas, 1974; Cuatrecasas and Anfinsen, 1971). In the description of Experiment 4-2, the purification of β-galactosidase, Fig. 4-16 shows the chemical steps involved in synthesizing the affinity matrix. The chemistry used to couple covalently the molecule of choice to an inert material will, of course, depend on the nature of the molecule to be coupled. A variety of "activated" dextran beads are available that permit different kinds of chemical coupling for synthesizing affinity chromatography media (Fig. 4-9).

GENE FUSION APPROACHES

Modern molecular biology techniques have been devised to take advantage of affinity chromatography to purify almost any protein. One common approach is the gene fusion technique, in which the gene encoding the protein of interest is fused to a second gene encoding a protein that is easily purified by affinity chromatography; this second protein is called the **"tag."** The vector that one uses for such cloning is a plasmid containing the tag sequence and is constructed with particular restriction sites that permit fusing the gene of interest to the tag gene in the same reading frame. This gene fusion thereby encodes a large protein consisting of the protein of interest fused to the tag protein (Fig. 4-10). Often the fusion vector is designed to introduce a short **"linker"** amino acid sequence between the protein of interest and the tag protein. This linker region can contain amino acid sequences that are cleavage sites for particular pro-

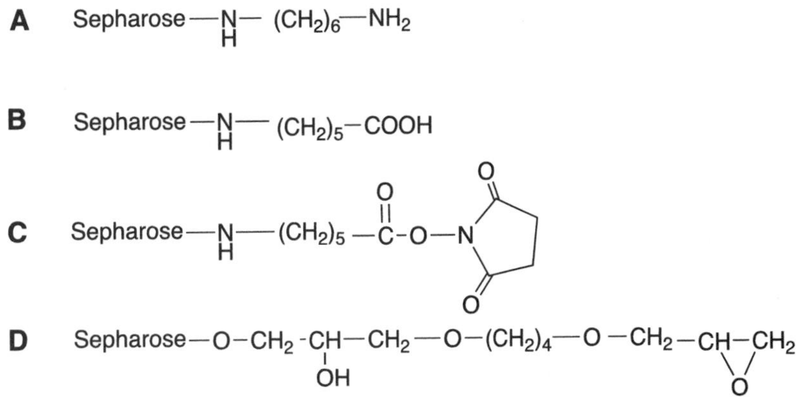

A   Sepharose—NH—$(CH_2)_6$—NH_2

B   Sepharose—NH—$(CH_2)_5$—COOH

C   Sepharose—NH—$(CH_2)_5$—C(=O)—O—N (succinimide)

D   Sepharose—O—CH_2-CH(OH)—CH_2—O—$(CH_2)_4$—O—CH_2—CH—CH_2 (epoxide)

FIGURE 4-9 Partial structures of *(A)* AH-Sepharose, *(B)* CH-Sepharose 4B, *(C)* activated CH-Sepharose 4B, and *(D)* epoxy-activated Sepharose 6B. Courtesy of Pharmacia.

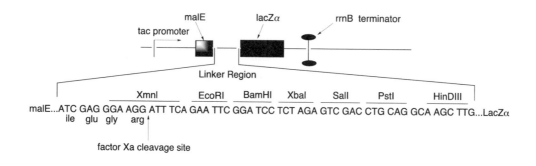

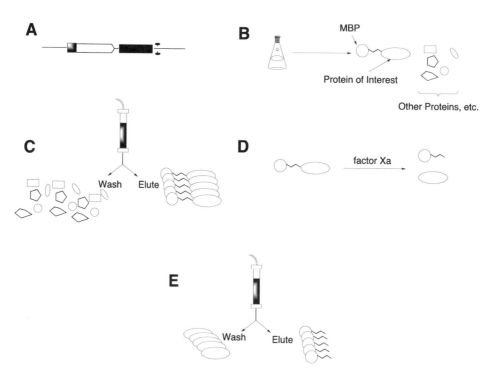

FIGURE 4-10 Strategy for construction of a gene fusion to *malE* and purification of the fusion protein. *(Top)* The *pmal* plasmid vector (New England Biolabs) for construction of gene fusions to the *malE* gene encoding the maltose-binding protein (MBP). The target gene is cloned into the linker region (sequence is expanded just below the vector schematic) just downstream from the *malE* gene and in the same reading frame, forming a fusion gene. The *malE* gene lacks its termination codon(s) so the downstream genes are translationally coupled. Note that the linker region contains a factor Xa cleavage site positioned upstream from the multiple cloning sites for inserting the target gene. The transcription of the fusion gene may be induced from the strong *tac* promoter with IPTG or lactose (Chapter 11), and the *rrnB* terminator downstream from the fusion gene provides efficient transcriptional termination, which is necessary to prevent the overexpression of downstream genes. The *lacZα* gene permits blue/white screening of putative clones (see Chapter 11). *(Bottom)* Strategy for expression and purification of the target protein. *(A)* The fusion gene is constructed and introduced into bacteria; *(B)* the bacterial culture is grown, the fusion gene is expressed, the cells are harvested, and a crude extract is made; *(C)* the crude extract is subjected to affinity chromatography on an amylose column (New England Biolabs), resulting in great purification of the fusion protein; *(D)* the purified fusion protein is cleaved with factor Xa protease, liberating the target protein from the maltose-binding protein; *(E)* the mixture is rechromatographed on the amylose column to separate the target protein from the MBP.

teases. In addition, the plasmid vector can be designed to have a strong **promoter** that provides a high level of **transcription** of the fusion gene upon induction (Fig. 4-10).

The strategy used to purify a protein of interest when fused to the **maltose-binding protein** of *E. coli* is illustrated in Fig. 4-10. A bacterial culture is grown until the cells reach midlogarithmic phase, and then the expression of the plasmid-borne fusion gene is induced. Several hours later, when the cells contain many copies of the fusion protein, the cells are harvested and broken open. The debris and particulate material are removed by centrifugation to clarify the extract, and the crude extract (supernate), containing several thousand different proteins, is passed over a column containing an immobilized maltose derivative. The maltose-binding fusion protein product binds to this column, but almost all of the other cellular proteins pass right through the column and are easily washed away (Fig. 4-10). After all of the proteins that do not bind the column tightly are washed away, the maltose-binding protein fused to the protein of interest can be eluted by adding buffer that contains maltose to the column. This results in an almost pure maltose-binding fusion protein. To obtain the desired protein of interest from the fusion protein, the fusion protein can be cleaved with the protease (e.g., factor Xa) that cuts in the linker region. After inactivation of the protease the digestion mixture can be passed again over the column that contains the immobilized maltose derivative. This column will retain the liberated maltose-binding protein tag, but the protein of interest, separated from the tag by protease digestion, flows right through the column. All that remains to be done then is to purify the protein of interest away from the inactivated protease. Many different proteins have been purified by this strategy. A similar strategy uses fusion to **glutathione *S*-transferase** instead of fusion to the maltose-binding protein.

METAL-CHELATE CHROMATOGRAPHY

A widely used method to purify many different proteins is metal-chelate chromatography. This method is based on the chelation of metals such as nickel. The most commonly used approach takes advantage of the fact that stretches of histidine residues bind tightly to metals such as nickel. The gene of interest is genetically manipulated to encode a polyhistidine tag, usually six to eight histidines in length, at either the N-terminal or the C-terminal end of the protein. The presence of a few extra histidine residues at the C-terminal or N-terminal end usually does not deleteriously affect the activity of the protein. The fusion protein containing the histidine tag is then passed through a column containing immobilized nickel to which it adheres; then it is specifically eluted from this column by imidazole, which competes with histidine for the immobilized nickel.

ANTIGEN-ANTIBODY (IMMUNOAFFINITY) CHROMATOGRAPHY

Another type of affinity chromatography that takes advantage of the great specificity and avidity of antigen-antibody interactions is immunoaffinity chromatography. Indeed, affinity chromatography is the method of choice to purify specific antibodies from serum. For this purpose, the highly purified antigen is coupled to the stationary phase. After adsorption of the antibody to the coupled matrix, elution of the specific antibody often requires harsh conditions, such as those resulting in denaturation of proteins. Fortunately, most antibodies can be renatured after elution to regain activity. If antibodies to a protein of interest are available, these antibodies may be coupled to an inert matrix to form an affinity medium. However, in many cases this approach is not very useful to obtain active enzymes, because the interaction of antibody and antigen is so strong that the very harsh conditions required to elute the protein of interest from the immobilized antibody destroy the desired activity. However, Thompson et al. (1992, 1994) have shown that it is possible to select monoclonal antibodies that bind proteins with great specificity but can nevertheless be made to release the protein under fairly gentle conditions. Thus, the method of immunoaffinity chromatography may be further exploited in the future.

RNA POLYMERASE CHROMATOGRAPHY

The strong protein-protein interactions between **RNA polymerase II** of eukaryotic cells and the many transcription factors that regulate its activity have been exploited by Burton et al. (1986) to purify these regulatory factors by affinity chromatography. This has allowed the discovery of several transcription factors that bind RNA polymerase II. Burton et al. purified RNA polymerase II and covalently coupled it to an inert support. Extracts of nuclear proteins were then passed over the affinity matrix, and after the proteins that did not adhere to immobilized RNA polymerase were washed away, proteins that bound tightly to RNA polymerase were selectively eluted by high salt.

POLY-(U) CHROMATOGRAPHY

The purification of mRNA from eukaryotic cells is a first step in the construction of cDNA gene libraries that are commonly used to clone genes. The purification of mRNA is best accomplished by affinity chromatography on immobilized poly-(U) or poly-(dT). The method takes advantage of the fact that mRNA, but not other forms of RNA or DNA, contains long tracts of poly-(A) at the 3′ end. A very tight interaction of mRNA with the stationary phase containing poly-(U) occurs under conditions that favor Watson-Crick base pairing of complementary strands of nucleic acid. After contaminants are washed away, the mRNA is specifically eluted by conditions that prevent Watson-Crick base pairing, such as the presence of formamide, or a low pH.

DNA-AFFINITY CHROMATOGRAPHY

Immobilized nucleic acids are also used to purify proteins by affinity chromatography. Alberts et al. (1968) developed the method of purifying DNA-binding proteins on DNA cellulose; their studies on the mechanism of replication of the DNA of bacteriophage T4 constitute one of the milestones of biochemistry and molecular biology. DNA affinity chromatography is still widely used today, with agarose constituting the inert medium of choice for immobilization of DNA. After washing the column with buffer, elevating the salt concentration can selectively elute proteins that bind tightly to DNA.

DNA SEQUENCE-SPECIFIC CHROMATOGRAPHY

Kadanaga and Tijan (1986) developed an extension of DNA-affinity chromatography. This approach permits affinity purification of proteins that bind particular DNA sequences. Short, double-stranded DNA oligonucleotides containing a particular nucleotide sequence are coupled to an inert material by means of a spacer arm. After passing the complex mixture of proteins through the column and washing thoroughly, elevating the salt concentration can cause proteins that tightly bind the immobilized DNA sequence to be eluted. This method has been used to purify many different proteins that bind with high specificity to particular DNA sequences.

DYE-BINDING CHROMATOGRAPHY

Many dyes generated by the textile industry are large molecules with considerable conformational flexibility and many side chains. The flexibility of these dyes permits them to adopt conformations that are similar to many small molecules and protein side chains. They are often good mimics for nucleotides, such as ATP or NADH. Thus, in many cases, these dyes bind to proteins and can serve as competitive inhibitors of enzymatic activity. Numerous dyes have been used in affinity chromatography, and commercially available reactive dye affinity media are available. Although the specificity of dye chromatography is rarely as great as true affinity chromatography, the ease of the technique and the ready availability of the chromatography media have made it a useful method for the purification of many proteins.

ION-EXCHANGE CHROMATOGRAPHY

PRINCIPLES OF ION-EXCHANGE CHROMATOGRAPHY

Ion-exchange chromatography separates molecules on the basis of their charged groups, which cause the molecules to interact electrostatically with opposite charges on the stationary-phase matrix. Therefore, the procedure is limited to purification of ionizable molecules. The stationary phase carries ionizable functional groups coupled to an inert matrix material. Because of the principles of electroneutrality, these immobilized charges are electrostatically associated with exchangeable **counterions** from the solution. Charged molecules to be purified compete with these counterions for binding to the charged groups on the stationary phase and are thereby retarded on the basis of their charge. Different types of molecules will bind to the matrix with affinities that depend on both the conditions used and the types and number of individual charged groups. These differences lead to resolution of various molecule types by ion-exchange chromatography. In a typical use of ion exchange chromatography of a protein, a mixture is applied to the column. After molecules that do not bind are washed away, conditions can be gradually adjusted, such as by increasing the concentration of a simple counterion or by altering the pH, to release the molecule of interest from the stationary phase. Molecules with different charges will elute at specific points in the chromatography as adjustments are made, and the protein of interest can be separated from many others.

Ion-exchange chromatography is named on the basis of the exchangeable counterion. When the stationary phase bears a positive charge and the exchangeable ion is an anion, the process is referred to as **anion-exchange chromatography**. When the stationary phase bears a negative charge and the exchangeable ion is a cation, the process is referred to as **cation-exchange chromatography**. We will not use cation-exchange chromatography as part of the experimentation in this book, but we will use anion-exchange chromatography with diethylaminoethane (DEAE) groups linked to a cellulose column matrix to purify *E. coli* alkaline phosphatase as described later in this book. In this section, we will discuss some of the principles and types of ion exchange chromatography.

PRACTICAL CONSIDERATIONS OF ION-EXCHANGE CHROMATOGRAPHY

Nearly any molecule that has an exposed ionizable group can be purified by ion-exchange chromatography. Whether an ionizable group is charged depends on the pH of the solvent and the pK_a of the ionizable group (e.g., the carboxyl groups of amino acids have pK_a values in the range of 2–5, the imidazole group of histidine has a pK_a of 6, the SH group of cysteine has a pK_a of 8.3, the α-amino groups of amino acids have pK_a' values of 9–11, the phenolic group of tyrosine has a pK_a of 10, the ε-amino group of lysine has a pK_a of ~10.5, and the guanidino group of arginine has a pK_a of 12.5). Proteins consist of many different amino acids, and the overall charge is caused by the composite effect of many different ionizable groups. The pH at which a protein is uncharged is called the **isoelectric pH**, and is abbreviated **pI**; the pI of most proteins is in the range of 5–9. Ion-exchange chromatography of proteins is usually performed at least 1 pH unit away from the pI of the protein of interest to assure that it is charged. When the pH for chromatography is below the pI, the molecule will be positively charged and you would use a cation-exchange resin; when the pH is above the pI, the molecule will be negatively charged and you would use an anion exchange resin. Since interactions of ion-exchange groups with proteins depend on the surface (accessible) charges of the latter, even a protein at its isoelectric pH may bind to the column matrix. The interactions in ion-exchange chromatography are often quite complex.

Chromatography with weak ion-exchange resins containing DEAE or carboxymethyl (CM) groups (Table 4-2) is useful for separating molecules that differ only slightly in charge. The stationary phase contains an excess of ionizable groups, so many that the molecules can be "captured" and "released" many times as they pass through the column. The equilibrium distribution of molecules between the solvent and the stationary phase is affected by the presence of charged groups on their surface (which can be controlled by the pH), and by the presence of alternative counterions. At a given pH and concentration of counterions, molecules will be characteristically retarded and will be eluted from the stationary phase in roughly an inverse order of charge. Elution at a fixed concentration of counterion is referred to as **isocratic** elution. Although all charged compounds will eventually be eluted by isocratic elution (because all molecules are in equilibrium), elution of molecules that are highly ionized may require a very large volume of elution buffer, especially if the alternative counterions in the elution buffer compete poorly with the molecule of interest for binding to the immobilized charged groups. By altering the mobile phase, e.g., increasing the counterion concentration or changing the pH, the equilibrium distribution of molecules bound to the column versus those in the mobile phase can be shifted, allowing the selective elution of bound molecules in fairly small volumes of elution buffer.

Increasing the concentration of the counterion will elute proteins as the counterions compete with the proteins for ions of opposite charge on the stationary phase (Fig. 4-11). For example, Na^+ and H^+ are counterions frequently used for cation-exchange chromatography, and Cl^- and OH^- are commonly used for anion-exchange chromatography. In practice, proteins are usually fractionated by allowing adsorption to the stationary phase at a low concentration of KCl or NaCl and then are eluted from the stationary phase by increasing the concentration of KCl or NaCl in the elution buffer.

TABLE 4-2		
Ion Exchange Groups Used in the Purification of Proteins		
FORMULA	**NAME**	**ABBREVIATION**
Strong anion		
$-CH_2N^+(CH_3)_3$	trimethylaminoethyl	TAM
$-C_2H_4N^+(C_2H_5)_3$	triethylaminoethyl	TEAE
$-C_2H_4N^+(C_2H_5)_2CH_2-CH(OH)CH_3$	diethyl-2-hydroxypropylamino-ethyl	QAE
Weak anion		
$-C_2H_4N^+H_3$	aminoethyl	AE
$-C_2H_4N^+H(C_2H_5)_2$	diethylaminoethyl	DEAE
Strong cation		
$-SO_3^-$	sulpho	S
$-CH_2SO_3^-$	sulphomethyl	SM
$-C_3H_6SO_3^-$	sulphopropyl	SP
Weak cation		
$-COO^-$	carboxy	C
$-CH_2COO^-$	carboxymethyl	CM

When the concentration of the counterions is continuously varied during the elution process, it is referred to as **gradient elution,** whereas when the concentration of counterions is altered in a stepwise fashion during elution, this is referred to as **step elution.** The pH of buffer adjacent to the ion-exchange groups on the stationary phase will generally be different from that of the bulk solvent. This is due to the **Donnan effect** in which the ionized groups either attract or repel protons. Anion-exchange groups repel protons because they are positively charged; thus, the pH of the buffer near the column matrix (where the protein is binding) is ~1 unit higher than that of the bulk buffer. This phenomenon can lead to "mysterious" losses in protein activity.

PROCEDURES FOR ION-EXCHANGE CHROMATOGRAPHY

Ion-exchange chromatography may be performed in bulk, on thin layers with glass or plastic plates coated with the stationary phase, or, more commonly, in chromatography columns. In both column and thin-layer formats, the molecule of interest is bound and released many times as it progresses down the length of the stationary phase. Unlike what occurs in gel-filtra-

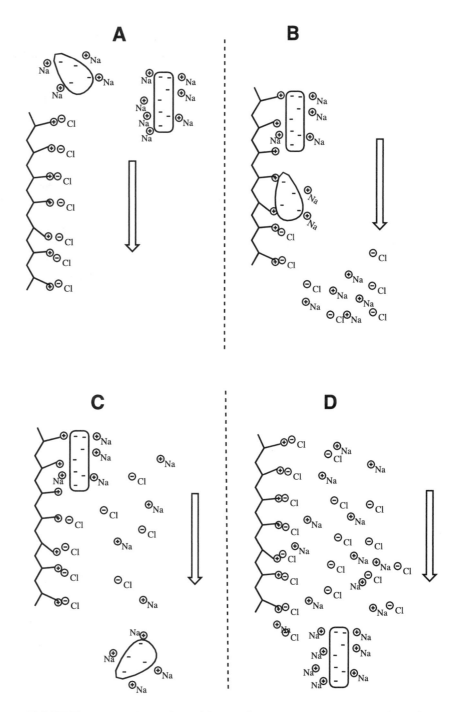

FIGURE 4-11 Illustration of ion-exchange processes occurring when negatively charged proteins are separated on an anion-exchange column. *(A)* Column is equilibrated with low [NaCl]. *(B)* Proteins are loaded at low [NaCl] and bind to the matrix, displacing Cl⁻ and Na⁺ ions formerly associated with the matrix and proteins. *(C)* As [NaCl] increases, the less acidic protein elutes. *(D)* At higher [NaCl], the more acidic protein elutes.

tion chromatography, sample molecules do not elute ahead of the buffer in which they are loaded. It is generally not important to load the sample in a small volume in ion exchange chromatography, because the conditions for applying the sample are usually chosen so that the molecule of interest binds strongly to the stationary phase. Moreover, during gradient elution various molecules in the applied samples become "focused" as they proceed down the column. This is because molecules near the top of the column are released when the salt gradient reaches a certain concentration at that position of the column. Molecules that are bound further down the column are not released until later when the higher salt concentration (and the already released molecules of interest) reaches them. For this reason, ion-exchange chromatography is often an excellent choice for an early chromatography step of a complex purification scheme; the sample can be applied in a large volume and the molecule of interest may be recovered in a relatively small volume. For the same reason, it is also often a useful step after a gel-filtration chromatography step.

Relatively simple devices can be used to apply a gradient of increasing counterion (or any gradient of two different solvents) to a chromatography column, and one such device is depicted in Fig. 4-12. As elution buffer from the stirred chamber is drawn into the column, buffer from the limit buffer

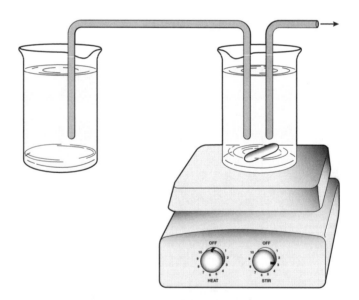

FIGURE 4-12   A simple gradient maker for column chromatography. The chamber on the left usually contains a high concentration of salt (limit concentration), while the stirred chamber initially has a low concentration of salt. As the eluant is removed from the stirred beaker to the column, the limit buffer is drawn over to the stirred beaker to maintain equal depths. Gradually, the salt concentration in the stirred beaker becomes higher, forming the gradient.

container is drawn into the mixing chamber, altering its buffer concentration. In practice, the mixing chamber is usually filled initially with buffer containing a low concentration of counterions, and the limit buffer container is filled with buffer containing a high counterion concentration. As the buffer from the mixing chamber is drawn into the column, the counterion concentration in the mixing chamber gradually increases. Altering the shapes of the stirred chamber and the limit buffer chamber may produce convex, concave, or linear gradients of the counterions.

A wide variety of different immobilized charged groups are used in ion exchange chromatography, and some of the more commonly used are listed in Table 4-2. In the purification of alkaline phosphatase, you will use the weak anion exchange group DEAE attached to a cellulose support. Strong ion exchange resins have ionizable groups that are usually completely ionized over the whole pH range used for protein purifications (i.e., they are strong acids or bases), while the weak ion exchangers are typically only partially ionized (i.e., they are weak acids or bases). The word "strong" does not refer to the strength of the binding but rather to whether the resin groups are sensitive to pH. The ionizable groups listed in Table 4-2 are available coupled to various inert supports, such as cellulose, dextran, agarose, and polyacrylamide.

QUESTIONS

1. A pH gradient can be used to elute molecules during ion-exchange chromatography. What type of gradient (high to low pH or low to high pH) should be used for anion- and cation-exchange chromatography?

2. How could you determine the exact salt concentration at which a protein eluted from an ion-exchange chromatography column?

3. Is the pH likely to be higher or lower than that of the bulk solvent near the ionizable groups of a cation-exchange resin?

4. At a given pH, will any proteins bind to both cation- and anion-exchange resins? Explain.

HYDROPHOBIC-INTERACTION CHROMATOGRAPHY

PURPOSES AND PRINCIPLES

Although we do not use hydrophobic-interaction chromatography in any of the laboratory exercises described in this book, it is an important tech-

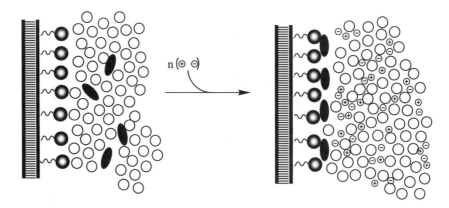

FIGURE 4-13 Adsorption of a protein to a hydrophobic chromatography matrix induced by increasing the concentration of salt.

nique for the purification of proteins and peptides, and we briefly discuss it here. When a nonpolar molecule is added to a polar solvent, the nonpolar molecule becomes surrounded by partially ordered polar solvent molecules, causing a decrease in entropy, which is energetically unfavorable. To minimize this decrease in entropy, nonpolar molecules tend to aggregate, thereby decreasing the exposed surface on which polar molecules can become ordered. Aggregation of nonpolar molecules is the phenomenon known as hydrophobic interaction. Changing the composition of the solvent to make it more or less polar also changes the hydrophobic interactions among nonpolar molecules present in the solvent. For example, adding high concentrations of **lyotropic ions**, such as ammonium sulfate, favors the interactions of hydrophobic surfaces of proteins, leading to aggregation and precipitation. This is the basis of the process of salting out proteins from solution by lyotropic ions (Chapter 6).

Proteins contain many different amino acids, some with polar and some with nonpolar side chains. Usually, the solvent-exposed surfaces of proteins are rich in amino acids with polar side chains, while the interiors of proteins are rich in amino acids with hydrophobic side chains. However, there are some amino acids with hydrophobic side chains on the surface of proteins. If a protein is designed to interact with other proteins or with lipid membranes, the interacting surfaces are likely to have patches with a higher proportion of hydrophobic side chains on the surfaces than is usually found with soluble proteins. The differences in the hydrophobicity of the surfaces of proteins are exploited in hydrophobic-interaction chromatography. The stationary phase is usually an inert material coupled to hydrophobic functional groups. Examples of hydrophobic functional groups used include ethyl, propyl, butyl, hexyl, octyl, and phenyl. A sample can be applied to the column in a buffer that is highly polar, which enhances the interactions of the hydrophobic surface patches of the proteins with the hydrophobic stationary phase. The principle is the same as that of salting out. The column can then be eluted with a solvent gradient of polar to nonpolar, such as a gradient of decreasing salt or a gradient of increasing ethylene glycol.

REVERSE-PHASE CHROMATOGRAPHY

A special type of hydrophobic interaction chromatography, known as **reverse-phase chromatography**, is used for the purification of peptides and a variety of small molecules. In this technique, the stationary phase usually consists of a long-chain hydrocarbon (C_8 or C_{18}) linked to an inert silica-based support. The chromatography is usually performed under conditions in which peptides are completely denatured, and thus, the primary separation will be on the basis of hydrophobicity. As the eluting aqueous solvent is made less polar by dilution with acetonitrile or methanol, the peptides (or other substances) are eluted in the order of their increasing hydrophobicity.

REFERENCES

Alberts, B. M., F. J. Amodio, M. Jemklins, E. D. Gutmann, and F. L. Ferris. 1968. Studies with DNA-cellulose chromatography. I. DNA-binding proteins from *Escherichia coli*. Cold Spring Harbor Symp. Quant. Biol. **33**:289.

Burton, Z. F., L. G. Ortolan, and J. Greenblatt. 1986. Proteins that bind to RNA polymerase II are required for accurate initiation of transcription at the adenovirus 2 major late promoter. EMBO J. **5**:923.

Cuatrecasas, P. 1970. Protein purification by affinity chromatography. J. Biol. Chem. **245**:3059.

Cuatrecasas, P., and C. B. Anfinsen. 1971. Affinity chromatography. Methods Enzymol. **22**:345.

Fulton, S., and D. Vanderburgh. 1996. The Busy Researcher's Guide to Biomolecular Chromatography. PerSeptive Biosystems, Framingham, Mass.

Kadanaga, J. T., and R. Tijan. 1986. Affinity purification of sequence-specific DNA binding proteins. Proc. Natl. Acad. Sci. USA **83**:5889.

Scopes, R. K. 1987. Protein Purification. Principles and Practice, 2nd ed. Springer-Verlag, New York.

Steers, E., Jr., and P. Cuatrecasas. 1974. β-Galactosidase. Methods Enzymol. **34B**:350.

Steers, E., Jr., P. Cuatrecasas, and H. B. Pollard. 1971. The purification of β-galactosidase from *Escherichia coli* by affinity chromatography. J. Biol. Chem. **246**:196.

Thompson, N. E., and R. R. Burton. 1994. Purification of recombinant human transcription factor IIB by immunoaffinity chromatography. Protein Exp. Purif. **5:**468.

Thompson, N. E., D. A. Hager, and R. R. Burgess. 1992. Isolation and characterization of a polyol-responsive monoclonal antibody useful for gentle purification of Escherichia coli RNA polymerase. Biochemistry **31:**7003.

SEPARATION OF BLUE DEXTRAN, CYTOCHROME *c*, AND POTASSIUM CHROMATE BY GEL-FILTRATION CHROMATOGRAPHY

MATERIALS

1. Column (15 mL)
2. Sephadex G-75 slurry
3. Elution buffer (20 mM sodium phosphate, pH 7.0)
4. 13×100 mm test tubes
5. Sample mixture containing blue dextran, cytochrome *c*, and potassium chromate

PROCEDURE

1. Pour a column (see above) so that the bed is ~13 cm in height. (The inner diameters of the columns used for this procedure in our laboratory measure 0.7 cm, so their radius is 0.35 cm.) Equilibrate the column in the elution buffer.

2. Add the mixture of blue dextran ($M_r \approx 2,000,000$, blue in color), cytochrome *c* ($M_r = 13,000$, orange or pink in color), and Na_2CrO_4 ($M_r = 162$, yellow in color) to the column, and collect fractions. Observe whether these molecules are separated during the chromatographic run. Collect 0.5- to 1-mL fractions during the run.

3. Determine the volume at which each of the three colored molecules in the sample elutes (visually estimate which tube has the peak fraction of each colored molecule). It may be useful to plot the estimated visual intensity of each color as a function of the tube number or volume of the column. This can help indicate whether the elution profile was symmetric.

QUESTIONS

1. What are the values for V_o, V_i, V_t, and V_g for your column?

2. Were the three colored molecules completely separated during your chromatography run? Explain.

PURIFICATION OF *E. COLI* β-GALACTOSIDASE BY AFFINITY CHROMATOGRAPHY ON *p*-AMINOBENZYL-1-THIO-β-D GALACTOPYRANOSYL AGAROSE

The *E. coli* enzyme, β-galactosidase, encoded by the *lacZ* gene of the *lac* operon, hydrolyzes lactose to yield glucose and galactose (Fig. 4-14). The enzyme is fairly robust, and you will conduct the purification and assay of the enzyme at room temperature. The β-galactosidase may be assayed by monitoring the cleavage of the chromogenic substrate *o*-nitrophenol-β-D-galactoside (ONPG), which when cleaved gives rise to the yellow compound, *o*-nitrophenolate at alkaline pH (Fig. 4-15). The concentration of *o*-nitrophenolate ions can be measured spectrophotometrically. This product is very similar to *p*-nitrophenol, which was described in the spectrophotometry experiments (Chapter 2) and will be discussed again in the purification of alkaline phosphatase (Chapter 7). This chromophore is commonly used in enzymatic assays.

In this experiment, you will use *p*-aminobenzyl-1-thio-β-D-galactopyranoside agarose, which is commercially available, as the affinity matrix. The galactopyranosyl group is a good substrate analog for the enzyme. Synthesis of this matrix is outlined in Fig. 4-16 and is a typical example of how you might prepare an affinity matrix.

You will evaluate the affinity chromatography step as a means of purification of β-galactosidase by determining the specific activity (units of activity per milligram of protein) of the solutions both before and after the

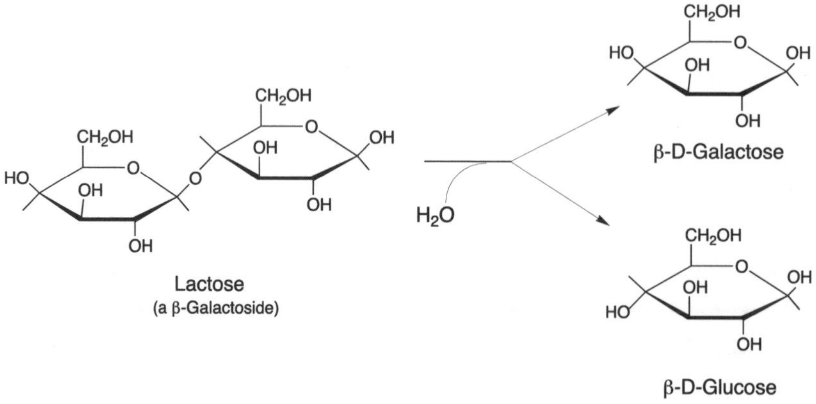

FIGURE 4-14   Reaction catalyzed by β-galactosidase.

chromatography step. We use the Bradford assay for determining the protein concentration (Chapter 3) and the activity assay described above.

MATERIALS

Chromatography

E. coli cell extract containing β-galactosidase

1. There are several sources that can be used. We will use an extract of cells containing a fusion of *lacZ* to the highly expressed *glnK* promoter of *E. coli* (strain MA100, containing a *glnK::lac* transcriptional fusion and a deletion of the *glnB* gene). The *glnK* promoter is regulated by nitrogen in *E. coli*, but in cells lacking the *glnB* product, the expression of *glnK* is constitutively high. Thus, by using this strain of bacteria, we can obtain cells containing high levels of β-galactosidase by growing them on rich broth media such as nutrient broth.

    To prepare the extract, cells containing β-galactosidase are harvested by centrifugation, washed in buffer B (20 mM Tris-Cl, pH 7.4, 10 mM $MgCl_2$), and resuspended in buffer B. The cells are disrupted by sonication on ice, cell debris is removed by centrifugation, and the resulting clarified extract is used. To save classroom

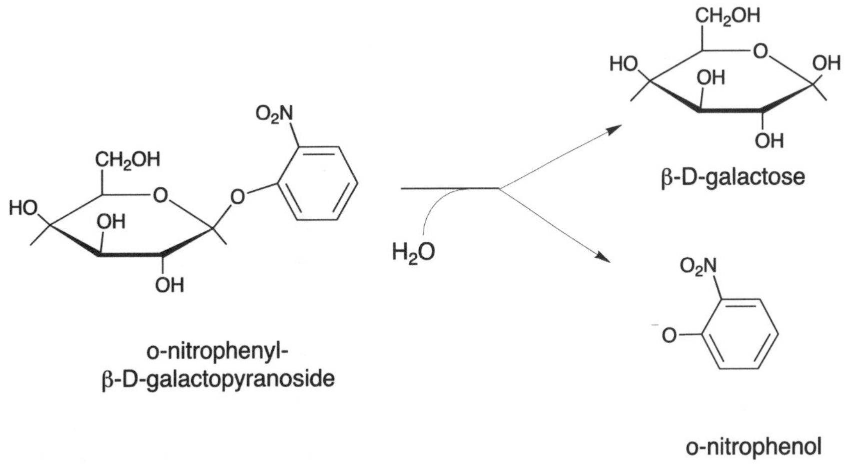

FIGURE 4-15 Reaction catalyzed by β-galactosidase used for spectrophotometric assay at 420 nm.

EXPERIMENT 4-2

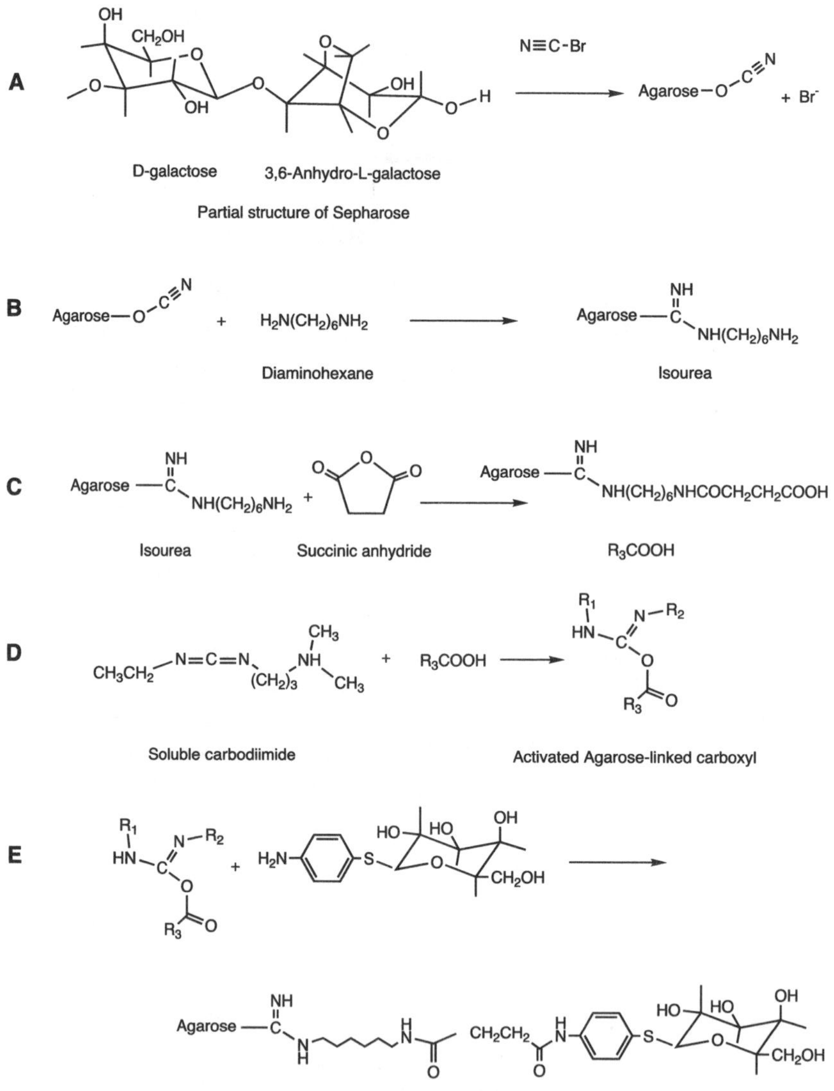

FIGURE 4-16 Synthesis of affinity resin for purifying β-galactosidase from *E. coli*. *(A)* Activation of Sepharose (Pharmacia) by reaction with cyanogen bromide at pH ~ 11. The hydroxyl groups react to form the species shown, as well as other products. *(B)* Reaction of diaminohexane with the activated agarose at pH ~ 9. *(C)* The isourea product is coupled to succinic anhydride at pH 6 to extend the arm and give a free carboxyl group. *(D)* This free carboxyl group is activated with the soluble diimide, 1-ethyl-3-(3-dimethylaminopropyl)carbodiimide (EDAC) at pH 4.7. *(E)* The *p*-aminophenyl thiogalactoside is included and reacts to give the desired resin .

time, instructors often prepare the clarified extract before the laboratory session, and each laboratory group is provided with some of the extract.

2. Buffer B: 20 mM Tris-Cl, pH 7.4, 10 mM $MgCl_2$

3. Buffer B + NaCl: Buffer B containing 1.5 M NaCl and 10 mM β-mercaptoethanol (βME)

4. Elution buffer 2: 0.1 M sodium borate, pH 10, 10 mM βME

5. Affinity resin: *p*-aminobenzyl-1-thio-β-D-galactopyranoside agarose

6. Elution buffer: 5 M NaCl

7. Stabilizer: 1.4 M βME

8. Glass chromatography column

Assay of β-Galactosidase

1. Z-buffer:
   60 mM sodium phosphate, pH 7.0
   10 mM KCl
   1 mM $MgSO_4$
   50 mM βME

2. Quenching and stopping solution
   1 M Na_2CO_3

3. ONPG: 4 mg mL^{-1}

Protein Determination

1. Bradford reagent (see Chapter 3)

2. BSA standard (see Chapter 3)

PROCEDURE

Chromatography

1. Take 1 mL of the *E. coli* extract and add 320 μL of 5 M NaCl and 9 μL of 1.4 M βME. This will bring the final concentrations of NaCl and βME to 1.5 M and 10 mM, respectively.

2. Save a 100-μL aliquot of this sample for use in the enzyme assays. The remaining 1.2 mL of this sample will be loaded onto the affinity column.

EXPERIMENT 4-2

3. Wash a 1-mL affinity column with 10 mL of buffer B + NaCl to equilibrate the column in this buffer. When the 10 mL is just at the top of the column, close the valve and gently load 1.2 mL of the cell extract from step 1 onto the column.

4. Set up a series of tubes in a rack under the column, so that you can collect 0.5-mL fractions.

5. Open the valve so that the sample runs into the column, and close the valve again as soon as the sample has entered the bed (see Fig. 4-6).

6. Gently add 10 mL of buffer B + NaCl to the top of the column, and open the valve again.

7. Collect 0.5-mL fractions until the level of the eluting buffer is just at the top of the bed, and close the valve again.

8. Gently add 10 mL of elution buffer 2 to the reservoir at the top of the column, and open the valve again.

9. Collect 10-drop fractions in clean test tubes. Collect at least 15 fractions.

β-Galactosidase Assay

1. Combine 2 mL of Z-buffer and 100 μL of ONPG (4 mg mL^{-1}) in a test tube.

2. Start the reaction by adding 10 μL of the sample to be assayed, and mix well.

3. After 10 min, add 0.5 mL of 1 M Na_2CO_3 and mix well.

4. Read the absorbance at 420 nm against a blank that received 10 μL of elution buffer 2 instead of sample.

Determine the following: Did the crude extract have β-galactosidase activity? Did the column retain the β-galactosidase activity? If it was retained, was the β-galactosidase activity eluted from the column by elution buffer 2? If it was eluted, in what fraction (tube) did this activity appear?

You will have many fractions to assay, and you will have several controls (unfractionated extract, elution buffer 2 alone) to perform. It would be wise to plan your assay so that many samples can be assayed side by side. The easiest way to do this is to set up a series of assay tubes and then to start the reactions at 30-s intervals by adding samples. Be sure to keep track of the time. Each assay should run only for 10 min before the addition of Na_2CO_3. This will require that you plan carefully and label reaction tubes before the assay. Your experiment will proceed smoothly if you prepare a sketch of a suggested protocol as follows:

Tube	1 2 3 4 5 6 7 8 9 10 11 12 13 14 15 16 17 18
Z-buffer	2 mL ⟶
ONPG	100 μL ⟶

a. Mix thoroughly

b. Add the following in 30-s intervals, mix, and incubate at room temperature:

Column fractions 1–15 : To tubes 1–15 (10 μL from each fraction). To tube 16, add 10 μL of buffer B wash. To tube 17, add 10 μL of the crude lysate. To tube 18, add 10 μL of elution buffer 2 (0.1 M sodium borate, pH 10, 10 mM βME).

c. After 10 min, add 0.5 mL of 1 M Na_2CO_3 to each tube at 30-s intervals. Mix and read on the spectrophotometer at 420 nm with tube 18 as the blank.

Protein Determination

Determine the total protein concentration in the starting extract and those fractions containing purified β-galactosidase by using the Bradford method.

QUESTIONS AND CALCULATIONS

1. Did the β-galactosidase activity remain associated with the bulk of cellular proteins or was β-galactosidase effectively purified by the procedure?

2. Was the 1-mL column of sufficient capacity to bind all of the β-galactosidase activity or should a larger column have been used?

3. Plot the results of the assays versus the fraction number (absorbance at 420 nm versus fraction number). For samples with very high β-galactosidase activity, how would you know if the assay was accurately reporting the level of activity (i.e., how would you know that the assay result is limited only by the amount of enzyme present)? Try to design some simple experiments to test the validity of the assays.

REAGENTS NEEDED FOR CHAPTER 4

✓ Blue dextran: Sigma D5751, 1 g

✓ Cytochrome *c*: Sigma C3006, 100 mg

✓ Potassium chromate: (K_2CrO_4) Sigma P0454, 100 g

✓ Chromatography columns: A number of economical columns may be used. We use BioRad Econocolumns 1 cm × 20 cm, product 737-1021. There are 5 columns per package.

✓ Sephadex G-75: Sigma G-75-120, 50 g, which is sufficient for 35 columns

✓ Sodium phosphate: Sigma S9638, 250 g

✓ 13 × 100 mL test tubes

✓ *E. coli* cells expressing β-galactosidase: A number of options are available. Wild-type cells (*lac*⁺) grown on lactose will work. We typically use a *lac* fusion strain in which expression of the *lac* operon is high on rich media. This simplifies the growing of the strain and provides a high yield of cells. *E. coli* strains can be obtained from the American Type Culture Collection (ATCC) for a nominal fee, or from the authors for the cost of shipping.

✓ Tris base: Sigma T3253, 1 kg

✓ Magnesium chloride ($MgCl_2$): Sigma M9272, 500 g

✓ Sodium chloride (NaCl): Sigma S7563, 1 kg

✓ β-Mercaptoethanol: Sigma M6250, 100 mL

✓ Sodium borate: Sigma S9640, 500 g

✓ *p*-Aminobenzyl-1-thio-β-D-galactopyranoside agarose: Sigma A-0414, 10 mL per 10-student group (can be reused many times)

✓ Potassium chloride: Sigma P9333, 500 g

✓ Magnesium sulfate: Sigma M7506, 500 g

✓ Sodium carbonate: Sigma S9140, 500 g

✓ *o*-Nitrophenyl-β-D-galactopyranoside (ONPG): Sigma N1127, 5 g

5 Gel Electrophoresis of Proteins

Process of Electrophoresis

Polyacrylamide Gels

SDS-Polyacrylamide Gel Electrophoresis (SDS-PAGE) of Proteins

Detection of Proteins in SDS-Polyacrylamide Gels

Applications of SDS-PAGE

Simple (Nondenaturing, Native) Gel Electrophoresis

Experiments 5-1 to 5-2

Reagents Needed for Chapter 5

PROCESS OF ELECTROPHORESIS

When an electric field is placed across a solution containing ions, a current develops in which anions move toward the anode and cations move toward the cathode. A solution that contains few ions (e.g., pure distilled water or benzene) would have very high resistance and would carry very little current. Electrophoresis is the process by which charged molecules are separated in an electric field because of their differential mobilities. Factors affecting the mobility of a molecule in an electric field include the charge of the molecule, q, and the voltage gradient of the electric field, E, which together provide the force to move the ions, and the frictional resistance of the supporting medium, f, which impedes their movement. **Mobility** is defined as the rate of migration traveled with a voltage gradient of 1 V/cm and is measured in square centimeters per second per volt. For practical reasons, one usually determines relative mobilities of proteins, R_f, as the ratio of the distance each protein migrates to that of a small anionic dye (the tracking dye). The high charge-to-mass ratio of the dye causes it to migrate near to the electrophoretic front and ahead of the proteins.

$$R_f \propto \frac{qE}{f} \qquad (5\text{-}1)$$

In the experiments described in this book (Experiments 5-1 and 5-2 and Chapter 7), we use electrophoresis to resolve the individual components of mixtures of proteins. Electrophoresis is also used in other experiments (11-1, 11-2, 11-3, 11-4, 12-1, 12-2, 12-3, 12-4, and 12-5) in which we resolve the individual components of mixtures of nucleic acids.

Since electrophoresis experiments involve the use of considerable voltages and thus are inherently dangerous, please review the laboratory safety section of Chapter 1 and any notes from safety lectures given in earlier laboratory sessions before proceeding with the laboratory exercises. In addition, familiarize yourself with the apparatus and check with your instructors on how to use it safely.

Proteins have a net charge at any pH other than their isoelectric point **pI** (see Ion Exchange Chromatography in Chapter 4); thus, when placed in an electric field, proteins will migrate toward the electrode of the opposite charge. For most proteins, the pI is in the range of 3-10, with the majority of proteins having a pI of <8. Thus, at pH 8 or higher, a majority of proteins have a net negative charge and will migrate in an electric field toward the anode (positive electrode). In order to prevent the loss of resolution caused by diffusion and vibrational and convectional disturbances, and to retain the separation of different proteins after completion of the experiment, electrophoretic separations are performed using a support medium. Ideally, the support medium should be strong, hydrophilic (to prevent hydrophobic interactions between the proteins in the sample and the support medium), uncharged, and stable over a range of temperatures, pH, and osmolarity; and it should have a carefully controlled and adjustable porosity. The pore size of the supporting medium is important because it is a major contributor to the frictional coefficient f. The sieving effect of the support medium, as well as the charge-to-mass ratio of the molecules, fractionates the molecules in the sample. In sodium dodecyl sulfate-polyacrylamide gel electrophoresis (SDS-PAGE), this sieving effect is the principal means of fractionation (see below). This permits the separation of molecules of identical charge to mass ratios, but of different sizes (Stokes radius). The sieving effect in gel electrophoresis is different from that in gel-filtration chromatography (Chapter 4). In gel filtration, molecules are excluded from the interiors of the gel beads according to their size, so that the largest molecules elute first. In gel electrophoresis, the gel matrix is a mesh-like substance rather than a bead, and therefore, it acts as a sieve, where frictional forces decrease the mobility of larger molecules more than those of smaller molecules. Therefore, large molecules travel more slowly than small molecules.

Various substances have been used as the supporting medium (matrix) for electrophoretic separations. Historically, paper and starch gels have played an important role in the development of the technique of protein electrophoresis, although these support media lack many of the desirable properties listed above. Currently, the supporting medium of choice for almost all protein electrophoresis applications (and many nucleic acid electrophoresis applications) is polymerized **acrylamide**, which is in the form of a gel and has all of the above mentioned desirable properties. The porosity of the polymerized polyacrylamide gel may be controlled by varying the percentage of acrylamide and/or the degree of cross-linking of the acrylamide chains (see below).

Precautions: Acrylamide, until it polymerizes, is a potent neurotoxin. Be sure to use gloves for all manipulations involving acrylamide. When weighing or manipulating the dry powder, wear a mask and carry out operations in a well-ventilated hood. Unused solutions should be polymerized or diluted greatly and rinsed carefully into the sink drain (clean out the sink before using it again). Spills should be soaked up promptly with paper towels, and the towels should be disposed of in an appropriate toxic-waste receptacle. After polymerization, acrylamide loses its toxicity but is difficult to remove from equipment.

POLYACRYLAMIDE GELS

Polyacrylamide gels are formed by the copolymerization of acrylamide ($CH_2=CH-CO-NH_2$), a water-soluble monomer, with a cross-linking agent to form a three-dimensional lattice (Fig. 5-1). The cross-linking agent of choice for most applications is **N,N′-methylene bisacrylamide (BIS)**, which is shown below.

$$(CH_2=CH-CO-NH-CH_2-NH-CO-CH=CH_2)$$

BIS contains two double bonds, which in polymerization reactions cross-link with adjacent chains of polyacrylamide. Acrylamide, if polymerized in the absence of a cross-linking agent, forms only linear polymers, resulting in viscous aqueous solutions rather than a gel. The polymerization reaction occurs by a free radical chain mechanism; the free radicals are generated either by photolysis of a labile compound (e.g., riboflavin) or by chemical decomposition of a labile compound. We describe the latter method of radical generation with **ammonium persulfate** and **tetram-**

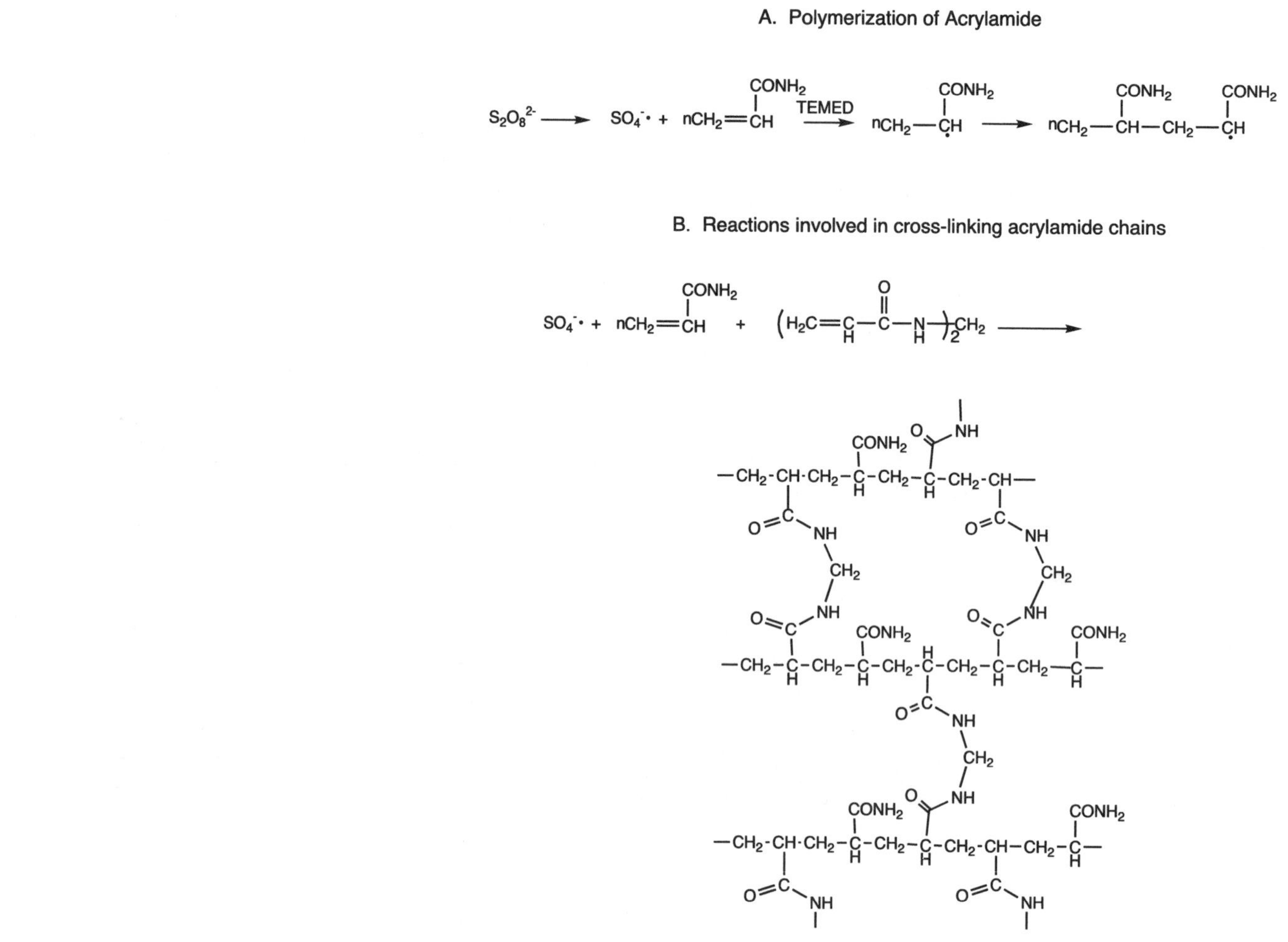

A. Polymerization of Acrylamide

B. Reactions involved in cross-linking acrylamide chains

FIGURE 5-1 Polymerization of acrylamide and bisacrylamide.

ethylethylenediamine (**TEMED**, $(CH_3)_2N-CH_2-CH_2-N(CH_3)_2$). **Ammonium persulfate** is the di-sulfate ester of hydrogen peroxide ($^-O_3S-O-O-SO_3^-$), and readily homolyzes into unstable $\cdot SO_4^-$ radicals. TEMED is a tertiary amine that reacts with these radicals to form TEMED free radicals, which in turn react with acrylamide to induce polymerization as shown in Fig. 5-1.

The average size of the pores in a polyacrylamide gel can be controlled by varying the amount of monomer used or by varying the degree of cross-linking, with higher degrees of cross-linking resulting in narrower pores. For practical purposes, however, the degree of cross-linking is generally kept constant, and the percentage of the monomer is varied to make gels of different porosity. For a gel of a given composition, the pores have a statistical distribution of pore sizes. Relatively small proteins will migrate in the gel with minimal impediment, whereas the migration of larger proteins will be retarded because they cannot pass through all the pores in the gel. Very large proteins may not enter the gel at all. Therefore, only proteins that are within a particular size range (molecular weight range, or, more accurately, Stokes radius range) will be separated on any given gel. In Table 5-1, the percentage of monomer used for polymerization and the molecular-size ranges of proteins that are separated from one another in such a gel are given. A gradient gel is a special form of gel in which the degree of porosity is varied continuously from the bottom to the top by creating a gradient in the percentage of the acrylamide. Gradient gels can separate proteins having a wide range of M_r.

SDS-POLYACRYLAMIDE GEL ELECTROPHORESIS (SDS-PAGE) OF PROTEINS

SDS-PAGE PRINCIPLES

As noted above, the separation of proteins by electrophoresis in polyacrylamide gels is influenced by both the charge of the proteins at the chosen pH and the size of the proteins and the attendant frictional resistance as they migrate in the electric field. Thus, if these two characteristics compensate for one another, proteins of different charge and size may occasionally migrate at the same rate in an electrophoresis experiment. To simplify the analysis of mixtures of proteins, it is possible to make the separation rely on only the size of the polypeptide chains. This is achieved by denaturing the proteins with the detergent **sodium dodecyl sulfate** (SDS, Fig. 5-2). Dodecyl sulfate binds strongly to proteins and causes the folded proteins to become denatured into extended "rods" coated with SDS (Fig. 5-3). On average, one dodecyl sulfate molecule will be present for every two amino acids. Because each dodecyl sulfate molecule has two negative charges at pH values used for electrophoresis, the net charge of the coated polypeptide chains will be much more negative than that of uncoated chains. Furthermore, the charge-to-mass ratio will be essentially identical for different

TABLE 5-1	
Fractionation Range for Polyacrylamide Gels	
PERCENT ACRYLAMIDE	OPTIMUM M_r RANGE
5–12	20,000–150,000
10–15	10,000–80,000
>15	<15,000

Polyacrylamide gels also contain a fixed proportion of bisacrylamide (5% of acrylamide).

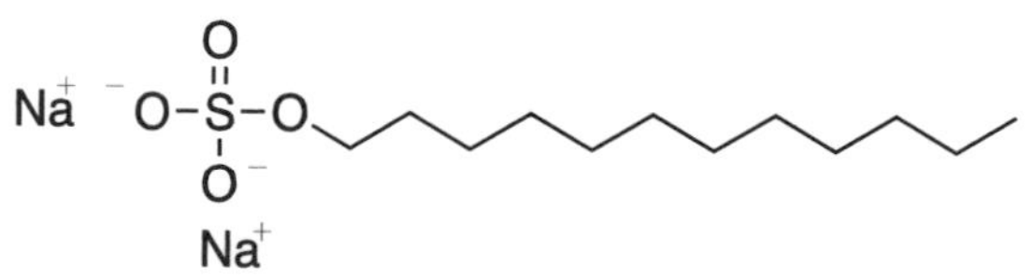

FIGURE 5-2   Structure of Sodium Dodecyl Sulfate (SDS).

proteins because the SDS coating dominates the charge. Thus, the separation of the denatured, detergent-coated polypeptide chains will be due almost exclusively to the size of the polypeptide chains.

ELECTROPHORESIS DEVICES

A variety of electrophoresis devices have been used for the separation of proteins on polyacrylamide gels; the most popular formats involve gels that are shaped like tubes or slabs (Fig. 5-4). **Slab gels** have the advantage of permitting the side-by-side comparison of different protein mixtures as illustrated in Fig. 5-5 and are used most frequently in biochemistry and molecular biology. We use slab gels in the experiments described in this book.

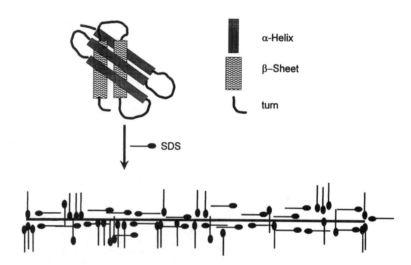

FIGURE 5-3   Denaturation of proteins by dodecyl sulfate. The native protein *(top left)* consists of two central β sheets surrounded by three α helices connected by turns. Amino acid side chains are not shown. The polar heads of the SDS molecules are represented by ●, while the hydrophobic tails are depicted as straight lines. When treated with SDS, the protein unfolds and the side chains are coated with SDS. Polar and nonpolar side chains and portions of the backbone result in an amalgam of different interactions with SDS.

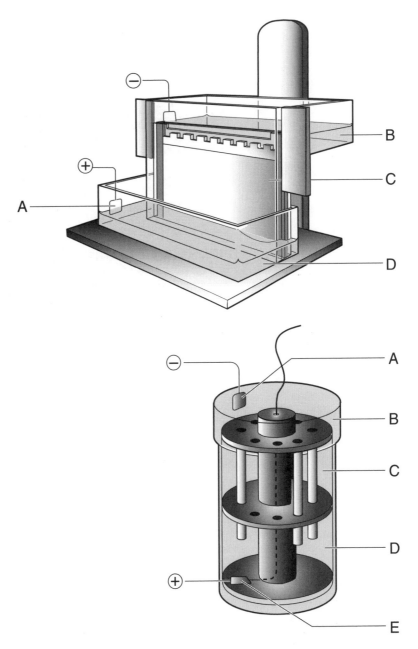

FIGURE 5-4 Formats for polyacrylamide electrophoresis. *(Top)* Slab gel apparatus. *(A)* Anode in the lower buffer chamber. *(B)* Upper buffer chamber with the cathode immersed (–). *(C)* Gel mounted vertically between glass plates. *(D)* Lower buffer chamber. The protein bands descend vertically and separate according to their mobility as shown in Fig. 5-5. *(Bottom)* Tube gel apparatus. *(A)* Cathode. *(B)* Upper buffer chamber. *(C)* Glass tubes containing the polyacrylamide gel and making contact with the two buffer chambers. *(D)* Lower buffer chamber. *(E)* Anode. This type of apparatus is also used for isoelectric focusing for the first direction in two-dimensional gels (Fig. 5-10).

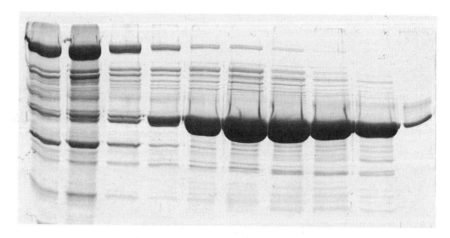

FIGURE 5-5 An example of a Coomassie-stained slab SDS-PAGE. This gel was used to scan fractions eluting from a Bio-Gel A-0.5M gel-filtration column for the presence of the protein NRI, a dimer of identical 51.5-kDa subunits. After allowing for the void volume of the column, an aliquot from every third fraction was run, from left to right on the gel. A prominent band corresponding to NRI is observed in samples 4-9. Gel courtesy of Quan Son.

The slab-shaped gels are formed outside the apparatus between two glass plates that are separated from one another by 1.5-mm-thick Teflon spacers (Fig. 5-6). The open space between the glass plates at the bottom of the gel sandwich is sealed in a bed of molten agarose as shown in Fig. 5-6. After the agarose gels, the liquid polyacrylamide gel is poured between the glass plates and allowed to polymerize. The **stacking gel** is poured on top of the **resolving gel**, and a comb is inserted at the top to cast the sample wells as shown. Then, before the gel sandwich is put into the electrophoresis device, the agarose plug at the bottom can be peeled away.

In order to achieve a good separation of different proteins in a mixture, as in gel-filtration chromatography, it is essential that the proteins be applied to the gel in very small volumes. Because it is often not practical to uniformly load very small volumes of the sample on an electrophoresis gel, a special type of gel known as a stacking gel is cast directly on top of the larger resolving gel within the glass plates. The stacking gel has properties that cause the proteins in the sample to be concentrated into a narrow zone at the top of the resolving gel, permitting effective and reproducible separation of the SDS-coated proteins by size. The stacking gel is polymerized with a small percentage of acrylamide and bisacrylamide to ensure high porosity, and is buffered with Tris-HCl buffer at pH 6.8, whereas the resolving gel contains a higher percentage of acrylamide and is cast in Tris-HCl at a higher pH (8.8). The upper (and lower) reservoir buffers usually used in the electrophoresis process contain Tris at pH 8.3 with glycine as the counterion. The protein mixture in Tris at pH 6.8, along with a

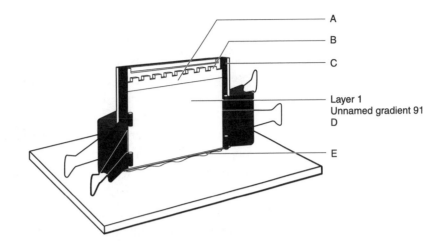

Layer 1
Unnamed gradient 91
D

FIGURE 5-6 Assembly for pouring polyacrylamide gels: *(A)* Stacking gel, *(B)* Comb, *(C)* Teflon spacers (1.5 mm), which separate the glass plates and seal the sides, *(D)* glass plates supporting the resolving gel, and *(E)* agarose to seal the bottom of the plates. The Teflon spacers (1.5 mm) separate the two glass plates and seal the sides. The plates are sealed at the bottom with agarose on a glass plate.

tracking dye (to permit monitoring of the electrophoresing front), is loaded onto the surface of the stacking gel in small wells formed by casting the stacking gel in the presence of a Teflon comb (see Figs. 5-4 and 5-6). Because the stacking gel is very porous, there is little difference in the mobility of proteins of different size as they migrate through the stacking gel. The principles of stacking are given below.

STACKING PRINCIPLES

When glycine from the upper reservoir enters the low pH (6.8) of the stacking gel, it will principally be in the neutral zwitterionic form, with only a fraction (1%) in the negative glycinate form. This prevents glycine from being an effective carrier of current. The Cl^- ions remain effective current carriers at pH 6.8 and migrate rapidly toward the anode. During this electrophoresing in the stacking gel, the Cl^- ion concentration becomes depleted at the cathode end of the gel (top), forming an increasing concentration gradient toward the anode (bottom). The SDS-coated protein molecules and the dye, which have charge-to-mass ratios greater than that of the glycine but less than that of Cl^-, must now migrate to carry the electrophoresis current behind the Cl^- and ahead of the glycine. As electrophoresis continues, protein molecules reaching the resolving gel will become greatly retarded, allowing trailing protein molecules to catch up, ensuring that the volume of the protein sample "presented" to the resolving gel will be much smaller than the volume of the sample initially loaded onto the stacking gel. This movement of proteins and dye in the stacking gel also creates an ion gradient, so that eventually the glycine must carry the current behind the proteins. This has the effect of concentrating (stacking) the proteins in a thin band sandwiched between the Cl^- ions and the glycine molecules at the interface between the stacking and resolving gels. It should be noted that for the stacking gel to function properly, it must be at the appropriate pH (6.8) where the glycine counterion is zwitterionic with only slight anionic character.

When the stacked bands enter the higher pH of the resolving gel, glycine becomes anionic, carrying a higher charge-to-mass ratio than that of the proteins. Now the newly formed glycinate ions move faster than the proteins, with mobility approaching that of the Cl^- ions. A new concentration gradient forms, requiring that the proteins and tracking dye carry the current in the trail of the Cl^- and glycinate ions. The proteins move according to their relative mobilities and are separated by the sieving effect of the resolving gel according to size. The high mobility of the tracking dye assures that it will migrate faster than the proteins. When the tracking dye reaches the end of the gel, the electrophoresis is terminated so that the proteins do not run off the end of the gel.

In order for SDS-polyacrylamide gels to resolve proteins based on size, it is important that all of the proteins be uniformly coated with SDS. It has been empirically shown that this can be achieved with a concentration of SDS of 0.1% (w/v). This concentration of SDS is therefore included in the electrophoresis buffer and in the stacking and resolving gels. The protein sample is usually mixed with a solution containing SDS at a final concentration of 1% (w/v). The sample is then heated to facilitate the denaturation of the proteins and the coating of the polypeptide chains with dodecylsulfate. This treatment works with the vast majority of proteins; however, some proteins are remarkably resistant to denaturation and/or to binding dodecyl sulfate. In particular, extremely acidic proteins (negatively charged) do not bind the detergent well and are not resolved strictly by their size in SDS-PAGE.

The presence of detergent and the resulting unfolding of polypeptide chains disrupt most of the interactions that hold multimeric proteins together, resulting in monomeric polypeptide chains. However, disulfide cross-links between polypeptide chains are not destroyed by heating in the presence of SDS, and thus polypeptide chains so linked will remain covalently bound. To eliminate these cross-links, the reducing agent β-mercaptoethanol is often added to the sample along with SDS.

DETECTION OF PROTEINS IN SDS-POLYACRYLAMIDE GELS

DETECTION TECHNIQUES

There are several methods for visualizing the presence of protein bands after their electrophoresis in SDS-polyacrylamide gels. The most commonly used technique is to stain the proteins with tightly binding dyes. A number of dyes have been used for this purpose; the most commonly used dye is **Coomassie Brilliant Blue R-250**, a dye closely related to that used in the Bradford assay for total protein concentration (Chapter 3). To stain the polypeptide bands, the gel is carefully removed from between the glass plates and placed into a small staining tray. The protein bands are fixed and most of the dodecyl sulfate is removed by soaking the gel in a solution that contains dilute acetic acid and methanol. The gel is then soaked in a solution containing Coomassie Brilliant Blue R-250 dissolved in dilute acid and methanol. Often, acceptable results are obtained by fixing and staining simultaneously in a single solution containing the dye. To visualize the protein bands, the usual practice is to stain the gel until it is uniformly blue and then remove the stain that is not adhering to proteins by soaking the gel in a solution of dilute acid and methanol. An example of a gel stained by this method is shown in Fig. 5-5. Such stained gels can be scanned in a densitometer to quantify the relative amounts of the various protein bands. It should be noted that just like in the Bradford assay (Chapter 3), individual proteins bind to the dye distinctly, and thus in the absence of specific knowledge on the binding of each protein to the dye, quantification of stained bands is only approximate.

A more sensitive but also more laborious and expensive way of visualizing protein bands in polyacrylamide gels involves staining the proteins with silver. This method is based on the chemistry used in photographic development. The gel is soaked in a dilute solution of methanol to fix the proteins to the gel. This is then incubated with an acidic solution of silver nitrate, which reacts with protein sites. An image is formed by reduction of ionic silver to its metallic form by formaldehyde at alkaline pH (sodium carbonate). Development is stopped by acidifying the solution with acetic acid. Various proteins react differently with silver nitrate, so that quantification with this method is also only approximate as with the other methods of protein detection. On average, the method is 10- to 100-fold more sensitive than staining with Coomassie Brilliant Blue R-250.

If the protein bands are labeled with radioactive isotopes of sufficient energy, they may be directly detected by autoradiography. For example, it is possible to make proteins that contain ^{35}S by synthesizing them in vitro in translation extracts containing $[^{35}S]$-methionine, or by producing them in vivo in organisms that are fed the labeled amino acid. In these cases, the radiolabeled protein bands may be detected by placing the dried gel directly against a piece of x-ray film in the dark. Decay of the radioactive compound will expose the x-ray film, revealing the presence of the protein bands. The autoradiography technique has also been used to track the results of phosphorylation or dephosphorylation of proteins in experiments with protein kinases and phosphatases, since in these cases it is possible to follow the incorporation or release of ^{32}P from proteins. In some cases, a low-energy isotope such as ^{14}C or ^{3}H will be incorporated into proteins. In these cases, direct autoradiography will not work, since the radioactive decay does not release a particle of sufficient energy to escape the gel efficiently and strike the film surface. To allow the detection of these low-energy emissions, the gel may be soaked in commercially available fluorophores. These fluors absorb low-energy emissions by intimate contact and produce light, which is recorded by exposure of the x-ray film.

If an antibody to the protein of interest is available, the protein may be detected in SDS-polyacrylamide gels by reaction with its specific antibody. This technique will be covered below in the discussion of immunoblotting.

RECORDING AND DRYING SDS-PAGE RESULTS

A record of the results of SDS-PAGE may be obtained by photographing the stained gels. This is generally the method of choice for gel images that are intended for publication. However, photographing gels is expensive and often not necessary for standard laboratory notebook purposes. For these less important gels, it is often suitable to dry the stained gel directly, with the dried gel itself serving as the record of the experiment. Several methods have been developed to dry gels. The standard method is simply to place the gel onto a piece of white filter paper (Whatman 3 mm), cover the top of the gel with plastic wrap, and dry the gel in a gel dryer, which provides vacuum and heat. After drying in this manner, gels tend to curl up, so it is a good idea to place the dried gels under a heavy book for a few days to keep them flat. Alternatively, the gel can be dried on a piece of cellophane in either a gel dryer or a special brace that holds the gel on the cellophane while it slowly dries at ambient conditions.

APPLICATIONS OF SDS-PAGE

ROUTINE ANALYSIS OF THE PURITY OF PROTEIN SAMPLES

Because of its simplicity, SDS-PAGE is commonly used for the routine estimation of protein purity and quantity. For example, SDS-PAGE is commonly used to track the course of purification of proteins for which there is no convenient enzymatic assay. Even when there is a good enzymatic assay available, it may be useful to examine the protein samples on SDS-PAGE during protein purification because the electrophoresis methods permit you to visualize the contaminants. You may be able to distinguish whether the contamination is due to small amounts of many different proteins, or alternatively, to a fairly large quantity of a particular protein. In the latter case, use of the gel electrophoresis method might permit you to establish conditions for the optimal separation from the bothersome contaminant. Since the identities of the contaminants are usually not known, the electrophoretic method is the only simple way that one can conveniently monitor the behavior of the contaminant in various fractionation procedures. Since the mobility of polypeptides on SDS-PAGE is almost exclusively due to size, it is not possible to detect contaminants of the same size as the protein of interest. However, this is seldom a problem, as the resolving power of the technique can often distinguish differences of ~1% in molecular mass.

Because many proteins are composed of more than one kind of polypeptide subunit, the presence of multiple bands on SDS-PAGE does not necessarily indicate enzyme impurity. However, in such cases, there should be a logical relationship between the quantity of protein in each band. If the multiple polypeptides are part of a single oligomeric enzyme, the ratio of different subunits should remain constant during the purification process.

DETERMINATION OF THE MOLECULAR MASS OF POLYPEPTIDE CHAINS

As discussed above, the differing mobilities of nearly all proteins in SDS-PAGE are due to their size. Because of this property, SDS-PAGE is frequently used to estimate the molecular mass of polypeptide chains. To calibrate the electrophoresis experiment, a number of polypeptides of known molecular mass are run on the gel simultaneously with the unknown polypeptide, and the relative mobility R_f of each species is determined. The relative mobility is calculated by dividing the distance the protein migrates by the distance the tracking dye migrates. The tracking dye is a small, charged, colored molecule that runs much faster than proteins and approximates the movement of small ions in the electric field. Since the tracking dye is colored, it permits the visualization of the electrophoretic front and the calculation of R_f even if adjacent gel tracks do not run exactly the same. (For this analysis, it is important not to run the tracking dye off of the gel or you will not be able to calculate R_f.) It is useful to include at least five different proteins of known molecular mass that span the molecular mass of your protein. A plot of the relative mobility of each protein versus the log of the molecular weight will be linear, and the molecular weight of the unknown polypeptide can be determined from its position on this plot (Fig. 5-7). Alternatively, the distance migrated versus log molecular weight may be plotted, but the resulting curve is only linear in its central region, and the analysis requires that the dye front be at the same position for each gel track.

IMMUNOBLOTTING (WESTERN BLOTTING)

One of the most common techniques used in molecular biology today is immunoblotting, also known as "Western blotting." This technique is used to detect and quantify proteins that react with a specific antibody. For example, the technique can be used to determine the concentration of the antigenic protein in cells that are in particular physiological states, or it can be used to determine if a given antigen is present. In this technique, an SDS-PAGE fractionation of the proteins in a sample is carried out. The gel is then removed from the gel sandwich and placed flat onto a nitrocellulose membrane. A set of absorbant pads is used to support the gel and membrane as a sandwich, and the assembly is held in a supporting clamp. An electric field transverse to the sandwich forces the proteins to migrate out of the gel and onto the membrane to which they adhere. Because the membrane has free sites remaining, it is coated with a mixture of nonspecific proteins to block these free sites. Often milk is used for this purpose. The membrane is then soaked in a solution containing an antibody to the protein of interest

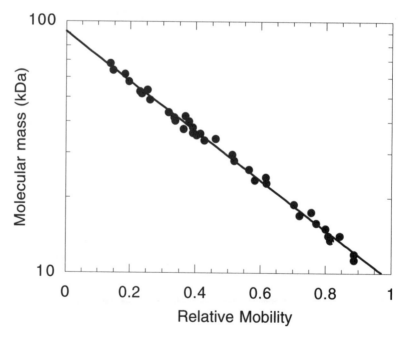

FIGURE 5-7   Plot of relative mobility versus log molecular mass in kilodaltons. Data obtained using a series of known proteins (Weber and Osborn, 1969).

(this antibody is called the **primary antibody**). Since all the protein-binding sites on the membrane are blocked, the antibody can adhere to the membrane only if it interacts with its specific antigen. After the nonadherent antibody is rinsed off, the presence of the antibody may be detected in one of several ways. The most frequently used way is to introduce a second antibody (called the **secondary antibody**) that will react with any antibody from the same biological source as the primary antibody. For example, if the primary antibody was from a rabbit, the secondary antibody could be a goat antibody raised against rabbit immunoglobulin. The secondary antibody is usually coupled covalently to an enzyme that catalyzes a chromogenic reaction, and the presence of the adherent secondary antibody can be assayed by immersing the membrane into the substrate of the coupled enzyme. This provides very sensitive detection of the antigenic protein.

PURIFICATION OF DENATURED PROTEINS

Proteins resolved by SDS gel electrophoresis are coated with detergent and unfolded. Thus, they usually lack enzymatic activity. For some applications, however, denatured protein is useful, and for these applications, SDS-PAGE may be used as a preparative method. For example, when samples are being prepared for injection into animals in order to raise an antibody, it is often useful to do a final purification by SDS-PAGE and then either elute the protein from the gel or simply inject the mashed-up protein band, gel and all, into the animal. The folklore is that the gel acts as an adjuvant in the animal and actually stimulates the production of antibody. Another use is to isolate a pure band of protein for N-terminal sequence analysis. This sequence can then be used for designing oligonucleotide primers for cloning the gene or for verifying that a protein band is the expressed product of a known gene.

In most cases in which an activity is sought, you would not use SDS-PAGE as a preparative method. However, occasionally, SDS-PAGE is the method of choice for the purification of small concentrations of proteins, such as bacterial sigma factors. Sigma factors are small, very acidic proteins and are components of the transcriptional apparatus. After separation by SDS-PAGE, the gel is weakly stained, and the desired protein bands are excised. The bands are then chopped up finely and soaked in solutions of high ionic strength to elute the proteins from the gel. The gel bits are removed by centrifugation, the proteins in the supernate are precipitated with 80% acetone, and the precipitate is collected by centrifugation and washed in 80% acetone. These steps get rid of most of the SDS. The precipitate is then dissolved in a solution containing a strong denaturant, such as guanidine-HCl, which completely denatures proteins. The sample is dialyzed against successively decreasing concentrations of guanidine-HCl and eventually against buffer lacking denaturant. During these dialysis steps, the sigma factors refold and regain activity. In principle, other proteins may also be refolded in this way.

SIMPLE (NONDENATURING, NATIVE) GEL ELECTROPHORESIS

As noted above, SDS-PAGE involves the denaturation of proteins. An alternative method is native or nondenaturing gel electrophoresis, also called simple electrophoresis, in which detergent is not used and the proteins retain their native structure and activity. Simple electrophoresis may be carried out in tubes or in slab gels. We use the latter in the laboratory exercise described in this book. The resolving gels will be overlaid with a stacking gel, just as in SDS-PAGE, but the gels and the electrophoresis buffer will lack detergent. The migration of proteins in nondenaturing gels is due to both the net charge and the size of the protein. Usually, a pH of 8–9 is chosen for the resolving gel, depending on the protein to be visualized. Protein bands are visualized by using dyes, as with SDS-PAGE, after fixing the proteins to the matrix with acetic acid and methanol.

Although the resolving power of nondenaturing gel electrophoresis is not as great as that obtained with SDS-PAGE, there are several applications in which the technique is useful. For example, the technique permits the detection and identification of an enzyme by staining for its activity. In the simplest case, when the enzyme catalyzes a chromogenic reaction, the appropriate colored product will be sufficiently brilliant to visualize directly after immersing the gel in substrate. We detect alkaline phosphatase by this technique in one of the laboratory exercises described in this book. In other cases, it is possible to couple the enzymatic reaction to be assessed with a second reaction that yields a colored product (see Enzymatic Methods in Chapter 9).

Nondenaturing gel electrophoresis has also proven useful in the study of enzyme structure and function. Covalent modifications of proteins that alter the charge and/or the mass of the protein can frequently be detected by simple gel electrophoresis. For example, the covalent uridylylation and de-uridylylation of the homotrimeric *E. coli* PII signal transduction protein were monitored by native gel electrophoresis (Fig. 5-8). Although adding a uridine monophosphate (UMP) moiety to the PII protein increases the molecular mass only slightly, each covalent modification adds negative charges to the protein, causing a change in the charge-to-mass ratio and hence to the electrophoretic mobility. The native gel electrophoresis method revealed the intermediates of PII as the trimeric protein acquired three uridylyl groups.

It is quite easy to purify small quantities of an oligomeric enzyme by native gel electrophoresis and then to determine whether the enzyme is composed of different subunits by SDS gel electrophoresis. First, you resolve the enzyme on a native gel and identify the band exhibiting the activity. This band is excised and subjected to SDS electrophoresis to separate the subunits by the difference in their mass. In Fig. 5-9 (top), the native gel electrophoresis of dimeric forms of the signal transduction protein NRII and a fusion protein containing NRII and the *E. coli* maltose-binding protein (MBP) are shown, along with a hybrid dimer formed

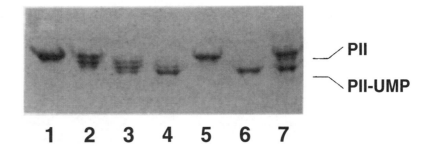

FIGURE 5-8 Use of native gel electrophoresis to monitor the reversible uridylylation of the PII protein of *E. coli.* Highly purified PII protein (lane 5) was uridylylated by purified uridylyltransferase. A time course of the uridylylation reaction is shown in lanes 1–4. Purified PII-UMP is shown in lane 6, and a mixture of purified PII and PII-UMP is shown in lane 7. Since the PII protein is a trimer of identical subunits and uridylylation occurs at a unique site within each subunit, there are four possible states corresponding to 0, 1, 2, or 3 uridylyl groups per trimer. (From Atkinson et al., 1994.)

in vitro from these types of monomers. After excision of the different dimer bands, the enzyme was denatured and fractionated by SDS-PAGE to observe the number and size of the different subunits (Fig. 5-9, bottom).

PORE LIMIT (GRADIENT GEL) ELECTROPHORESIS

The rate of migration (mobility) of proteins in nondenaturing gel electrophoresis is determined by the charge of the proteins at the chosen pH and the frictional resistance of the supporting medium. Thus, the separation is not caused by size alone, and the technique is not usually suitable for the determination of molecular weight. A special variety of nondenaturing gel electrophoresis, known as pore limit electrophoresis, is suitable for the determination of the molecular mass of native oligomeric proteins. The technique is similar to the simple electrophoresis technique discussed above, except that the resolving gel is cast with a polyacrylamide gradient from 30% at the bottom of the gel to 3% at the top. In this technique, there is no need for a stacking gel, and the pH of the system is usually sufficiently alkaline to ensure that most proteins move toward the anode. Proteins migrate through the gel until they reach a point where the pore size of the medium is too small and they can proceed no further. Thereafter, the individual proteins become focused at the point where the pore size becomes limiting. By including standards on the gel, the molecular mass of the unknown proteins can be estimated.

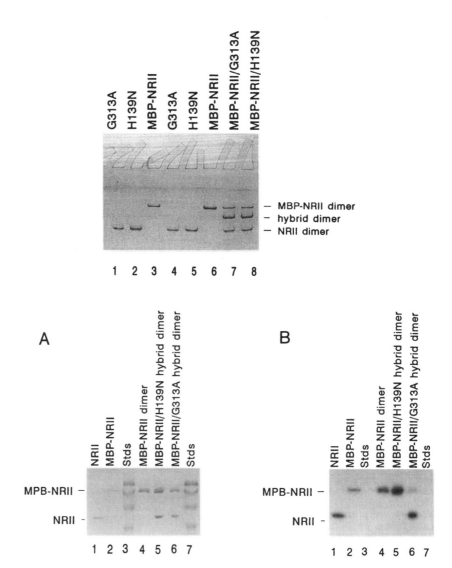

FIGURE 5-9 Use of native gel electrophoresis to purify small quantities of an enzyme and use of SDS-PAGE to characterize the enzyme. *(Top)* Nondenaturing gel electrophoresis showing various dimeric forms of the signal-transducing kinase NRII of *E. coli.* Wild-type NRII is a homodimer of 37.5 kDa subunits. MBP-NRII is a homodimer of 77.5 kDa subunits, consisting of MBP fused to NRII (See Fig. 4-10) The hybrid dimers are heterodimers containing one MBP-NRII and one NRII subunit. The various forms were phosphorylated with γ-^{32}P-ATP before running the gel, separated on the gel, detected by Coomassie Brilliant Blue staining, photographed, and excised. *(A)* SDS-PAGE analysis of the dimers excised from the gel shown above. The gel was stained with Coomassie Brilliant Blue R-250. Note that the hybrid dimers contained one of each kind of subunit. *(B)* Autoradiography of the gel shown in panel A revealed which of the subunits were phosphorylated. (From Ninfa et al., 1993.)

ISOELECTRIC FOCUSING

In simple electrophoresis, the mobility of proteins is due to their charge, which is pH dependent. At its isoelectric pH (pI), a protein is not charged and thus will not move in an applied electric field. This property is exploited by the technique of isoelectric focusing (IEF), which separates proteins by their pI values.

For analytical purposes, the procedure is usually performed in gels cast within thin tubes. A pH gradient is set up within the IEF gel by the inclusion of polymeric buffer components known as ampholytes. These, like proteins, have many positive and negative charges and various pI values. To perform IEF, mixtures containing thousands of different ampholytes are used; if the pI of the protein of interest is known, a set of ampholytes with the pI centered on this pH gradient may be used. The ampholytes are included in the gel upon casting, and the gel is subjected to an applied electric field for several hours. The cathode is in contact with a weak base, and the anode is in contact with a weak acid. During the "pre-running" phase, each of the ampholytes migrates to a position where the pH equals its pI, establishing a pH gradient across the gel. The sample is then introduced, and electrophoresis is continued until the net current is near zero. At this point, each of the components will have migrated to the position in the gel where the pH is equal to its pI, and thus, there are no species present with a net charge.

IEF is useful for determining the pI of proteins. IEF is capable of very high resolution and can be used to detect small changes in proteins, such as covalent modifications and limited proteolysis. However, its use is limited by the high cost of the ampholytes and by the requirement for special equipment. Nevertheless, in combination with SDS-PAGE in the two-dimensional electrophoresis method discussed in the next section, IEF has played an extremely important role in the identification of numerous proteins.

TWO-DIMENSIONAL GEL ELECTROPHORESIS

Two-dimensional gel electrophoresis is the combination of IEF and SDS-PAGE. Typically, a tube IEF gel is run, and the resulting tube gel is then placed horizontally across the top of the stacking gel of a slab SDS-polyacrylamide gel. The proteins are thus separated by their pIs in the first (IEF) dimension and according to their molecular mass in the second (SDS-PAGE) dimension. Since both of these techniques are of very high resolution, the cumulative resolution obtained when both are combined is outstanding. Two-dimensional electrophoresis of total cell proteins from various organisms can resolve hundreds of different proteins. An example of the use of the technique is shown in Fig. 5-10.

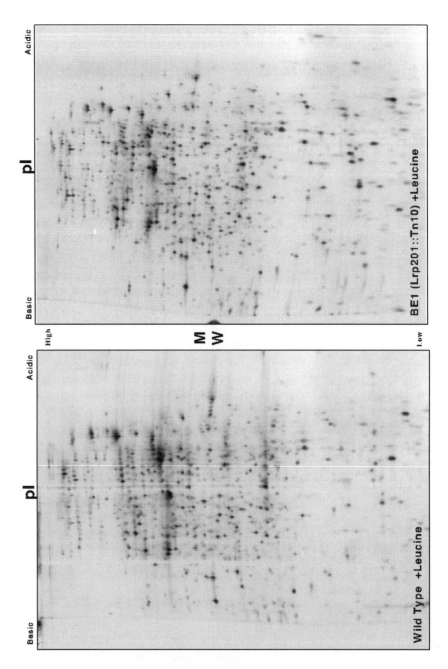

FIGURE 5-10 Two-dimensional electrophoresis. The separation was by IEF tube gel in the vertical dimension and by SDS-PAGE in the horizontal dimension. Higher-molecular-weight proteins are at the left. The top gel was the total cytoplasmic proteins from an *E. coli* mutant lacking the leucine responsive regulatory protein (LRP), while the bottom gel was the total cytoplasmic proteins from wild-type *E. coli*. Differences in the two patterns reveal which proteins are regulated (for expression) by the LRP. This image was provided by Travis Tani, Department of Biological Chemistry, University of Michigan.

Two-dimensional electrophoresis has played an important role in the identification of proteins whose level of expression is affected by a physiological condition or by a mutation. For example, the proteins that are induced by heat shock treatment were discovered by using the two-dimensional method. Several methods have been used to identify the proteins that are visualized as "spots" after two-dimensional electrophoresis. If a point mutation in a known gene is observed to result in a different mobility for a particular protein spot, it is likely that this gene encodes the protein in the altered spot. Moreover, the protein spots may be excised from the gel and subjected to amino acid analysis and N-terminal sequencing. Occasionally, the identity of a protein spot may be gleaned from this information, especially if there are other supporting data. Finally, the gene encoding a protein spot may be cloned by performing protease digestion of the protein in a spot, purifying the peptides and sequencing 10–20 residues from their N-termini, designing oligonucleotides that may encode these sequences, and identifying the gene from a gene library by hybridization against the derived oligonucleotides (see Chapter 11).

The technique of two-dimensional gel electrophoresis has been used to catalogue a large fraction of the proteins of *E. coli*. This work was performed in the laboratory of Frederick Neidhardt of the Department of Microbiology and Immunology at the University of Michigan Medical School (Van Bogelen et al. 1992). These investigators used plasmid clones that each contained a small portion of the *E. coli* genome. They used a technique that permits the specific labeling of proteins that are encoded by extrachromosomal elements (such as plasmids) and enumerated the proteins encoded by each of hundreds of different plasmids. Since the genetic identity of the plasmids was known, this painstaking effort resulted in a catalogue of most of the proteins encoded by *E. coli*. This catalogue is known as the **gene-protein index**.

REFERENCES

Atkinson, M. R., E. S. Kamberov, R. L. Weiss, and A. J. Ninfa. 1994. Reversible uridylylation of the *Escherichia coli* PII signal transduction protein regulates its ability to stimulate the dephosphorylation of the transcription factor Nitrogen Regulator I (NRI or NtrC). J. Biol. Chem. **269:**28,288.

Hames, B. D., and D. Rickwood. 1990. Gel electrophoresis of proteins, 2nd ed. IRL Press, Oxford.

Laemmli, U. K. 1970. Cleavage of structural proteins during assembly of the head of bacteriophage T4. Nature **227:**680.

Ninfa, E. G., M. R. Atkinson, E. S. Kamberov, and A. J. Ninfa. 1993. Mechanism of autophosphorylation of *Escherichia coli* Nitrogen Regulator II (NRII or NtrB): Transphosphorylation between subunits. J. Bacteriol. **175**:7024.

Shapiro, A. L., E. Vineula, and J. V. Maizel. 1967. Molecular weight estimation of polypeptide chains by electrophoresis in SDS-polyacrylamide gels. Biochem. Biophys. Res. Commun. **28**:815.

Van Bogelen, R., P. Sankar, R. L. Clark, J. A. Bogan, and F. C. Neidhardt. 1992. The gene-protein database of Escherichia coli: edition 5. Electrophoresis **13**:1014.

Weber, K., and M. Osborn. 1969. The reliability of molecular weight determinations by dodecylsulfate-polyacrylamide gel electrophoresis. J. Biol. Chem. **244**:4406.

DETERMINATION OF MOLECULAR WEIGHT BY SDS-PAGE

OBJECTIVES

You will be provided with a set of protein standards of known molecular weight and with several protein "unknowns." Your objective is to determine the molecular mass of the unknowns by comparison with the standards. Your instructor will supply the identity and molecular mass of the standards and may suggest what percentage acrylamide gel should be most useful for characterizing the unknowns.

If you carried out the affinity chromatography experiment of Chapter 4, extra lanes on the gels may be used to examine the crude extract and affinity-purified β-galactosidase.

Unpolymerized acrylamide is a potent neurotoxin!

Be sure to review the precautions in this chapter before proceeding with the experiment.

METHOD

Stock Solutions

1. 30% Acrylamide/bis monomers:
acrylamide	29.2 g
bis	0.8 g
H_2O	100 mL
    Filter and store in dark (brown bottle) at 4 $^\circ$C.

2. 1.5 M Tris-HCl, pH 8.8:
Tris base	18.15 g
H_2O	50 mL
    Adjust pH to 8.8 with 1 N HCl (~24.3 mL).
    Bring to 100 mL with H_2O.

EXPERIMENT 5-1

3. 0.5 M Tris-HCl pH 6.8:
    Tris base   3 g
    H_2O 40 mL
    Adjust to pH 6.8 with 1N HCl.
    Bring to 50 mL with H_2O.

4. 10% SDS:
    SDS   10 g
    Bring to 100 mL with H_2O.

5. 10% Ammonium persulfate (APS):
    ammonium persulfate   1 g
    Bring to 10 mL with H_2O (make fresh, shelf life 1 week).

6. 4× Resolving gel buffer:
    1.5 M Tris-HCl, pH 8.8   20 mL
    10% SDS   1.6 mL
    H_2O 18.4 mL

7. 4× Stacking-gel buffer:
    0.5 M Tris-HCl, pH 6.8   20 mL
    10% SDS   1.6 mL
    H_2O 18.4 mL

8. 1× Electrophoresis buffer:
    Tris base   12 g
    glycine   57.6 g
    SDS   4 g
    H_2O 4 liters (no need to adjust the pH)

9. 4× Sample loading buffer:
    4× stacking-gel buffer   10 mL
    10% SDS   18 mL
    β-mercaptoethanol   2 mL
    glycerol   20 mL
    0.01% w/v bromophenol blue

10. 50/10 fixing and destaining solution:
    50% vol/vol methanol
    10% vol/vol acetic acid

11. Staining solution:
    Coomassie Brilliant Blue R-250   1 g in 500 mL of 50/10

12. 10/10 destaining solution:
    10% vol/vol methanol
    10% vol/vol acetic acid

PREPARING THE RESOLVING GEL

To make resolving gels of the appropriate acrylamide concentration, see Table 5-2. The amounts listed are enough to permit the casting of two minigels.

To initiate polymerization, add 3 μL of TEMED, mix gently, and add to the gel mold (Fig. 5-6). Overlay the solution *gently* with water, and allow to polymerize. You should see the gel-H_2O interface disappear and then reappear when the gel is polymerized.

PREPARING THE STACKING GEL

1. Prepare the stacking-gel solution according to Table 5-3. The amounts are enough for two minigels.

2. Remove all H_2O that was overlaid onto the resolving gel. This can be done by turning the gel sandwich upside down and using a Kimwipe to remove the bead of H_2O that remains on the edge of the glass. Do not touch the surface of the polymerized resolving gel.

3. Add the TEMED to the stacking-gel solution and immediately pour the stacking gel onto the top of the polymerized resolving gel.

4. Insert the comb into the wet stacking gel, and allow the gel to polymerize at ambient temperature.

TABLE 5-2
Resolving-Gel Solution

MATERIAL	10% GELS	12 % GELS	15% GELS
H_2O	3.25 mL	2.7 mL	1.92 mL
4× resolving buffer	2 mL	2 mL	2 mL
30% acrylamide	2.7 mL	3.2 mL	4 mL
10% APS	80 μL	80 μL	80 μL
TEMED[a]	3 μL	3 μL	3 μL

[a]Add immediately before pouring.

EXPERIMENT 5-1

TABLE 5-3	
Stacking-Gel Solution	
MATERIAL	**VOLUME**
H_2O	4.6 mL
4× stacking-gel buffer	2.0 mL
30% acrylamide	1.3 mL
10% APS	48 μL
TEMED[a]	5 μL

[a]Add immediately before pouring.

5. When the stacking gel is polymerized, the comb may be removed gently, and the gel sandwich can be loaded into the electrophoretic apparatus.

LOADING THE SAMPLES

1. Mix three volumes of sample with 1 volume of 4× sample loading buffer.

2. Heat the samples in a heating block to 90 °C for 5 min.

3. Inject 10 or 20 μL of the heated sample mixture into the wells of the gel. Be sure to include the standards in two or more lanes on each gel.

ELECTROPHORETIC SEPARATION

1. Apply power at 100–150 V until the tracking dye (usually bromophenol blue) has reached the top of the resolving gel. The power may then be increased to 200 V, if desired.

2. When the tracking dye has reached the bottom of the resolving gel, turn off the power, remove the gel sandwich, disassemble the glass plates, carefully remove the stacking gel with a razor blade and discard, and place the resolving gel in a small tray.

3. Immerse the gel in the staining solution until it is uniformly blue.

4. Decant the staining solution back into the bottle provided (this can be reused many times) and immerse the gel in 50/10 destaining solution.

5. When the gel is mostly destained, decant the 50/10 solution and immerse the gel in 10/10 destaining solution.

6. For the purposes of this laboratory exercise, measure the migration of proteins and tracking dye with a ruler. Calculate the mobility of each protein (mobility = (distance the protein migrated)/(distance the tracking dye migrated)). Plot mobility versus log molecular mass for the standard proteins, and then determine the molecular mass of the unknowns from this plot.

QUESTIONS

1. What is the deduced molecular mass of the unknown proteins?

2. How would you know whether an enzyme is monomeric or is oligomeric and consists of only one type of subunit?

3. How would you modify the experiment to deduce whether the subunits of a multimeric protein are held together by disulfide bonds?

EXPERIMENT 5-2

IDENTIFICATION OF ALKALINE PHOSPHATASE BY SIMPLE (NONDENATURING) GEL ELECTROPHORESIS

OBJECTIVE

We use simple gel electrophoresis, which does not involve the denaturation of proteins, to separate alkaline phosphatase from other proteins. The protein with alkaline phosphatase activity will be identified by an activity stain (shown in Fig. 5-11), while in other samples, the presence of all proteins will be detected using Coomassie Brilliant Blue R-250 staining. By comparison of the results from these two different staining methods, you should be able to identify which protein in the mixture had alkaline phosphatase activity.

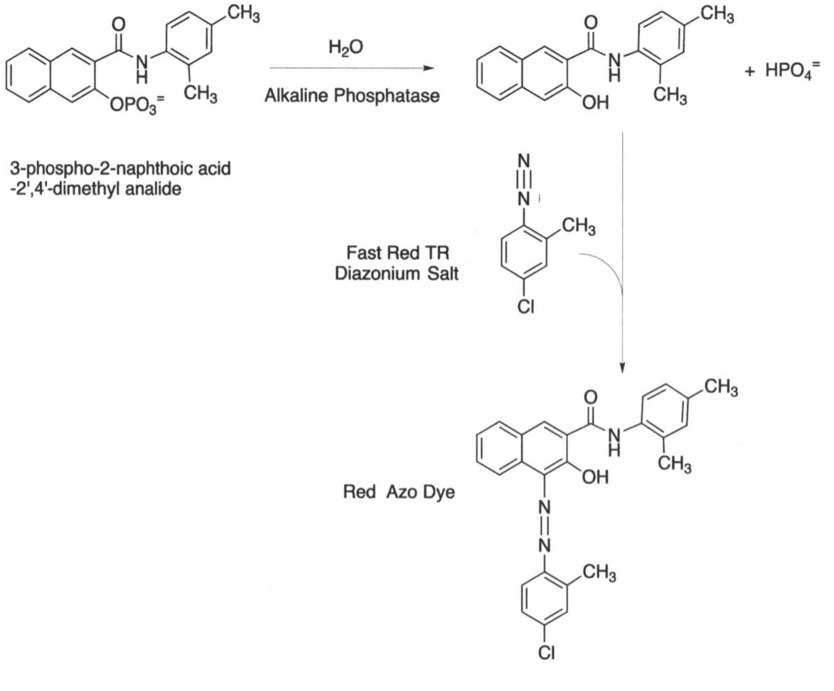

FIGURE 5-11 Reactions used for the staining of alkaline phosphatase in nondenaturing gels.

METHOD

Gel Electrophoresis Stock Solutions

1. Resolving gel: 4× 1.5 M Tris-HCl, pH 8.7. (This pH will vary depending on the proteins that you are trying to resolve.)

2. Stacking gel: 4× 0.5 M Tris-HCl, pH 8.0. (Note that stacking is less efficient at this pH, but it works.)

3. Electrophoresis buffer:

Tris base	40 g
glycine	45.5 g
H_2O	5 liters

4. Sample buffer:

4× stacking-gel buffer	10 mL
glycerol	20 mL
bromophenol blue, 0.01% w/v	
H_2O	20 mL

Alkaline Phosphatase Staining Solution

A mixture of 4-chloro-2-methylbenzenediazonium salt with 3-phospho-2-naphthoic acid-2´,4´-dimethylanilide in Tris buffer. This is sold as a kit for staining gels (Sigma Fast Red TR/naphthol AS-MX alkaline phosphatase substrate, product F-4523).

PROCEDURE

1. Cast and overlay the resolving gels as in the SDS-PAGE process. The correct amounts of acrylamide and other components are listed in Tables 5-2 and 5-3 (except that all buffer solution should lack SDS for the simple electrophoretic experiment). When the resolving gel is polymerized, cast a stacking gel by using the proportions of each component listed in the tables for Experiment 5-1.

2. Your instructors will provide you with a mixture of proteins, one of which has alkaline phosphatase activity. Load the mixture onto the gel twice in two widely separated lanes. After electrophoresis, you will slice the gel in half between these two lanes to provide duplicate samples. You may also run standards provided by your instructors on each half of the gel.

EXPERIMENT 5-2

3. Electrophoresis conditions are similar to those used in the SDS-PAGE experiment. After the tracking dye has reached the bottom of the gel, stop the electrophoresis and disassemble the gel sandwich. Slice the gel in half with a razor blade.

4. Stain and de-stain one of the two halves of the gel with Coomassie Brilliant Blue R-250, as in Experiment 5-1.

5. Stain the other half of the gel for alkaline phosphatase activity. To do this, place the gel in the staining solution containing Tris buffer and Fast Red TR/naphthol AS-MX. Carefully watch the staining reaction to observe appearance of the red band. When this band is clearly visible, stop the reaction by pouring off the staining solution, and rinse the gel in H_2O. The staining solution may be reused several times.

6. Place the two gel halves next to one another, and identify which of the Coomassie-stained bands corresponds to alkaline phosphatase.

REAGENTS NEEDED FOR CHAPTER 5

✓ Acrylamide/bisacrylamide 30% solution (29/1): Sigma A3574, 100 mL

✓ TEMED: Sigma T9281, 25 mL

✓ Ammonium persulfate: Sigma A3678, 10 g

✓ SDS: Sigma L3771, 100 g

✓ Coomassie Brilliant Blue R-250: Sigma B0149, 10 g

✓ Glycine: Sigma G8898, 1 kg

✓ Tris base: Sigma T8404, 1 kg

✓ HCl

✓ β-Mercaptoethanol: Sigma M7154, 25 mL

✓ Glycerol: Sigma G8773, 1 liter

✓ Bromophenol blue: Sigma B8026, 5 g

✓ Methanol

✓ Acetic acid

✓ Alkaline phosphatase staining kit (Fast Red TR/naphthol AS-MX alkaline phosphatase substrate tablets set): Sigma F-4523, 1 kit

✓ Alkaline phosphatase *(E. coli)*: Sigma P4069, 25 units

✓ Protein standards and unknowns: A variety of standards are available. Sigma M3913 (low range, 6500–66,000) or M4038 (wide range, 6500–205,000) is adequate. We typically use side fractions of proteins purified in our laboratory for the unknowns. If these are not available, carbonic anhydrase (Sigma C2273, MW 29,000) or fructose-6-phosphate kinase (Sigma F0387, MW 84,000) is suitable. We also encourage students to run the crude extracts and affinity-purified β-galactosidase from experiment 4-2.

6 Overview of Protein Purification

INTRODUCTION

In the experiments for Chapter 7, you will purify alkaline phosphatase from *E. coli* by using a series of steps prescribed in this book. Before undertaking that project, it is useful to consider how you might design a protein purification strategy for a protein that has never been purified before. Proteins are purified for many uses: clinically as pharmaceutical agents, as specimens for studies of either protein activities and their regulation or of structure/function relationships in proteins, as tools in industrial processes (grain milling, as detergents or emulsifying agents, etc.), as reagents for research (for synthesizing a particular biological substrate, as restriction endonucleases, as protein molecular weight standards, etc.), and as tools for diagnostic studies. For most of these purposes, it is necessary to develop and apply a practical purification procedure before the desired protein is useful.

WHY DO YOU WANT THIS PROTEIN?

Before initiating protein purification, you should consider the requirements of your intended application. Five important factors to consider are the following:

1. How much of the purified protein do you need?

2. How homogeneous does the protein need to be?

3. Do you need to recover the protein in an active form or in the native conformation?

4. What is the least expensive way you can satisfy your requirements?

5. How long and how much effort will it take to achieve your requirements?

You may have to balance these issues. For example, the least expensive way to obtain a purified protein may involve a considerably higher expenditure of time and/or effort and thus may not be practical.

For some applications, you may not need very much of the purified protein nor need it to be active or in its native conformation. For example, you may want to raise an antibody to the purified protein by injecting it into an animal. In that case, the key aspect of the purification would be to obtain a small amount of the protein in a pure enough form so that most of the antibodies raised in response to the injected sample are specific to the protein of interest. For this type of application, you can sometimes use material extracted from an SDS-polyacrylamide gel band as the purified antigen.

For other applications, you may need a considerable amount of an enzyme that is highly active but not necessarily very pure. For example, as you will learn in Chapter 11, microorganisms contain restriction endonucleases, which hydrolyze DNA at specific sequences. These enzymes are used in research applications for gene cloning and for studying gene structure and function; many restriction enzymes are available commercially. The purified restriction endonucleases obtained from commercial sources frequently are not very pure. The only purity requirements for these enzymes are that they must be free of contaminating restriction endonucleases and other activities that affect DNA, such as endonucleases and exonucleases, ligases, etc. Indeed, often a different protein, such as BSA, is added to the purified restriction endonuclease as a stabilizing agent. In a similar way, the enzymes, alcohol dehydrogenase, pyruvate kinase, and lactate dehydrogenase, which we will use in the laboratory exercises demonstrating enzymatic methods of analysis and coupled assays (Chapter 9), do not have to be pure. It is sufficient if these enzymes are free from each other, creatine phosphokinase, and other enzymes that would confuse the experiments.

In other circumstances, it is necessary to have a very pure protein. For example, if you were trying to ascertain the activities of a protein, such as demonstrating that a particular activity is due to the protein of interest, you would want to have a very pure enzyme. The study of enzyme regulation or structure/function usually requires a homogeneous enzyme. Crystallization of a protein as a prelude to structure determination may require highly purified enzyme, since high-quality crystal forms generally occur only in the absence of contaminating proteins.

Regardless of the application and the attendant purity, activity, and quantity requirements, optimizing the purification of a new protein involves the same intellectual steps: (1) development of a suitable assay procedure, (2) selection of the best source material, (3) solubilization of the desired protein, and (4) development of a suitable series of fractionation steps, concentration procedures, and storage procedures.

DEVELOPMENT OF A SUITABLE ASSAY PROCEDURE

In order to purify a protein, you must have an assay available for this protein so that you can monitor how it partitions during various fractionation steps. To be practical, it must be possible to perform the assay rapidly on many samples, and it must reliably indicate the amount of the desired protein present at the various stages of purification. In addition, the assay used should not significantly deplete the overall yield of the enzyme from the preparation. Some of the strategies used and problems frequently encountered will be mentioned in this section.

Losses invariably occur during fractionation steps. If you had to perform 10 fractionation steps to purify an enzyme and each step had an ~80% efficiency, after the final step you would have <20% of the starting material.

ASSAYS OF ENZYME ACTIVITY

As we learned in our affinity purification of β-galactosidase (Chapter 4), a colorimetric enzymatic assay may be conveniently applied during protein purification. Our assay was specific, very sensitive, cheap, and rapid, and could easily be applied to many samples. These are important characteristics that you should look for in an assay. Indeed, as you will see in our purification of alkaline phosphatase, a very rapid qualitative assay such as a "spot test" can be quite useful, even if it is not as rigorously quantitative as the continuous assay.

Sometimes the enzyme you are attempting to purify cannot be easily assayed directly but can be coupled to other enzymes. For example, if you were to try purifying creatine phosphokinase, a coupled assay would probably be the assay of choice. You will develop a coupled assay for this enzyme in Chapter 9.

More serious problems arise when the activity that is measured requires more than one protein from that preparation or requires both a protein and a cofactor. A fractionation step might separate the two components needed for activity, such that no single resulting fraction has any activity at all, and it would appear that you had lost all of your activity. To ascertain whether a critical factor was separated, you might try to combine aliquots of the various fractions to see if activity is regained.

ASSAYS THAT MEASURE THE DESIRED PROTEIN DIRECTLY

Another approach is to measure the presence of the desired protein directly in the various fractions. The weakness of such assays is that they reveal nothing about the activity of the protein, and thus, you could spend some time purifying the protein only to find out that activity was lost at an unknown point along the way. Nevertheless, such assays (as described below) may be useful if the procedure is quick, and the activity proves to be stable.

In the Ninfa and Ballou laboratories, we often purify proteins from genetically manipulated bacteria that overexpress the protein of interest. Sometimes the overexpression is such that the desired protein constitutes 1% or more of all soluble protein in the crude extract. In that case, we can easily see the protein band on an SDS-polyacrylamide gel, and gel electrophoresis can be used to monitor the presence of the desired protein in various fractions. In other cases, the protein may not be present in sufficient quantity to be visualized on gels but may be immunologically detected. If you are fortunate enough to be purifying a protein with a cofactor such as a heme or a flavin, the protein will be colored, so that it can often be detected directly.

Another approach is to "tag" the protein of interest with a radioactive or fluorescent label, so that the presence of this label can be monitored in protein fractions. For example, many proteins that are integrated into the cell membrane serve as receptors for ligands such as other proteins, peptides, or nucleic acids. If the ligand can be produced in radiolabeled form, it may be possible to covalently cross-link the labeled ligand to the receptor and track the purification of the ligand-receptor complex by monitoring the presence of the labeled ligand in the separated fractions.

TIME, TEMPERATURE, AND YIELD

Proteins in living cells vary in stability. Many proteins are "turned-over" rapidly as a result of the action of proteases that digest the proteins into peptides and, eventually, into amino acids. Thus, to avoid proteolysis, especially in the early stages of protein purification, it is usually desirable to proceed with the fractionation of the protein mixture as rapidly as possible. The action of proteases may be retarded by the addition of one or more protease inhibitors (such as phenylmethylsulfonyl fluoride, PMSF) to the crude mixture, but even this often does not completely solve the problem of proteolysis. Also, many enzymes are unstable for other reasons during protein purification. For example, they may stick to the walls of the vessels, aggregate, become oxidized, or interact in unfavorable ways with other components of the crude protein mixture. Thus, depending on the stability of the protein you are trying to purify, assays that take a lot of time might not be very useful for tracking the progress of the purification. For example, assays of biological activity (i.e., those that depend on observation of responses of whole organisms or tissues to a protein or enzyme) are often quite impractical. In Chapter 7, we will purify alkaline phosphatase, one of the most stable enzymes known. Indeed, we chose to purify this enzyme because the class periods do not permit us to carry out complete purifications quickly. The leisurely pace of our purification scheme would not be useful for most enzymes.

Keeping the extract cold helps to retard the inactivation of most proteins in crude extracts. Thus, most manipulations are performed in the cold room or on ice. It should be noted, however, that not all proteins are resistant to cold—in fact, some enzymes are irreversibly inactivated by cold. Thus, it helps to know some information about the stability and temperature resistance of the protein that you are trying to purify.

In general, all of the above problems can be minimized if the entire purification is carried out in as short a period as possible. A key criterion for development of a good purification protocol is to make the whole series of steps proceed quickly.

SELECTION OF THE BEST SOURCE MATERIAL

If a choice of starting tissue is available, choose the source that has the best combination of the following qualities: it contains the greatest amount or most stable protein, it is least costly, or it is most convenient to use. For example, even if hamster liver has about twice as much of the desired protein per gram of tissue as bovine or porcine liver does, the bovine or porcine source might be better, because very large amounts of these tissues can be easily obtained. If you need to conduct more than one purification, as is likely, it is highly desirable to have a source of tissue that reproducibly has a sufficient quantity of the protein of interest or one of the other desirable characteristics listed above. If you are isolating the protein from tissue obtained at a slaughterhouse, it may be worthwhile to demonstrate that several different batches of the same tissue all have the protein of interest in similar quantity. Otherwise, you must identify a batch of the tissue sample that has a sufficient quantity of the protein and store as much as possible of this tissue batch for future purification.

In addition to having a good amount of the protein, it is desirable that the protein be easily and reproducibly extractable from the tissue source; it is also helpful if no components are present that are bothersome contaminants or could otherwise impede the purification process. It does not do much good to have a tissue source rich in the desired protein, if the desired protein cannot be successfully solubilized from the tissue, if the tissue is rich in proteases that digest the protein of interest, or if the tissue contains a component that is a bothersome contaminant that cannot be easily removed from the protein of interest by the fractionation steps you are using. Often, you may want to reevaluate the source of tissue as you learn more about the protein during the initial purification attempts.

HYPER-EXPRESSION OF THE DESIRED PROTEIN(S)

If the gene for the desired protein is available, it may be possible by genetic manipulation to achieve significant overexpression of the desired protein. Many different "expression systems" for the overproduction of proteins in bacteria are available. Unfortunately, proteins from eukaryotes are subject to more complicated transcription, covalent modifications, and post-translational processing events than probably will occur in bacteria. In eukaryotes, the genes are not usually contiguous, but have both introns (portions not expressed) and exons (portions expressed); in the eukaryotic cell the exons are ligated together in the form of the mRNA that will translate the proper gene product (the protein). You can usually obtain the appropriately constructed gene by isolating the mRNA and using reverse transcriptase to synthesize the cDNA that contains the expression message. Even if you can express the proper peptides from the complementary DNA (cDNA) if subsequent protein processing is required, the bacterially produced protein may not be suitable. Another option in that case is to overexpress the desired protein in tissue culture cells by using expression systems based on baculovirus (an insect virus) vectors (see Chapter 11).

GENETICALLY TAGGING THE DESIRED PROTEIN(S)

In addition to permitting the overexpression of proteins, genetic manipulations may also be used to tag the desired protein with a peptide sequence that promotes the export of the protein from the cell, or permits affinity purification. (This is discussed in Chapter 4.) For example, adding a "signal sequence" to the N-terminal end of the protein may often bring about export of proteins from bacteria. The signal sequence serves as an export message for the cellular secretory apparatus, which cleaves off the sequence and secretes the desired protein outside the cell and thus away from most contaminating proteins. In this case, the culture medium would serve as the starting tissue source.

Genetic manipulations can also be used to add tags that permit affinity purification of proteins. For example, the *E. coli* MBP may be fused to the protein of interest. The fusion protein can then be purified by affinity chromatography on maltose columns, taking advantage of the great affinity of the MBP for maltose (Fig. 4-10). It is possible to engineer such constructs so that they contain sites for cleavage by specific proteases between the tag and the protein of interest. Then the tag can be removed by brief treatment with the specific protease. The tag can be easily separated from your protein by passing the solution over the affinity column again. This time the tag will stick, but your protein will pass directly through the column. Another system that can tag your protein places a polyhistidine stretch of amino acids (polyhistidine binds nickel) on one end of the protein. A special nickel affinity column (it costs more than five cents) is used to bind your protein. Then your protein can be eluted with imidazole. Often the polyhistidine tag does not affect the function of your protein. If it does, it may be possible to engineer a specific protease site by which it can be removed after purification. Other systems include fusion of glutathione-*S*-transferase (which binds glutathione) with the protein of interest or fusion of biotinylation sites on the protein. In the latter case, the protein of interest will be biotinylated in vivo and can be purified by affinity chromatography on columns containing immobilized streptavidin. (See also Chapters 4 and 11.)

SOLUBILIZATION OF THE PROTEIN

In order to purify the protein, you must solubilize it for subsequent purification steps. This involves liberating the protein from the intact cells. The key considerations are the efficiency of the extraction process and whether preferential extraction of the protein from the cells yields significant purification.

In our purification of alkaline phosphatase, we take advantage of the fact that this enzyme is located in the **periplasm** (the gel space between the cell wall and the plasma membrane) from which it can be selectively released without significant contamination by the vast number of cytoplasmic proteins. Although the osmotic shock treatment that we will use does not have 100% yield, the gains in purification by this gentle extraction method make it the method of choice. Essentially, after the extraction, we are left with the relatively easy task of purifying alkaline phosphatase away from the relatively small number of other periplasmic proteins.

Eukaryotic cells contain many compartments, such as various membranes, the nucleus, the endoplasmic reticulum, the Golgi apparatus, mitochondria, chloroplasts (in plants), lysosomes, etc. Occasionally, it can be observed that a particular subcellular compartment is especially rich in the protein of interest, and in these cases, it is worth exploring whether fractionation of subcellular compartments may serve as a useful purification step. Various types of **differential centrifugation**, including **density gradient centrifugation**, can be used to obtain subcellular compartments. Usually, when working with eukaryotic tissue it is advantageous to separate the organelles to obtain the specific source. Voet and Voet (1995), Scopes (1987), Harris and Angel (1989), Cooper (1977), and Deutscher (1990) describe many schemes for obtaining rather pure fractions of particular organelles. For example, it is possible to isolate chloroplasts from plant cells and observe that the specific activity of the photosynthetic apparatus is greatly enriched in this organelle, as compared with that in the whole-cell extract. Alternatively, the contents of subcellular compartments may be obtained by selective lysis.

Unless the protein of interest is exported from the cell or can be released from the surface of the cell, it will be necessary to disrupt the cells to release the protein. Various **cell disintegration techniques** are used. Of course, the protein of interest must be able to withstand the treatment used for solubilization. The most gentle techniques involve cell lysis by freezing and thawing of cells, osmotic disruption, digestion of cell walls with enzymes, or chemical solubilization, such as extraction with toluene or treatment with detergents. Also, homogenization by grinding cells may provide a fairly gentle extraction. If these techniques are not suitable, homogenization in a blender or grinding with abrasives such as sand or alumina may be useful. Finally, more vigorous cell disintegration techniques may have to be used. These include lysis at very high pressure (French press), disruption by ultrasound (sonication), or disruption in a "bead mill" in which vibration with glass beads is used to rip the cells open.

Regardless of the cell disruption method used, it is important that the method can be reproducibly performed and that the sample can be maintained at the desired temperature during the disruption process. Since some of the extraction techniques generate heat, various devices are used to dissipate the heat generated during tissue disintegration.

The purification of membrane proteins presents unique problems, since in many cases, the membrane-spanning parts of these proteins must be kept in a hydrophobic environment to prevent aggregation and maintain activity. The solubilization of membrane proteins often involves the use of nonionic detergents such as octyl-glucoside to "coat" the membrane-spanning parts of the protein. Successive chromatographic steps must then be performed in the presence of the solubilizing detergent. Many different gentle detergents have been used to purify membrane proteins.

INITIAL STEPS OF PURIFICATION

CLARIFICATION OF THE EXTRACT

Once the cells are disrupted, the resulting crude extract usually has to be cleaned up before starting the fractionation steps. This is because many of the extraction methods release small membrane micelles, particulate and aggregated matter, nucleic acids, and other troublesome materials that will foul up successive steps. The most common clarification method involves moderate- to high-speed centrifugation of the crude extract, similar to the method we use in the purification of alkaline phosphatase. The addition of protamine sulfate or streptomycin sulfate to 2% (wt/vol) results in precipitation of nucleic acids and ribosomes, so such an addition before centrifugal clarification is often very useful.

In some cases, the crude extract will contain considerable particulate material, and this is often removed by filtration with cheesecloth or paper filters. Care must be exercised not to introduce air or cause foaming of the extract during these filtration steps, since this can often denature proteins.

DIFFERENTIAL SOLUBILITY

Techniques that take advantage of the different solubilities of individual proteins may be used with a threefold advantage: they provide some fractionation of the proteins in the crude extract, they concentrate the fraction, and they serve to clarify the crude extract, enhancing the effectiveness of high-resolution chromatographic steps to follow. The most common techniques used are salting-in and salting-out, isoelectric precipitation, precipitation by acetone or alcohols, and denaturation by heat.

Proteins are least soluble and often precipitate, when the pH of the solution equals their pI, the pH at which the net charge on the protein is zero. This property may be used to precipitate selectively the protein of interest or to precipitate many other contaminating proteins leaving the protein of interest in solution. However, this technique is most useful when the protein of interest has a pI that is very different from those of the majority of cellular proteins, i.e., when the protein we desire is exceptionally acidic or basic. Of course, if enzymatic activity is sought, the stability of this activity at the conditions to be used must be verified.

Proteins can be "salted-in" and "salted-out" of solution by varying the ionic strength of the solution, and often this can be an excellent purification step, particularly when the protein of interest differs significantly in solubility from most other proteins in the crude extract. The unique solubility properties of the *Klebsiella aerogenes* Nac transcription factor provide an easy method for purifying this protein. Nac is insoluble in solutions containing <50 mM KCl but is completely soluble in solutions containing 150 mM KCl. Thus, considerable purification of Nac can be achieved by making a crude extract at 150 mM KCl, clarifying the extract by high-speed centrifugation, and then dialyzing against buffer containing 50 mM KCl. Nac selectively precipitates during this dialysis and can be easily collected by centrifugation. The precipitated Nac can then be solubilized in 150 mM KCl and clarified. Two cycles of such treatment yield essentially homogeneous preparations of Nac that are highly active.

For many proteins, salting out with the lyotropic agent ammonium sulfate is a very useful fractionation step (see hydrophobic-interaction chromatography in Chapter 4). This step may also be used to clarify crude extracts. Ammonium sulfate is highly soluble in water, and frequently has stabilizing effects on proteins; thus, the salting-out technique is gentle. As before, to be most useful as a fractionation step, the solubility properties of the protein of interest must differ from those of the majority of the contaminating proteins in the extract. Since salt precipitation depends on both the concentration of protein in the crude extract and the temperature, these factors must be carefully controlled for the procedure to be reproducible. Additions of ammonium sulfate to solutions cause a decrease in the pH; thus, it is also advisable to have excess buffering capacity present and frequently monitor the pH.

A typical procedure for ammonium sulfate fractionation would be to bring the crude extract to a concentration of ammonium sulfate just below that necessary to precipitate the protein of interest; those proteins that do precipitate are removed by centrifugation. Then the ammonium sulfate concentration is increased to a level that precipitates most of the desired protein in a reasonable period of time. The precipitated protein is recovered by centrifugation and suspended in a small volume of buffer lacking ammonium sulfate. Alternatively, if the protein of interest is salted out by very low ammonium sulfate concentrations, the crude extract may be brought to this concentration and the precipitated protein collected by centrifugation. The former technique is useful for the purification of proteins that are soluble at elevated concentrations of ammonium sulfate, while the latter method is useful for the purification of proteins that are precipitated by ammonium sulfate at low concentration.

We will use the technique of heat denaturation during our purification of alkaline phosphatase. Since most cellular proteins are irreversibly denatured by heating at 80 °C, this is a powerful technique for the purification of the small number of proteins that remain soluble and active after such treatment.

Finally, proteins may be selectively precipitated by alcohols or by acetone. These techniques are not widely used because they denature or inactivate many proteins, and the use of large volumes of organic solvents is undesirable. However, for those proteins that survive the treatment, they can be powerful fractionation steps and have the additional advantage of helping to clarify the extract.

DEVELOPING A SERIES OF HIGH-RESOLUTION CHROMATOGRAPHIC STEPS

GENERAL PRINCIPLES

In Chapter 4, we discussed at length some of the various types of chromatography commonly used in the purification of proteins. At this point, we comment briefly on the methods used to establish which types of chromatography are most useful for the purification of a protein of interest. Important factors to consider are the following: (1) the capacity and yield of the chromatographic step, (2) dilution of the sample resulting from a chromatographic step, (3) the speed of the step, (4) the retention of activity throughout the step, and (5) the cost of the required materials and equipment. A helpful guide to chromatographic procedures written by Fulton and Vanderburgh (1996) can be obtained from the World Wide Web at http://www.pbio.com.

We have already commented on the importance of having a high yield for each chromatographic step. Usually, a single chromatographic step will not result in a homogeneous protein preparation, and thus a series of different steps will be required. Steps that result in good purification, but considerable loss of material, are useful only if the starting material is inexpensive and easy to obtain or small amounts of the protein are needed. Similarly, the capacity of each chromatographic step must permit the fractionation to be performed at reasonable expense, within a reasonable time frame, and without excessive dilution of the sample.

In the affinity chromatography laboratory exercise (Experiment 4-2) described in this book, we use an affinity chromatography matrix linked to an immobilized substrate analog for the purification of a small quantity of *E. coli* β-galactosidase. That laboratory exercise is intended to illustrate the process of examining the feasibility of a chromatographic step for the purification of an enzyme. It is strongly advised that similar small-scale feasibility studies be conducted for each proposed step before attempting to purify large amounts of the desired protein. It is important that the material used for the small-scale experiments be as similar as possible to the samples that will be used in the large-scale purification. Even though this has been carefully controlled, it is sometimes observed that chromatographic steps function differently after scaling up. For example, in some cases, the binding of the protein to the stationary phase appears stronger after scale-up, such that harsher elution conditions may be required to elute the protein.

WHY YOU "NEVER FOLLOW A SEAL ACT WITH A SEAL ACT"

Usually, it is best to assemble a series of chromatographic steps that separate molecules according to their different molecular properties. As they used to say in Vaudeville, "never follow a seal act with a seal act." It would be questionable to follow an ion exchange step with another ion exchange step that achieves separation based on the same property of the protein. Contaminating proteins that copurify with the protein of interest in the first ion exchange step are likely to also copurify with the desired protein in the second step. It would be better to follow an ion exchange step with a gel-filtration step (separating according to molecular mass), a hydrophobic interaction chromatography step (separating according to hydrophobicity), a chromatofocusing step (separating according to pI), or an affinity chromatography step (separating according to biological specificity).

GRADIENT ELUTION FROM CHROMATOGRAPHIC COLUMNS

Although in our laboratory exercises we use a single concentration of eluting buffer to elute proteins from chromatographic columns, often a more effective method of elution is to use a gradient of the eluting substance (usually salt). These gradients may be formed by using simple devices or sophisticated ones as described in Chapter 4. Shallow gradients have the advantage of providing the best separation of the desired protein from contaminants differing slightly in property, but have the disadvantage of diluting the sample. Steep gradients provide less effective fractionation, but dilute the sample less and may even concentrate it (Chapter 4). The choice as to the best gradient to use depends on the effectiveness of the other chromatographic steps to be used.

Often proteins are eluted from chromatographic media in a broad peak that is spread out over a large volume of elution buffer. This is commonly seen with hydrophobic interaction chromatography and dye affinity chromatography (Chapter 4). In such cases, unless a suitable method for the concentration of the sample is available, the step may not be useful.

In other situations, proteins are eluted from chromatographic steps in fairly sharp peaks with "shoulders" overlapping with the peaks of eluted contaminating proteins. In this case, it is important to consider whether yield or purification is more important. If yield is essential, the whole peak of eluted protein must be taken despite the presence of many contaminants. However, if purity is the main requirement, fractionating the eluate into many small fractions and collecting only the most pure of the fractions of the protein of interest may be most advantageous. The "side fractions" of the eluted protein may be kept separately and recycled through the procedure at a later time.

The number of fractions collected after elution should be carefully considered. If your assay for the protein of interest is difficult, time consuming, or expensive, then fractionation of the sample into many fractions of small volume would not be appropriate. However, the fractionation power of the column is lost if only a few large-volume eluted fractions are taken. As a rule, it is seldom useful to collect fractions that are >10% or <1% of the bed volume. Since buffer is usually the least expensive element used in protein purification, within the above limits, it is wise to collect many fractions until the behavior of the desired protein in the chromatographic step is well characterized.

METHODS USED TO CHANGE BUFFER AND CONCENTRATE PROTEIN SAMPLES

Almost invariably, you will have to change the buffer that the sample is in during protein purification. For example, ion exchange chromatography often requires that the protein sample be applied at low ionic strength, followed by elution at higher ionic strength. Thus, the resulting sample is in a solution of high ionic strength. If a second chromatographic step to be used requires application of the sample at low ionic strength, the salt in the sample will have to be removed.

Many proteins are labile when dilute; therefore, to maintain the activity of the desired protein, it may be necessary to concentrate the sample at some point in the purification process. In our purification of alkaline phosphatase, we illustrate both of these principles. After heat denaturation, the sample will be concentrated by ammonium sulfate precipitation. Since the next step, chromatography on DE52, requires application of the sample at low ionic strength, ammonium sulfate will be removed from the resuspended protein by dialysis against a buffer of low ionic strength. These steps (ammonium sulfate precipitation followed by dialysis) are effective and widely used. The limitation of the dialysis method is that it is slow.

Other methods used to change the buffer include gel-filtration chromatography, as noted in Chapter 4. Gel filtration of small samples can be used to change over the buffer rapidly by using small columns that exclude the protein of interest, as noted in Chapter 4. Because gel filtration may also be a useful purification step, it may be useful to use a large gel-filtration column with a resolving range that includes the desired protein in place of dialysis steps. Gel filtration is slow, just like dialysis, but has the added advantage of providing protein fractionation. As noted in Chapter 4, gel filtration results in some further dilution of the sample.

Another way of changing over the buffer and concentrating the sample is to apply the sample to a small chromatographic column and elute with a small volume of elution buffer that completely removes the protein of interest. For example, in the purification of the *E. coli* sigma factor σ^{54}, the penultimate step is gel filtration on a large column. The eluted peak fraction from this column, usually ~50 mL in volume after pooling, is then applied to a small (2-mL) phenyl-Sepharose hydrophobic interaction column. This column has high capacity for the σ^{54} protein and retains all of it. The sigma factor is then eluted with 2 mL of buffer containing 50% ethylene glycol. Thus, a 25-fold concentration of the protein is quickly achieved. Also, the eluting buffer happens to be the optimal storage buffer for this protein, so the eluted samples need only be aliquoted and put into the freezer. Similarly, ion exchange columns may be used to concentrate dilute protein samples.

Finally, concentration of the sample and buffer exchange may be obtained by using various filtration devices. Most of these devices use vacuum or centrifugation to force the sample against a membrane that retains the protein of interest but lets buffer pass through. After several cycles of concentration and dilution into the desired buffer, the exchange of buffer is often sufficiently complete. Losses from this method result from sample sticking to the membrane. In addition, these methods tend to be expensive and, if a large starting volume must be processed, may be excessively slow.

A LOGICAL SERIES OF STEPS

Each new protein purification will present its own set of challenges. However, a logical series of steps may be used to achieve effective purification with high yield in a minimal amount of time. A few general principles to follow are the following:

1. It is best to place a step that provides high capacity and high purification early in the scheme. This will ensure that additional steps providing lower resolution are most effective. For example, affinity chromatography steps are often excellent as a first chromatography step.

2. Gel filtration ought to be used at a point in the process in which a buffer exchange would be needed anyway, preferably late in the purification scheme.

3. The sequence of steps should be arranged so that material from one step may be used immediately in the next step with a minimum of manipulation. For example, hydrophobic-interaction chromatography is frequently used after ion-exchange chromatography. This is because the ion exchange step will often result in a sample that has elevated salt concentration, and a high salt concentration is usually useful in applying proteins to hydrophobic interaction columns.

STORAGE OF THE PURIFIED PROTEIN

If you are purifying a large amount of a protein, it makes sense to be sure you know how to store it so that activity is not lost. In initial experiments, try several different storage conditions to realize suitable conditions for storage.

Proteins are typically stored at low temperatures, such as at –80 °C in ultralow freezers, or in liquid nitrogen. Cryogenic agents are frequently added to stabilize the protein during freezing and thawing. Commonly used cryogenic agents include salts, glycerol, ethylene glycol, and dimethyl sulfoxide. The usefulness of each of these should be checked on a small scale. Since the presence of various contaminants may affect the stability of the purified protein, you may have to conduct a complete small-scale preparation to obtain purified material with which to test storage conditions.

Some proteins are sensitive to cold and survive only at room temperature. This is unusual, but should be kept in mind.

Other stabilizing agents that might be examined for inclusion are small molecule substrates or products, or another protein such as BSA, which can partially coat the walls of storage containers so that surface denaturation of your protein is minimized.

Whatever conditions you find suitable for storing your protein, it is a good idea to store it in aliquots in which the tubes are numbered. The tube number of the material used should then be recorded in the laboratory notebook for each experiment involving the purified protein. Then, if there is a "shelf life" problem such as loss of activity upon freezing and thawing of a tube, this will be quickly apparent from your notebook.

PROTEIN PURIFICATION TABLE

The protein purification table is intended to help you focus your attention on the purification process and identify steps that need improvement. A table should be compiled for each protein purification undertaken. Furthermore, the purification table serves as a convenient summary of each preparation and can be referred to by laboratory mates who are also using the protein and need to know how it was obtained. As you learn more about the biological properties of your protein, information from the purification table and notebook will take on increased significance. A sample purification table is given on the next page. Feel free to copy it for your experiments.

PROTEIN PURIFICATION TABLE

Protein: _____

Date Begun: _____ Investigator: _____

Protein Assay: _____

Starting Material: _____ Weight: _____

Definition of Protein Activity: _____

Assay Procedure: _____

PROCEDURE	FRACTION	VOL (ML)	UNITS ML	TOTAL UNITS	PROTEIN MG/ML	TOTAL PROTEIN (MG)	UNITS MG PROTEIN	YIELD (%)	PURIFI- CATION	REMARKS

REFERENCES

Cooper, T. G. 1977. The tools of biochemistry. John Wiley & Sons, Inc., New York.

Deutscher, M. P. 1990. Guide to protein purification. Methods in enzymology, vol. 182. Academic Press, Inc., New York.

Fulton, S., and D. Vanderburgh. 1996. The busy researcher's guide to biomolecular chromatography. PerSeptive Biosystems, Framingham, Mass.

Harris, E. L. V., and S. Angel. 1989. Protein purification methods: a practical approach. IRL Press, Oxford.

Scopes, R. K. 1987. Protein purification: principles and practice, 2nd ed. Springer-Verlag, New York.

Voet, D., and J. G. Voet. 1995. Biochemistry. 2nd ed., chapter 5. John Wiley & Sons, Inc., New York.

7

Isolation and Characterization of the Enzyme Alkaline Phosphatase from *Escherichia coli*

Objectives

Introduction and Basic Principles

Experiment 7-1

Characterization of Purified Alkaline Phosphatase

Appendix 7-1. Assay of Alkaline Phosphatase

Reagents and Equipment Needed for Chapter 7

OBJECTIVES

This chapter illustrates some of the principles of protein purification, enzyme and protein assays, and enzyme characterization. You will receive instructions for a complicated series of procedures to isolate and investigate properties of alkaline phosphatase. The procedures used include column chromatography, ammonium sulfate fractionation, heat denaturation, spectrophotometry, gel electrophoresis, and others. Some of the principles of steady-state enzyme kinetics will be illustrated in the experiments described in Chapter 8, which will use the purified alkaline phosphatase produced in the exercise.

INTRODUCTION AND BASIC PRINCIPLES

ALKALINE PHOSPHATASES

Phosphatases are a group of enzymes that catalyze the hydrolysis of phosphate monoesters by the following reaction:

$$R–O–PO_3H^- + H_2O \rightarrow R–OH + H_2PO_4^-$$

Alkaline phosphatases (orthophosphoric monoester phosphohydrolases, EC 3.1.3.1) are the phosphate hydrolases that have maximum activity at a relatively high pH (>7.0). These enzymes are widespread in nature, occurring in many eukaryotic and prokaryotic cells. We describe the extraction and purification of the major alkaline phosphatase from the bacterium, *Escherichia coli*. The physiological role of this enzyme in *E. coli* is to cleave phosphoryl groups from a wide variety of phosphorylated compounds, thereby providing the cell with a source of inorganic phosphate (P_i). The enzyme is found in the **periplasmic space**, that is, between the outer membrane and the cytoplasmic membrane of the cell. The outer membrane is permeable to many phosphorylated compounds because it contains channels formed by proteins known as porins. The inner membrane, however, is not permeable to most phosphorylated compounds. Alkaline phosphatase in the periplasmic space liberates P_i, which is then bound by a small carrier protein known as the P_i-binding protein. The binding protein delivers phosphate to a specific high-affinity transport system, known as the Pst system, which transports the phosphate across the cytoplasmic membrane.

Levels of alkaline phosphatase are low in wild-type *E. coli* cells grown in phosphate-rich media; however, if these cells are grown in a medium lacking P_i, expression of the alkaline phosphatase gene is activated, and the cells produce large amounts of the enzyme. For our purpose, we use an *E. coli* mutant in which control of phosphatase production is defective, so that the cells produce elevated amounts of the enzyme under all culture conditions. This kind of mutant is said to be **constitutive** for alkaline phosphatase expression. These mutant cells are the source of the enzyme for the experiments described in this book.

Alkaline phosphatase of *E. coli* is a zinc-containing enzyme with two metal atoms per molecule. The enzyme is dimeric ($M_r = 86,000$) with two identical monomers (Schlesinger and Barrett, 1965). The dissociated monomeric subunits have no activity. The pH optimum for activity is 8.0, and the isoelectric point (pI) is 4.5. As noted above, the enzyme in *E. coli* is secreted into the periplasmic space, i.e., the space between the outer membrane and the cytoplasmic membrane of the bacterium. Since there are relatively few proteins secreted into the periplasmic space, as compared with the vastly larger number of proteins within the cell, we obtain substantial purification of alkaline phosphatase by selectively releasing the periplasmic proteins from the cell. The cell wall consists of the cytoplasmic membrane coated with a cross-linked peptidoglycan mesh. This mesh is a polymer of *N*-acetylmuramic acid and *N*-acetylglucosamine units linked glycosidically and attached to short peptides that are cross-linked to provide mechanical strength analogous to reinforced concrete. The peptidoglycan is also attached to lipid moieties that anchor it to the plasma membrane. This mesh gives the cell the strength necessary for prevention of swelling from osmotic pressures that might result when concentrations of metabolites in the environment change. The outer membrane is very sensitive to osmotic shock. We treat the cells with lysozyme to partially fragment the peptidoglycan mesh and subject them to osmotic shock to liberate the periplasmic proteins into the medium. Most other components of the cell remain enclosed by the cytoplasmic membrane. This process forms vesicles called **spheroplasts** (Neu and Heppel, 1965).

E. coli alkaline phosphatase is somewhat unusual, because unlike most other enzymes, it is very heat stable. We use this heat stability to purify the enzyme by heat denaturing other proteins in the solution. The denatured proteins are insoluble and can be removed by centrifugation. In general, the heat stability of enzymes varies greatly, depending on the enzyme of interest. To help decide whether heat denaturation will be useful for a purification of a particular enzyme, it is necessary to determine the heat stability of the enzyme you wish to purify. For enzymes such as alkaline phosphatase, this is an excellent step, whereas for many other enzymes it is not useful. The partially purified enzyme, after concentration by salting-out with ammonium sulfate, is further purified by ion-exchange chromatography on DEAE cellulose. This purification process is adapted from the method originally described by Garen and Levinthal (1960). You will check the purity and molecular size of the enzyme by gel electrophoresis. In the exercises described in Chapter 8, you will determine the steady-state kinetic parameters of alkaline phosphatase.

METHODS OF PROTEIN PURIFICATION MEASUREMENT

As discussed in Chapter 6, a number of methods are available for the purification of enzymes. Enzymes are almost always proteins, and therefore, the main aim of enzyme purification is to remove all other proteins present in the crude enzyme preparation (the nonprotein substances are comparatively easy to remove from proteins). To check the purification of the enzyme, you must measure both the enzyme activity and the protein concentration at each stage of purification. As previously discussed for purification of β-galactosidase, the purity of the enzyme is determined by its **specific activity**, i.e., activity of the enzyme per milligram of protein. It is obvious that the higher the specific activity, the purer the enzyme. Therefore, during purification, the specific activity of the enzyme should increase until it becomes constant, which usually indicates that the enzyme is quite pure. Homogeneous alkaline phosphatase has a specific activity of 50 U/mg of protein at 25 °C and pH 8.0.

Another useful parameter to measure during enzyme purification is the **total activity** of the enzyme. For each stage of purification, there will invariably be some loss of enzyme, and therefore, total activity is measured at each step to determine the loss (or yield) of the enzyme during that particular step. As mentioned in Chapter 6, you may have to make a choice between different purification methods, choosing the method that gives either high specific activity or high yield.

The **enzyme activity** is usually determined by measuring the appearance of product or disappearance of substrate from the reaction catalyzed by the enzyme. Enzymes are catalysts, and therefore, the **rate** of the reaction will be proportional to the amount of enzyme present in the reaction mixture. However, the rate of the reaction catalyzed by an enzyme also varies with a number of other factors, such as temperature, substrate concentrations, rate of reverse reaction, etc. Therefore, the reaction conditions for the assay of an enzyme should be defined strictly. For enzyme assays, the "**initial velocity**," i.e., the reaction rate extrapolated to zero time, is always determined to minimize any reverse reaction (see Chapter 8) or product inhibition. The temperature at which the assay is performed should also be defined and controlled because the rates of chemical reactions vary with temperature (typically a change of temperature of 10 °C causes a change in rate of twofold). The enzyme activity is generally measured at high substrate concentrations so that the activity does not change with slight changes in the substrate concentration (yielding a state referred to as **zero-order kinetics**, see Chapter 8) and so that the velocity of the reaction is close to V_{max}.

Enzyme activity is defined as the amount of product formed per unit of time at constant temperature. The internationally accepted enzyme unit is "micromoles of product formed per minute (μmol/min)." In Appendix 7-1, methods for the measurement of alkaline phosphatase activity are described in detail.

As mentioned above, for determining the purity of the enzyme (the specific activity), the amount of protein must also be measured. As discussed in Chapter 3, there are several methods available for the quantification of proteins. In this laboratory exercise, you may use either the Lowry or Bradford methods to determine the protein concentration in our starting material and samples during each stage of purification.

REFERENCES

Garen, A., and C. Levinthal. 1960. A fine structure genetic and chemical study of the enzyme alkaline phosphatase of *E. coli*. I. Purification and characterization of alkaline phosphatase. Biochem. Biophys. Acta **38:**470.

Neu, H. C., and L. A. Heppel. 1965. The release of enzymes from *Escherichia coli* by osmotic shock during the formation of spheroplasts. J. Biol. Chem. **240:**3685.

Schlesinger, M. J., and K. Barrett. 1965. The reversible dissociation of the alkaline phosphatase of *Escherichia coli*. J. Biol. Chem. **240:**4284.

PURIFICATION OF ALKALINE PHOSPHATASE

Compared with most other enzymes, *E. coli* alkaline phosphatase is remarkably stable. Nevertheless, the enzyme solution should be kept on ice except where otherwise indicated. Between laboratory periods, it should be stored in the refrigerator, since these extracts are a good growth medium for bacteria and fungi.

At each step of the purification, it is necessary to remove an aliquot for analysis. Monitor the progress of the enzyme purification by assaying the aliquots from each step of the purification for enzyme activity (micromoles per minute) and protein content (milligrams per milliliter). The enzyme activity should be determined at each stage of purification (stages 1–4) on the day of purification with the kinetic assay (Appendix 7-1, Method 1). The protein content for samples at each step of preparation, however, can be measured together (along with the standard) on the last day of the purification process.

 Do not forget to measure the volume of the enzyme preparation at each stage before you go on to the next step!

The volume is conveniently measured by filling a test tube with water from a serological pipette to the same level as that in the tube containing the enzyme solution and noting the volume from the pipette.

FIRST DAY

MATERIALS

1. Hydrated *E. coli* K-12 cells, 10 mL in 0.2 M Tris-HCl (pH 8.0)
2. Lysozyme, 10 mg/mL
3. DNase I, 50 mg/mL (10,000 U/mL)
4. 0.1 M EDTA
5. 1 M $MgSO_4$
6. Polycarbonate tubes, 50 mL
7. Dialysis tubing, 1 in $\times$ 12 in

8.   Dialysis buffer (10 mM Tris-HCl, pH 7.4, containing 10 mM $MgSO_4$)

9.   High-speed centrifuge

10.   0.2 M Tris-HCl (pH 8.0)

11.   1 mM *p*-nitrophenyl-β-D-phosphate (PNPP) in buffer (Tris, pH 8.0)

12.   Standard (50 μM *p*-nitrophenol (PNP) in buffer)

PROCEDURE

1.   Place 10 mL of the cell suspension in a 50-mL polycarbonate centrifuge tube (and warm the tube with your hands). Make sure that this tube is clearly labeled with your names.

2.   Add 0.1 mL of lysozyme solution. This is an enzyme that hydrolyzes part of the cell wall, (i.e., the bond between *N*-acetyl muramic acid and *N*-acetylglucosamine), liberating the phosphatase from the periplasmic space.

3.   Add 0.1 mL of a DNase solution. This enzyme is added to hydrolyze any DNA that might be liberated from broken spheroplasts. The spheroplasts (the portions of the cell bounded by the plasma membrane ideally contain all of the DNA) are fragile, so that if broken would exude DNA, making the extract so viscous that centrifugation might be difficult.

4.   Incubate the extract for 2 min at room temperature, and then add 0.1 mL of 0.1 M EDTA. EDTA is ethylenediaminetetraacetic acid, also called ethylenedinitrilotetraacetic acid:

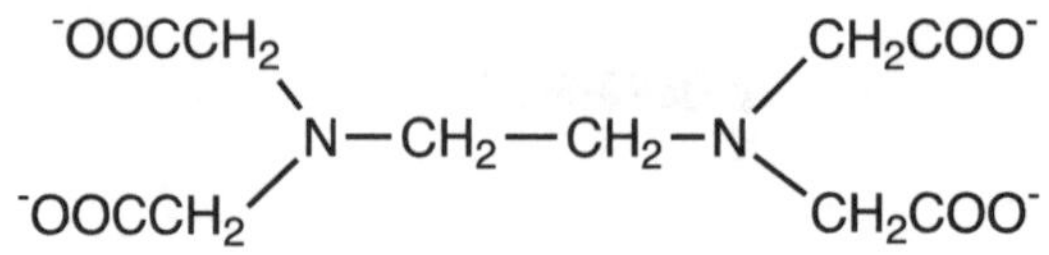

$$^-OOCCH_2 \diagdown \atop ^-OOCCH_2 \diagup N - CH_2 - CH_2 - N \diagup ^{CH_2COO^-} \diagdown_{CH_2COO^-}$$

EDTA binds tightly to divalent cations, such as Ca^{2+} and Mg^{2+} to form a soluble complex. The complex is cyclic, involving all four carboxyl groups and the adjacent nitrogen atoms. This type of multidentate binding is called **chelation**; the complex itself is a **metal chelate**.

5. Mix in the EDTA by gentle swirling to avoid breaking the fragile spheroplasts.

6. After 10 min, gently add 0.1 mL of 1 M magnesium sulfate and mix gently by swirling. This binds to the excess EDTA. EDTA is added to remove bound Ca^{2+} from the cell wall, thus softening the wall and facilitating its hydrolysis by lysozyme. However, EDTA should not be allowed to remain in contact with the enzyme for a prolonged period of time, because it might remove the Zn^{2+}, which is essential for catalytic activity, from the enzyme.

7. After 20 min, centrifuge the tube for 20 min at 12,000g in the Sorvall centrifuge that is set to 4° C. Refer to the centrifugation section in Chapter 1. The speed required to achieve 12,000g can be calculated by the following formula:

$$rpm = 299\sqrt{\frac{g}{radius}}$$

The "relative centrifugal force" is g, and we are aiming for 12,000. The "radius" refers to the distance in centimeters between the center of the rotor and some point in your sample. The rotor in the centrifuge (Sorvall® SS-34) is an "angle head" rotor, in which the centrifuge tubes are held at a fixed angle to the ground, not vertically or horizontally. Therefore, the value of the radius will vary in different parts of your liquid. For Sorvall centrifuges, the estimation of the g-force is obtained on a chart furnished from the Sorvall Co., which is a subsidiary of DuPont, Newtown, Conn. This is calculated on the basis of the maximum radius (10.8 cm) of the rotor, i.e., the bottom of the centrifuge tube. The formula reduces to

$$rpm = 91\sqrt{g}$$

When you remove your tube from the rotor, avoid shaking the tube or you might resuspend the pellet!

8. Remove the supernate to a fresh tube and measure the volume. We refer to this as the stage 1 enzyme. Take out 0.5 mL for enzyme assay and protein estimation, and store in an Eppendorf tube. (Use 0.05–0.1

EXPERIMENT 7-1

mL for the assay of enzyme activity on day 1. The remainder of this aliquot should be stored at 4 °C until the purification is complete, after which the protein content of all fractions can be measured at the same time.) Using a Pasteur pipette, transfer the remaining stage 1 enzyme to a dialysis bag, being careful not to puncture the bag.

DIALYSIS TECHNIQUE

9. While you are waiting during the above steps, prepare a section of dialysis tubing. This tubing has minute pores through which small molecules can pass. Therefore, dialysis allows exchange of small molecules while retaining the large molecules ($> 10,000\ M_r$) inside the sac. The process of dialysis was described in Chapter 1 (Fig. 1-9). We will cut the tubing into short lengths and soak it in water to wash out the glycerol and some of the impurities. If you let the tube dry out, microscopic cracks can develop, allowing the tube to leak.

    Keep the dialysis tubing submerged in a beaker of distilled water until you are ready to fill it with the enzyme solution. When you are ready, drain any water that may be inside the dialysis tubing, tie off one end with a pair of knots, and fill the bag with the enzyme solution (stage 1). Then tie another pair of knots at the top end of the bag in the same manner as before. *Leave plenty of room in the tubing for expansion.* It should be like a half-deflated balloon, *not* an overstuffed sausage.

NOTE✔

The dialysis bag should be wetted with distilled water at all times!

10. After loading your sample into the bag and closing the bag with double knots, tie a string to the top and attach a unique label. Bring this assembly to a large beaker. It should contain 10 mM Tris, pH 7.4, in 10 mM $MgSO_4$. Your instructor will replace the liquid with fresh buffer twice before your next laboratory period so that your sample can equilibrate fully with the proper buffer. The liquid inside the dialysis bag is called the **retentate**, and the liquid outside, which we are discarding in this experiment, is called the **dialysate**.

SECOND DAY

MATERIALS

1. Dialyzed stage 1 enzyme
2. Polycarbonate centrifuge tubes, 50 mL
3. Rubber stopper, size 8
4. Powdered ammonium sulfate
5. Sorvall centrifuge with SS-34 rotor
6. Tris-HCl buffer (10 mM, pH 7.4) containing 10 mM $MgSO_4$
7. Dialysis tubing 1 in $\times$ 12 in
8. 0.2 M Tris-HCl (pH 8.0)
9. 1 mM PNPP in buffer
10. Standard 50 μM PNP in buffer
11. H_2O bath at 80 °C

HEAT DENATURATION STEP

11. At the start of the laboratory period, retrieve your dialysis bag from the dialysis buffer tank and transfer the contents to a 50-mL polycarbonate centrifuge tube. The transfer is a bit difficult because the bag contents may be under pressure and liquid can spurt out uncontrollably as the bag is opened. Hold the bag vertically with two fingers just below the second knot from the top. Try to minimize compression of the bag by holding the cellulose close to the knot. With your other hand, make a very small nick just under the knot with scissors. This will let out trapped air without loss of liquid, provided you do not drop the bag. Now invert the bag over the centrifuge tube to let most of the contents out. To remove the remainder, squeeze out as much as possible.

    If your dialysis bag is under a great deal of pressure (i.e., it looks like a hot dog), it may be advisable to hang the bag in a suitable sized beaker and then, holding the bag from the knot at the top, reach into

EXPERIMENT 7-1

the beaker with a pointed scissors and carefully cut the bottom of the bag. This will enable the beaker to capture the retentate as it squirts out of the hole. The last bits of enzyme can be squeezed out. Then, transfer your enzyme fraction to a polycarbonate tube.

12. Cover the centrifuge tube with a size 8 rubber stopper (loosely at first) and heat the tube in a 75–80 °C H_2O bath for 15 min.

13. Cool the tube rapidly by running cold tap water on the outside.

14. Centrifuge at 10,000 rpm for 15 min.

15. Carefully remove the supernate after centrifugation

Be very careful not to touch or disturb the pellet!

16. Transfer the supernate to a clean 50-mL polycarbonate centrifuge tube.

17. Estimate the volume of liquid and remove 0.6 mL for assay of protein and enzyme. This is the stage 2 enzyme.

Stage 2 enzyme often contains more apparent enzyme activity per milliliter than stage 1 enzyme. This is probably due to the removal by dialysis or heat of an endogenous alkaline phosphatase inhibitor.

AMMONIUM SULFATE PRECIPITATION STEP

Ammonium sulfate is a highly water-soluble salt that is valuable in producing differential precipitation of proteins. The high ionic strength of ammonium sulfate solutions causes the more hydrophobic proteins to associate, resulting in precipitation. This type of association is described in Chapter 4 under Hydrophobic Interaction Chromatography and in Chapter 6 under Initial Steps of Purification. Proteins will usually precipitate within a small range of ammonium sulfate concentration, depending upon the particular protein. Thus, ammonium sulfate is often used to concentrate enzymes by first precipitating them by addition of an appropriate concentration and then resolubilizing them in a minimum volume of buffer. For example, one might add enough ammonium sulfate to produce 30% saturation, meaning a final concentration equal to 30% of the maximum ammonium sulfate that could dissolve at that temperature. This calculation is generally based on the solubility at 0 °C, and we follow that recipe, although the precipitation is done at room temperature. In the experimental procedure, we use ammonium sulfate primarily to concentrate the enzyme

EXPERIMENT 7-1

rather than to remove other proteins. You can check this point for your preparation when you determine the specific activity before and after this step.

18. Weigh out ammonium sulfate, using 0.603 g for every 1 mL you now have in your polycarbonate tube. This produces a concentration that is "90% saturated," if you were to cool it to 0 °C. Add the crystals slowly, stirring the liquid with a glass rod until all the crystals have dissolved. The enzyme should precipitate. Let the solution sit for 10–15 min so that full precipitation can take place.

NOTE✔

Adding ammonium sulfate, tends to acidify your solution. Therefore, it is advisable to check the pH often as you add ammonium sulfate and add appropriate base to maintain the desired pH. Alternatively, if you are not going to a very high percent saturation, you can make up a solution saturated in $(NH_4)_2SO_4$ and adjust the pH. Then add an appropriate amount to your enzyme-containing solution to achieve the desired final percent of saturation. This will prevent you from unwittingly acidifying your solution.

19. Now centrifuge the tube in the Sorvall as before, but at 18,000 rpm for 30 min.

HAZARD!

As in all centrifugation, you should make sure that there is no solid matter in the rotor tube cavity! If the centrifuge tube should press (at a tremendous g force) against a lump, the tube is likely to shatter at that point. This can cause severe damage to enzymes, tubes, and even the centrifuge. If someone has spilled solid matter into the cavity, wipe it out with a wet towel.

Ordinarily, precipitated protein can be spun down readily; one does not need the high centrifugal force that we describe in this activity. However, with ammonium sulfate precipitations, the salt increases the density of the solution, so that a higher speed is often necessary. The speed of sedimentation depends on the difference in density between the precipitate and the liquid. For example, a cork stopper or a precipitate having density the same or lower than that of the suspending liquid will never sediment, no matter how high a speed you use. Since ammonium sulfate raises the density of the buffer appreciably, the rate of sedimentation can be annoyingly slow, and this is compensated by spinning more rapidly.

EXPERIMENT 7-1

20. Carry your pelleted protein carefully to your laboratory table and discard the clear supernatant liquid. You might observe a white flocculent material floating on the surface of the liquid. This material most likely is lipid, which has a low density.

21. Add 0.5 mL of Tris-HCl buffer (10 mM, pH 7.4) containing 10 mM $MgSO_4$, and gently resuspend the pellet. Try to avoid introducing bubbles, since exposure to air and the formation of bubbles can lead to denaturation of proteins. Now transfer this suspension to a new dialysis bag. Use three 0.5-mL rinses of buffer to complete the transfer.

22. Dialyze as before until the next laboratory period. Make sure that you label your dialysis samples. The purpose of the second dialysis is to remove the ammonium sulfate.

THIRD DAY

MATERIALS

1. Dialyzed stage 3 enzyme

2. DEAE cellulose in Tris-HCl (pH 7.4, 5 mM) containing $MgCl_2$ (5 mM)

3. Buffer A, the column buffer, 5 mM Tris-HCl (pH 7.4) with $MgCl_2$ (5 mM)

4. Buffer B, the eluting buffer, 5 mM Tris-HCl (pH 7.4) with $MgCl_2$ (5 mM) and NaCl (125 mM)

5. 0.2 M Tris-HCl (pH 8.0)

6. 1 mM *p*-nitrophenyl phosphate (PNPP) in buffer

7. 10 N NaOH

8. Standard 50 μM *p*-nitrophenol (PNP)

9. Bradford reagent for protein estimation

10. 10 mM HEPES buffer with 5 mM $MgCl_2$, dialysis buffer

Record the volume of your retentate and take out 0.3 mL for the enzyme and protein assay. This is the stage 3 enzyme.

DEAE-CELLULOSE COLUMN CHROMATOGRAPHY

DEAE cellulose is an anion-exchange packing; it may be useful to review the principles behind the ion-exchange chromatography by rereading the section Ion-Exchange Chromatography in Chapter 4. Proteins contain anionic groups caused by aspartic acid and glutamic acid residues, which have anionic side chains. Thus, proteins can bind to the DEAE groups if the pH of the medium is below the pK of the DEAE group (pK = 10) so that the column amines are positively charged.

The proteins must be applied to the column in a low-ionic- strength buffer at a pH above the pI of the protein. Then the proteins are eluted with a buffer containing a higher concentration of ions. Under such conditions, the protein molecules are displaced by anions in the buffer.

Alkaline phosphatase binds very weakly to DEAE cellulose. It is important to ensure that the DEAE cellulose is fully equilibrated with the low-ionic-strength buffer before applying your enzyme. Equilibration will require considerable washing.

QUESTIONS

1. Why must the starting buffer be at a pH above the isoelectric pH of the protein of interest?

2. Why is it necessary to dialyze the enzyme after ammonium sulfate precipitation?

PROCEDURE

23. Prepare the column as you did in the chromatography experiments (Chapter 4). Pack it with a slurry of DEAE cellulose, leaving the stopcock at the bottom fully open. This helps make the packing more compact. Add the slurry in portions, allowing each portion to settle somewhat before adding the next. When you have a bed ~5 cm high, you are ready to proceed.

24. Use buffer A (5 mM Tris-HCl, pH 7.4, and 5 mM $MgCl_2$) to pack and wash the column. After the column matrix has settled, use some of this to rinse the walls above the packing, so that there is a nice flat surface of packing.

25. Remove the excess buffer above the packing, even if some particles are floating in it, so that the top of the packing is exposed but not dry.

EXPERIMENT 7-1

If you have a surplus of DEAE cellulose, return it to the reagent table.

26. Prepare about fifteen 13×100 mm test tubes with each one marked at the 2-mL level.

27. Add your stage 3 enzyme to the column, and collect *all* effluent liquid in 2-mL fractions in the marked test tubes.

28. When all the sample has just entered the column, rinse the rest of the enzyme solution with 2 mL more of the same buffer into the column bed. When this has percolated in, add 8 mL of the same buffer and collect 2-mL fractions.

29. When all of buffer A is eluted out, carefully add, without disturbing the surface, ~16 mL of buffer B and continue elution. Buffer B is the same buffer, but it also contains NaCl (0.125 M), which will elute the enzyme.

Do not confuse buffers A and B!

30. Collect more 2-mL fractions until the buffer B is completely drained out.

31. Keep the 2-mL fractions from the column and assay each one for both phosphatase activity and protein concentration. To speed up the identification of fractions containing enzyme activity, you may want to devise a miniaturized qualitative assay. You can perform a spot test for the enzyme as follows: place on a Parafilm sheet assay buffer (+ PNPP) into a row of 10-µl drops, one for each column fraction. Add 5 µl from the first fraction to the first drop. Repeat for subsequent fractions with the remaining drops. Column fractions with AP activity will cause the drop to turn yellow.

32. When you have identified which fractions have enzyme activity, carefully measure both the activity and the concentration of protein for each. Protein concentration can be measured in fractions that lack the enzyme activity as well.

33. Mix the contents of each tube well, and use 50 µL from each tube to estimate the enzyme activity (Method 2 or the fixed-time method) and another 50 µL to estimate the amount of protein by the dye-binding method (Chapter 3). If the colors seem very dark (eye estimation), dilute an aliquot of the fraction fivefold with buffer A and use 50 µL

of the diluted sample to measure the protein and/or enzyme. Remember to consider this dilution in your calculations.

34. Record the fraction numbers that contain phosphatase activity. Plot on graph paper the protein concentration on the left-hand *y*-coordinate and the phosphatase activity on the right-hand *y*-coordinate, each versus column fraction. You may use different symbols or colored lines to distinguish one graph from the other. A comparison of both graphs should reveal the separation brought about by the DEAE-cellulose chromatography.

35. Pool the fractions containing high phosphatase activity into one tube and mix gently. You might have a tube or two that show only a trace of enzyme; it is not worth pooling these with the rest, since this would mainly result in a dilution of your enzyme.

36. Estimate the total volume in the pooled fractions and assay aliquots of these for protein and enzyme activity by quantitative procedures (Lowry or Bradford method for the protein and the rate assay for the enzyme).

This is the stage 4 enzyme, our last purification step.

37. Cover the enzyme sample with Parafilm, and keep it refrigerated so that it can be used in the next experiment.

Remember to assay the enzyme activity on each day of purification to check the progress of purification. Protein values can be measured simultaneously all in 1 day so that you can run one standard curve for the assay of proteins in all the fractions. Save aliquots from each purification stage and assay the proteins on day 3 or 4 of the laboratory period.

Calculate for each stage the enzyme activity per milliliter (milliunits per milliliter), protein concentration (milligrams per milliliter) and specific activity of the enzyme (units per milligram). For each stage, also calculate the total activity (milliunits) and the percent yield relative to the stage 1 enzyme (stage 1 = 100%). The purification achieved (increase in the specific activity) at each step relative to stage 1 should also be calculated. Using these values, fill in the purification table (Chapter 6).

38. Save all of your enzyme aliquots from different purification stages. Especially preserve the stage 4 enzyme for steady-state kinetic studies (Chapter 8), electrophoresis, and possible structure analysis.

EXPERIMENT 7-1

CHARACTERIZATION OF PURIFIED ALKALINE PHOSPHATASE

A pure enzyme has a unique set of physical and chemical properties. These include a defined specific activity, kinetic parameters (V_{max}, K_m, etc., see Chapter 8), molecular weight, subunit composition, amino acid composition, and amino acid sequence. The protein that you have purified will be characterized with regard to its homogeneity by SDS-PAGE. You will also identify which band of protein observed in simple electrophoresis has enzymatic activity, by using an activity stain. These procedures are described in Chapter 5. The kinetic parameters of alkaline phosphatase will also be determined as the last part of this series of experiments and compared with the values reported in the scientific literature. The kinetic studies that we perform are described in Chapter 8.

APPENDIX 7-1   ASSAY OF ALKALINE PHOSPHATASE

Enzyme activity is generally determined by measuring the rate of either the formation of product(s) or the disappearance of substrate(s) under specified reaction conditions. For proper assay of enzymes, the conditions should be such that the catalytic rate is directly proportional to the amount of enzyme added. You will set up assays with this in mind in the development of the creatine phosphokinase assay (Chapter 9). The pH of the reaction mixture and the temperature of the assay should be kept constant during the reaction. Other variables that affect the enzyme activity should also be optimized.

It is important to measure initial velocities of enzyme-catalyzed reactions to assure that the observed rates are not limited by substrate depletion or product accumulation. Moreover, the substrate concentration is generally kept high ($3\text{–}4 \times K_m$) to minimize variations caused by small differences in substrate concentrations. To estimate the initial velocities, several methods can be used. If the reaction appears to be linear in the early portions of the progress curve, initial velocities can be estimated as shown in Fig. 7-1. Alternatively, you could try to fit the data to an equation using curve-fitting programs. If an acceptable fit is obtained, the derivative of the equation for time $t = 0$ may be used to estimate the initial velocity.

For the assay of enzymes, you should always measure a blank (or control) tube side by side with the actual assays to check for any nonenzymatic formation of the product or depletion of substrate. The most common control is the "no enzyme" blank; i.e., the reaction mixture contains everything except the enzyme. If possible, a "zero-time" blank in which the reaction is stopped just after adding the enzyme, should also be measured to check for any interference by the protein and/or other components present in the sample.

Alkaline phosphatase catalyzes the hydrolysis of all phosphomonoesters. We assay the enzyme by using a convenient method that involves an unnatural, chemically synthesized compound, *p*-nitrophenyl phosphate (PNPP), which is hydrolyzed by the enzyme to form inorganic phosphate (P_i) and *p*-nitrophenol (PNP) (equation 7-1). The ionic form of PNP (pK = 7.2) is colored, absorbing maximally at 410 nm, and thus the enzymatic reaction can be monitored by measuring the increase in color at 410 nm with a spectrophotometer. You already know the spectrum of PNP from the colorimetry experiment (Experiment 2-1).

$$\text{(7-1)}$$

pK = 7.2

We use a spectrophotometric assay by measuring the PNP formed either continuously (Method 1) or after a fixed time interval (Method 2).

METHOD 1: CONTINUOUS ASSAY OF ALKALINE PHOSPHATASE

MATERIALS

1. 1 mM PNPP in 0.2 M Tris-HCl (pH 8.0)

2. 0.2 M Tris-HCl (pH 8.0)

3. Enzyme solution appropriately diluted with 0.2 M Tris-HCl (pH 8.0) buffer (see below)

4. Standard PNP solution (50 μM) in pH 8.0 buffer

PROCEDURE

1. Pipette 1.5 mL of the substrate and 1.30–1.45 mL of buffer (depending upon how much enzyme you will add) into a 13 × 100 mm test tube.

2. By using this test tube as a blank, adjust your spectrophotometer to read zero absorbance at 410 nm.

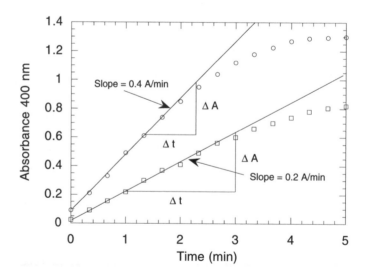

FIGURE 7-1 Typical absorbance versus time plot for the alkaline phosphatase activity assay. The data points represent the experimental values, while the solid lines represent the extrapolated lines for determination of the initial velocities.

3. Add 0.05–0.2 mL of enzyme solution to the test tube (final volume of 3.0 mL). Rapidly mix the liquids for a few seconds by sealing the tube with a square of Parafilm and inverting the tube several times.

4. Immediately place the tube back in the spectrophotometer, and read the absorbance at 20-s intervals. The absorbance reading operation is best done with a partner. One person reads the absorbance values, while the other person times and records the values. Record the readings in your laboratory notebook. To be sure that the enzyme assay is in the linear range, it should be performed at two levels of enzyme; the rate (ΔA/min) should be proportional to the amount of enzyme added.

NOTE✔

Your enzyme solution may be so active that the reaction goes too fast to record accurate readings (> 0.5 *A*/min). Start over with fresh solutions, but dilute an aliquot of the enzyme solution appropriately.

5. Stop taking readings when you reach an absorbance of ~0.6. If this takes more than 3 min, stop. You may have overdiluted your sample. If possible, your sample should be diluted so as to give 0.2–0.4 *A*/min.

   A standard solution of PNP is used so that the absorbance values can be converted to moles of substrate hydrolyzed.

6. Place 3 mL of this 50 µM solution in your cuvette, and read the optical density at 410 nm.

   From this value, calculate the molar absorptivity of PNP. You have previously performed this calculation in Experiment 2-3 on colorimetry. As you recall, the molar absorptivity of PNP can be calculated from your knowledge of the Beer–Lambert law: $\varepsilon = A/(\text{molarity})(\text{pathlength})$. Assume the pathlength of the liquid in your cuvette is 1 cm. In other words, your calculated value will be the absorbance of a 1 mM solution in a 1-cm cuvette.

   Most tables of molar absorptivity values are given for 1 M solutions, so test your understanding of the difference by converting your observed value to those units ($M^{-1} \cdot cm^{-1}$).

   The molar absorptivity of a solution of PNP should be determined every day you perform assays, since there may be some variation in the performance of the various spectrophotometers used in the laboratory and in the tubes that you use.

CALCULATIONS

Plot the absorbance versus time for each assay and use a ruler to draw straight lines through the points on your graph that describe the initial rate for each assay (see Fig. 7-1). The line may not go through the origin since some reaction may have taken place while you were mixing the enzyme with substrate. Calculate the slope of your straight line; the units are ΔA per minute (Fig. 7-1). Using the millimolar absorptivity of PNP (calculated as above), convert the ΔA per minute to millimolar per minute (i.e., rate of increase in concentration of PNP) from the Beer–Lambert law. Then calculate the amount of PNP formed (in nanomoles per minute) in the reaction mixture (3-mL volume). (Remember the relationship: concentration × volume = amount, e.g., micromoles × milliliters = nanomoles). As mentioned above, nanomoles per minute is defined as milliunits. Also calculate the enzyme activity as milliunits per milliliter of the enzyme. For example, if 0.1 mL of enzyme solution was added to 2.9 mL of the assay reagents and was found to produce 40 nmoles of PNP/min, 40 mU were present in the cuvette, and the activity of the enzyme solution is 40 mU/0.1 mL or 400 mU/mL.

METHOD 2: FIXED-TIME ASSAY OF ALKALINE PHOSPHATASE

The basic assumption in this procedure is that the enzyme activity is linear with time up to the point when the reaction is stopped. The advantage of the assay is that a large number of samples can be assayed at the same time and the spectrophotometric measurements can be done at a convenient pace. However, strict precautions should be taken that each tube is incubated for exactly the same time period, so that they can be compared with each other. It is also advisable to prove that the assay is linear for the time period you intend to use.

MATERIALS

1. 1.0 mM PNPP in 0.2 M Tris-HCl, pH 8.0

2. 0.2 M Tris-HCl, pH 8.0

3. 10 N NaOH

4. Enzyme solution

PROCEDURE

1. Pipette 1.5 mL of PNPP solution and 1.35 mL of buffer to 13 × 100 mm test tubes and mix.

2. Pipette 0.05 mL of the appropriately diluted enzyme, and mix quickly by inverting the tube after closing the opening with Parafilm. Note the time.

3. After exactly 3 min, add 0.1 mL 10 N NaOH, and mix quickly.

4. Read against an appropriate blank (no enzyme blank) at 410 nm. The color is stable for at least 1 h.

As mentioned above, the timing is important. For multiple assays, you should add the enzyme (and mix) and stop the reaction (and mix) at the same time intervals so that the total incubation time does not exceed 3 min. If the activity is very high (> 1.0 *A*/3 min), dilute an aliquot of the enzyme with buffer and run the assay again. Do not forget to measure the absorbance of the standard (50 µM) PNP solution.

CALCULATION

Calculate the change in *A* per minute and from that nanomoles per minute and nanomoles per minute per milliliter of enzyme as described above.

REAGENTS AND EQUIPMENT NEEDED FOR CHAPTER 7

Note: Use glass-distilled H$_2$O for all solutions

Hydrated *E. coli*: Sigma EC-1, 25 g of *E. coli* K12 cells are washed with 10 mM Tris-HCl, pH 8.0 and then centrifuged. The pellet is hydrated in 1000 mL of 0.5 M sucrose (20%) containing 30 mM Tris-HCl, pH 8.0 buffer. Recheck the pH of the final mixture, since sucrose may contain acid. Add extra buffer, if necessary. Each student pair needs 10 mL. For students, hydrate cells 48 h before use and store at 4 °C. To test the procedure, hydrate enough cells for 15 experiments and purify the rest to stage 1. Save 80% of the stage 1 enzyme for students who make mistakes. Use the remaining 20% of the stage 1 enzyme to check the rest of the purification protocol.

✓ Lysozyme 10-mg/mL of H$_2$O: Sigma L7001, 5 g (0.2 mL/student pair). Store the solution at 4 °C and on ice during the class period. Stable for 3–4 days at 4 °C.

✓ DNase I: Sigma D4527, from bovine pancreas, 10,000 E.U. (~5 mg)/mL H$_2$O. Need 0.1 mL/student pair. Store the solution as above for lysozyme.

✓ 0.1 M Sodium EDTA: Sigma E5314, 100 g. Make solution in water; need 0.1 mL/student pair.

✓ 1 M MgSO$_4$: Sigma M5921. Need 0.1 mL/student pair.

✓ Polycarbonate tubes (50 mL): 2 per student pair.

✓ Ammonium sulfate: Sigma A2939, 1 kg.

✓ Corex tubes, 15 mL: 2 per student pair. Also need two Sorvall SS-34® rotors with rubber adapters for the Corex tubes.

✓ Dialysis tubing: Sigma D9777, 50 ft x 1 in. Cut into pieces 12 in. long. Rinse it in distilled water, and keep wet and cold until given to the students.

✓ Dialysis buffer: 10 mM Tris-HCl, pH 7.4, containing 10 mM MgSO$_4$. Make 2 liters of 10× concentrate, and dilute with chilled distilled H$_2$O before use. (Tris base, Sigma 8524). Keep plenty of glass-distilled H$_2$O for this and other operations.

✓ DEAE cellulose (DE52):

Make up 1 M Tris-HCl buffer pH 7.4, and put an equal volume of DE52 slurry into it. Stir gently for 30 min. Check and adjust the pH to 7.4 (either with 1 M Tris base or 1 N HCl). (DEAE cellulose, Sigma D6418, 100 g).

Pour the entire slurry into a large column with a glass wool plug at the bottom (you can use a Buchner funnel if a large column is not available). Wash with 10 volumes of H_2O followed by 5 volumes of column buffer (5 mM Tris-HCl pH 7.4 + 5 mM $MgCl_2$). Do not let the column run dry.

> Check the pH of the effluent; it must be 7.4. If it is not, wash with more buffer.

> Suspend the cellulose beads in column buffer in a beaker, allow them to settle, and decant. Repeat two to three times to remove the fines. This will permit better flow rates.

> Note that it is essential that the equilibrated column is at low ionic strength (0.005 M) or alkaline phosphatase will not bind.

✓ Column buffer: 5 mM Tris-HCl

✓ Eluting buffer (buffer B): 5 mM Tris-HCl + 5 mM $MgSO_4$ containing 0.125 M NaCl (pH 7.4). Need ~20 mL/student pair. Make sure the students use correct buffers for loading, for both the first and second elutions. Mark all of these clearly and warn the students about these different buffers. (NaCl, Sigma S7653, 1 kg).

✓ Enzyme assay materials (needed every day of the experiment):

> 0.2 M Tris-HCl pH 8 (make 4 liters).

> 8.0 (371 mg/L). (PNPP, Sigma N3254, 1 g)

> Standard, 0.05 M PNP in 0.2 M Tris-HCl buffer pH 8.0 (7 mg/L). (PNP, Sigma 104-8, 1g)

> Parafilm (for spot tests and sealing tubes): Sigma D3172.

> Protein assay materials (needed on day 3): These are listed at the end of Chapter 3.

> Water bath: need two; check whether they will heat up to 80 °C.

> Sorvall centrifuges: check that with the SS-34 rotor, the centrifuge maintains 4 °C at 18,000 rpm for 30 min; the tubes must remain at 4 °C at the end of the run. If the temperature increases during the run, the instrument needs service.

> HEPES buffer: Sigma, H0891, 100 g.

> Bio-Rad columns and two-way stopcocks for DEAE-cellulose chromatography. (Use the same columns used for Chapter 4.)

> SDS-PAGE and simple (nondenaturing) gel electrophoresis materials: These are described at the end of Chapter 5.

8 Enzyme Kinetics

WHY USE STEADY-STATE KINETICS?

After purifying an enzyme, it is often valuable to characterize its steady-state kinetic properties by determining its K_m, V_{max}, and the inhibition constants for various substances, including its products. Such studies provide information about how the enzyme responds to substrates and inhibitors. It allows you to compare its kinetic properties to those of other enzymes catalyzing similar reactions, so that you might deduce whether they utilize similar mechanisms. From steady-state kinetic analyses, you can frequently determine the order of addition of substrates and release of products. Kinetic inhibition studies are important for evaluating the efficacy of pharmaceutical inhibitors as well as their mode of action on enzymes.

Enzymes such as racemases or epimerases catalyze single-substrate reactions. Hydrolases, such as proteases and phosphatases, use water as a second substrate. Because water is at such a high concentration (55.6 M), its concentration remains constant during the reaction, and you can treat the kinetics as if they were for a single-substrate reaction. For example, alkaline phosphatase (Chapter 7) is such a hydrolase for a number of phosphate monoesters, including *p*-nitrophenyl phosphate (PNPP). In this chapter, we discuss the treatment of kinetics for single-substrate reactions. The principles discussed are valid for multiple-substrate reactions, but in the latter, the analysis is somewhat more complex and requires more extensive data. For a general overview with examples, see Garrett and Grisham (1995). Kyte (1995) provides more detailed coverage and gives many examples with data obtained from research publications. For a comprehensive discussion, see Segel (1975), Cornish-Bowden (1995a), or Purich (1979), which are excellent laboratory references. For proper statistical treatment of data, see Cleland (1979) or Cornish-Bowden (1995b).

STEADY-STATE KINETICS PRINCIPLES

A simple one-substrate reaction is shown below.

$$\text{Substrate (S)} \rightleftharpoons \text{Product (P)}$$

When this reaction is catalyzed by an enzyme E, a reversible enzyme-substrate complex (ES) is formed, and this complex is converted to E + P as shown:

$$E + S \underset{k_{-1}}{\overset{k_1}{\rightleftharpoons}} ES \underset{k_{-2}}{\overset{k_2}{\rightleftharpoons}} E + P \tag{8-1}$$

The rate of formation of P is the velocity of the reaction at any time *t*. If the initial velocity *v* is measured (this is the rate of formation of P at $t = 0$), then $[P] = 0$. Therefore, the rate of the reverse reaction will be negligible and equation 8-1 simplifies to

$$E + S \underset{k_{-1}}{\overset{k_1}{\rightleftharpoons}} ES \overset{k_2}{\longrightarrow} E + P \tag{8-2}$$

There are two ways to derive kinetic equations appropriate for describing the above reactions:

1. By assuming ES is in rapid equilibrium with E and S (Michaelis-Menten hypothesis)

2. By assuming a steady-state condition, i.e., the rate of formation of ES equals the rate of breakdown of ES (Briggs-Haldane hypothesis)

For both of the conditions, [ES] is considered to remain essentially constant. The following derivation is made by assuming a steady-state condition. The simple equilibrium treatment of Michaelis and Menten is a special case of the general steady-state theory.

ASSUMPTIONS

1. $[S] \gg [E]$ so that the amount of substrate bound to the enzyme is negligible compared to the total amount of substrate. $[S]$ is assumed to remain constant during the reaction.

2. Only the initial velocity of the reaction is measured. Thus, the velocity measured is at $t = 0$, at which time $[P] = 0$ and the $[S] \approx [S]_{\text{initial}}$.

DERIVATION

Refer to equation 8-2. Suppose E_t = total enzyme added. Then the concentration of free enzyme $[E] = [E_t] - [ES]$ at any time during the reaction. Rate of *ES* formation

$$\text{d}[ES]/\text{dt} = k_1[E][S]$$

$$= k_1([E_t] - [ES])[S] \qquad (8\text{-}3)$$

Rate of ES breakdown

$$\text{d}[ES]/\text{dt} = k_{-1}[ES] + k_2[ES]$$

$$= (k_{-1} + k_2)[ES] \qquad (8\text{-}4)$$

Under steady-state conditions, the rates of formation and breakdown of ES are equal; therefore,

$$k_1([E_t] - [ES])[S] = (k_{-1} + k_2)[ES] \qquad (8\text{-}5)$$

After rearranging and defining K_m, we obtain

$$\frac{([E_t] - [ES])[S]}{[ES]} = \frac{k_{-1} + k_2}{k_1} = K_m \qquad (8\text{-}6)$$

$$\frac{[E_t][S]}{[ES]} - [S] = K_m$$

$$[ES] = \frac{[E_t][S]}{K_m + [S]} \tag{8-7}$$

The velocity of the catalyzed reaction at any time is

$$v = k_2[ES]$$

When the enzyme is saturated with substrate so that all of it exists as the *ES* complex, the velocity is maximum, V_{max} and

$$V_{max} = k_2[E_t]$$

Multiplying both sides of equation 8-7 by k_2 and substituting for v and V_{max} results in

$$v = \frac{V_{max}[S]}{K_m + [S]} \tag{8-8}$$

This is the **Michaelis-Menten** equation where K_m is the Michaelis constant and V_{max} is the maximal velocity.

THE SIGNIFICANCE OF K_m AND V_{max}

The K_m (Michaelis constant) is defined in equation 8-6 for the mechanism described by equation 8-2 as

$$K_m = \frac{k_{-1} + k_2}{k_1}$$

The K_m is the concentration of substrate that produces $V_{max}/2$ for the catalyzed reaction, which can be easily seen from equation 8-8. K_m has units of M^{-1}. The value of K_m is independent of the amount of enzyme used (or its purity) but does depend on the assay conditions, such as pH, presence of inhibitors or activators, temperature, ionic strength, etc. This is because (as shown by equation 8-2) K_m is made up of pure rate constants. Although K_m appears to be similar to a dissociation constant, further examination of equation 8-2 reveals that even in this simple case, $K_m \neq K_s$, where K_s is the dissociation constant for the enzyme and the substrate and is defined as

$$K_s = \frac{[E][S]}{[ES]} = \frac{k_{-1}}{k_1}$$

K_s is a direct measure of the binding affinity of the substrate for the enzyme (note that because it is the dissociation constant; the smaller the K_s, the higher the affinity). We discuss in more detail the nature of dissociation constants and how they can be measured in Chapter 10. In general, $K_m \neq K_s$ and therefore is not a direct measure of the affinity for substrate. Indeed, for more complex mechanisms, K_m for a given substrate will usually be considerably more elaborate than shown in equation 8-2 and will differ significantly from K_s. Nevertheless, there are instances where K_m will approach K_s. For example, in the simple case described by equation 8-2, if $k_{-1} \gg k_2$, then $K_m \approx K_s$ and K_m will be a reasonable measure of affinity. Although not a binding constant, K_m does indicate the apparent catalytic affinity of the enzyme for a given substrate in catalysis. It tells what concentration of substrate is required to give one half of maximal velocity with other substrates saturating. Using the Michaelis-Menten equation, you can calculate how much substrate will be required to reach any desired level of saturation during catalysis. K_m values have been widely used to compare the affinities of enzymes for different substrates.

V_{max} for a particular substrate is also a useful constant, but it should be realized that observed values for V_{max} (micromoles per minute) depend on the amount of enzyme used for the assay. However, for homogeneous enzyme, if V_{max} is defined as micromoles per minute per milligram of enzyme protein, it will indeed be a constant for a particular set of conditions (e.g., pH and temperature) for the enzyme. If the molecular weight is known and pure enzyme is available, one can calculate (from protein assays) how much enzyme is being used (in moles) in the enzymatic assay; then (for these conditions) V_{max} can be expressed as a fundamental constant for the enzyme that is called the maximum turnover number or the catalytic constant k_{cat}. The constant k_{cat} tells us how many moles of substrate per minute (or per second) will be converted to product by one mole of the enzyme under a particular set of environmental conditions when all substrates are saturating (i.e., at V_{max}). For the system in equation 8-2, $k_{cat} = k_2$.

In experiment 8-1, you will determine k_{cat} and K_m for your alkaline phosphatase preparation using PNPP as substrate. How might you estimate the purity of your enzyme preparation and determine how much enzyme you actually have in the assays you carry out? You need this information to determine a k_{cat}.

It should be noted that the initial velocity of the reaction is directly proportional to the amount of the enzyme added not only when $v = V_{max}$ (when $[S] = \infty$) but also at any other substrate concentration. (Can you deduce this from equation 8-8?) There is a common misconception that enzymes can be properly assayed only under "saturating" substrate concentrations. If $[S]$ remains constant during the assay, you can use any substrate concentration to assay the enzyme. However, for many enzymatic reactions, it is experimentally difficult to keep the substrate concentration constant during an assay when using subsaturating substrate levels. Therefore, when it is practical (if substrate is not too precious), it is usually advisable to keep the substrate concentration relatively high ($\geq 5\ K_m$) during the assay of enzymes. However, you must be careful not to have it so high that it inhibits the enzyme.

GRAPHICAL ANALYSIS

The Michaelis-Menten equation describes a rectangular hyperbola, which is illustrated in Fig. 8-1. When $[S]$ is very low, i.e., $K_m >> [S]$, equation 8-8 reduces to

$$v = \frac{V_{max}[S]}{K_m} \tag{8-9}$$

Thus, $v \propto [S]$, i.e., the reaction follows **first-order kinetics** with respect to S.

At high substrate concentrations $[S] >> K_m$, so that $K_m + [S] \approx [S]$; therefore, $v = V_{max}$ and the reaction rate will be independent of substrate concentration, i.e., the reaction follows **zero-order kinetics**. When the velocity v of the reaction is half that of V_{max},

$$v = \frac{V_{max}}{2} = \frac{V_{max}[S]}{K_m + [S]} \tag{8-10}$$

Thus, $K_m + [S] = 2[S]$ and $K_m = [S]$. Therefore, K_m can be determined graphically as shown in Fig. 8-1A. However, in this method, because it is difficult to determine precisely the value of V_{max} (value of v when $[S] = \infty$), the determination of K_m will also be imprecise.

Equation 8-8 can be manipulated to yield various equations that yield straight lines and facilitate the graphical extrapolation of V_{max} and K_m. The Lineweaver-Burk method uses the reciprocal form of equation 8-8:

$$\frac{1}{v} = \frac{K_m + [S]}{V_{max}[S]} = \left(\frac{K_m}{V_{max}}\right)\left(\frac{1}{[S]}\right) + \frac{1}{V_{max}} \tag{8-11}$$

A straight line is generated if $1/v$ is plotted against $1/[S]$ (Fig. 8-1B), where the y intercept is $1/V_{max}$ and the x intercept is $-1/K_m$. K_m and V_{max}

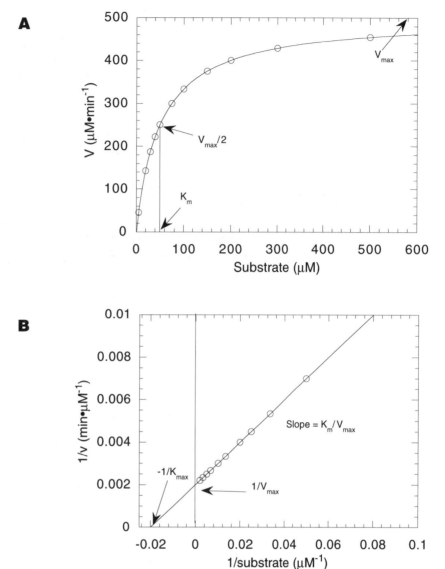

FIGURE 8-1 *(A)* Michaelis-Menten plot and *(B)* Lineweaver-Burk plot. K_m = 50 µM, V_{max} = 500 µM/min.

are easily determined by this method by extrapolating the straight line to its intercepts on the *x* and *y* axes.

There are several other methods of plotting equation 8-8 that also give straight lines. For example, in the **Hanes-Woolf** plot, straight lines are obtained by plotting [S]/*v* versus [S]:

$$\frac{[S]}{v} = \left(\frac{1}{V_{max}}\right)[S] + \frac{K_m}{V_{ma:}} \qquad (8\text{-}12)$$

The slope is $1/V_{max}$, and the y intercept is K_m/V_{max}.

In the **Eadie-Hofstee** plot, v is plotted versus $v/[S]$. In this case,

$$v = -K_m\left(\frac{v}{[S]}\right) + V_{max} \qquad (8\text{-}13)$$

Thus, the y intercept is V_{max}, the slope is $-K_m$, and the x intercept is V_{max}/K_m. Try these different methods of plotting using your data from Experiment 8-1, and show how you would determine the K_m and V_{max} from these plots. Evaluate the advantages and disadvantages of each plotting method.

Transformation of the data into linear forms not only makes it easier to calculate the K_m and V_{max} but also to determine visually whether the enzyme systems are following classical (Michaelis-Menten) kinetics behavior. If they are, the plots will be linear. In spite of their usefulness, these transformations will actually weight the data nonuniformly if you fit the transformed data with lines calculated by linear least-squares procedures. It is better to fit the untransformed data to the appropriate equations using nonlinear least-squares procedures, so that you do not arbitrarily weight the data. (Numerous computer graphics programs are available that are suitable for such fitting procedures.) Then replot the linear transformed data and the fit by one of these methods to obtain a better visual sense of the data. Note also that most data are imperfect and have scatter as well as experimental bias (especially if your controls are not well conceived). Be aware of this and examine your data carefully before reaching conclusions. Do not simply believe a computer output. Often you will need to use various other types of experiments and chemical intuition to confirm your suspicions derived from kinetic analysis.

COMPETITIVE, "NONCOMPETITIVE," AND "UNCOMPETITIVE" INHIBITORS

Generally, inhibitors that compete with substrates for the active site of the enzyme (thus increasing the effective K_m) are known as **competitive** inhibitors. Inhibitors that do not affect the formation of the enzyme-substrate complex but inhibit the enzyme by reacting at a position other than the active center are known as **uncompetitive** inhibitors. The reaction scheme in Fig. 8-2A is a simple model for reversible competitive inhibition (I is the inhibitor). The reaction scheme in Fig. 8-2B is a simple model for reversible uncompetitive inhibition.

Another type of inhibition occurs when the inhibitor binds to both E and the ES complex. This is usually called **noncompetitive inhibition** (a rather poor term) and is shown below in Fig. 8-3. This is drawn for the spe-

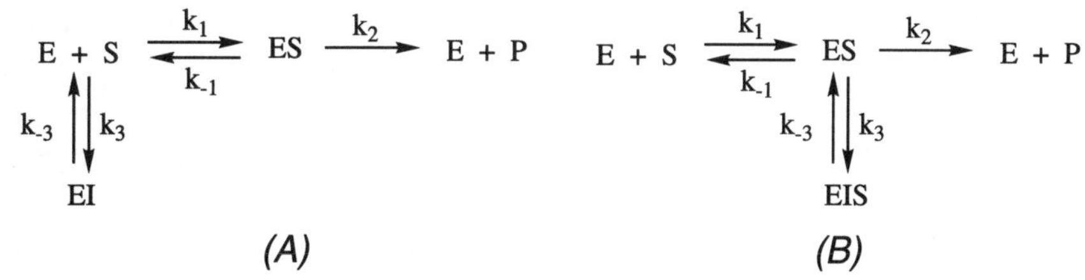

FIGURE 8-2   Schemes for (A) competitive and (B) uncompetitive inhibition.

cial case in which the inhibitor binding constant is the same for E and for ES (both terms use k_3 and k_{-3}). In this example, both EI and EIS are assumed to be inactive.

Michaelis-Menten plots (v versus S) for these examples are shown in Fig. 8-4.

For **competitive inhibition**, $K_i = k_{-3}/k_3$ (it is the dissociation constant for inhibitor), and it can be shown that

$$v = \frac{V_{max}[S]}{K_m\left(1+\dfrac{[I]}{K_i}\right)+[S]} \tag{8-14}$$

The reciprocal of equation 8-14 will be

$$\frac{1}{v} = \left(\frac{K_m}{V_{max}}\right)\left(1+\frac{[I]}{K_i}\right)\left(\frac{1}{[S]}\right)+\frac{1}{V_{max}} \tag{8-15}$$

Therefore, a plot (Lineweaver-Burk) of $1/v$ versus $1/[S]$ (Fig. 8-5A) will be a straight line with a y intercept $= 1/V_{max}$. V_{max} will be unchanged in the presence of an inhibitor, as you might expect, because it is competitive, and at infinite [S] and finite [I], the substrate will "win." The x intercept will be

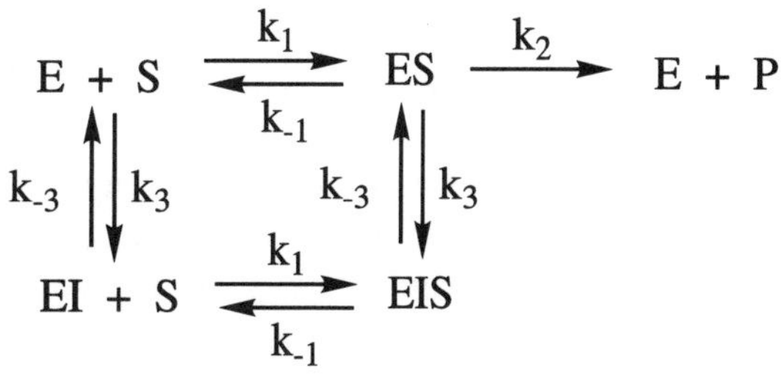

FIGURE 8-3   Kinetic scheme for noncompetitive inhibition.

at $-1/K_m(1+[I]/K_i)$. Thus, for competitive inhibition, only the slope will be affected and the observed K_m will be modified by $(1 + [I]/K_i)$.

For **uncompetitive inhibition**, in which the inhibitor binds only to the ES complex (Fig. 8-2B), the equation describing the velocity as a function of $[S]$ and $[I]$ is

$$v = \frac{V_{max}[S]}{K_m + [S]\left(1 + \dfrac{[I]}{K_i}\right)} \tag{8-16}$$

This is plotted in Fig. 8-4C. The reciprocal of equation 8-16 is

$$\frac{1}{v} = \left(\frac{K_m}{V_{max}}\right)\left(\frac{1}{[S]}\right) + \frac{1}{V_{max}}\left(1 + \frac{[I]}{K_i}\right) \tag{8-17}$$

When $1/v$ is plotted versus $1/[S]$, the slope is still K_m/V_{max}, but the intercept on the y axis is increased by the factor $(1 + [I]/K_i)$. The x intercept $= -(1+[I]/K_i)/K_m$. Double reciprocal plots of $1/[S]$ versus $1/v$ therefore yield parallel lines (equal slope) offset by the above factor (Fig. 8-5C). Thus, in uncompetitive inhibition, you will find that only the intercepts are affected.

For **noncompetitive inhibition** (special case shown in Fig. 8-3), it can be shown that

$$v = \frac{V_{max}[S]}{\left(K_m + [S]\right)\left(1 + \dfrac{[I]}{K_i}\right)} \tag{8-18}$$

This is plotted in Fig 8-4B. The reciprocal of equation 8-18 is

$$\frac{1}{v} = \left(\frac{K_m}{V_{max}}\right)\left(1 + \frac{[I]}{K_i}\right)\left(\frac{1}{[S]}\right) + \frac{1}{V_{max}}\left(1 + \frac{[I]}{K_i}\right) \tag{8-19}$$

Therefore, a plot of $1/v$ versus $1/[S]$ (Fig. 8-5B) will be a straight line with

$$y \text{ intercept} = \frac{1}{V_{max}}\left(1 + \frac{[I]}{K_i}\right)$$

and an x intercept: $-1/K_m$. Note that in noncompetitive inhibition the presence of the inhibitor I affects both the slope and the y intercept of Lineweaver-Burk plots. Although in Scheme 8-3 the affinities of the inhibitor for E and ES are the same (both use k_3 and k_{-3}), in most real circumstances these affinities will be different. Therefore, the equations will not be symmetric, so that in double reciprocal plots, lines plotted for different concentrations of inhibitor may not intersect on the x axis. This is often called **mixed inhibition**.

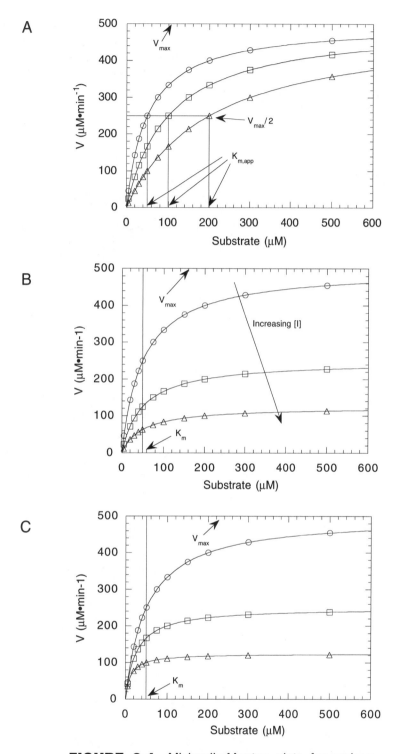

FIGURE 8-4 Michaelis-Menten plots for various types of inhibition. $K_m = 50$ μM, $V_{max} = 500$ μM/min, $K_i = 100$ μM; circles, [I] = 0; boxes, 200 μM; triangles, [I] = 500 μM. *(A)* Competitive inhibition, *(B)* noncompetitive inhibition, and *(C)* uncompetitive inhibition.

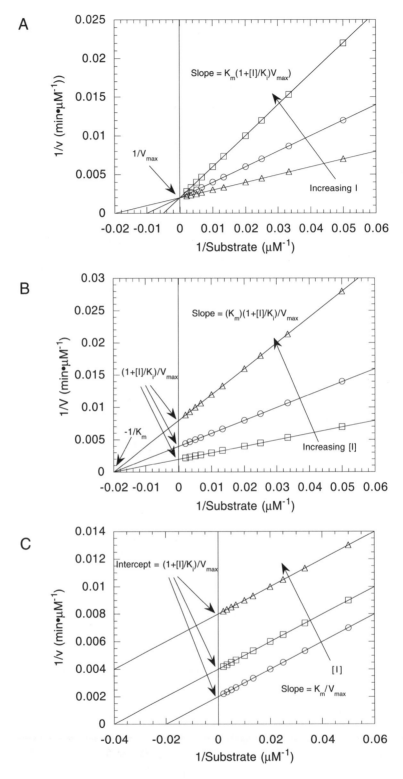

FIGURE 8-5 Lineweaver-Burk plots for various types of inhibition: *(A)* competitive, *(B)* noncompetitive, and *(C)* uncompetitive. Same conditions as in Fig. 8-4.

Noncompetitive and uncompetitive are clearly poor terms, since they are synonymous. Nevertheless, the terms are widely used and you need to keep in mind that uncompetitive inhibition implies that the inhibitor binds only to ES so that in Lineweaver-Burk plots only the intercepts are affected, whereas noncompetitive inhibition implies binding to both E and ES, so that both slope and the *y* intercept are affected. In contrast, competitive inhibition (binds only to E) gives plots with only the slope affected by the presence of inhibitor.

A practical way to determine competitive inhibition constants is to carry out a series of assays at two or more different substrate concentrations while varying the inhibitor concentration. Then, for each of the different substrate concentrations, make a plot of $1/v$ versus $[I]$, which is called a **Dixon plot**. This is shown in Fig. 8-6A. The intersection of the two lines in the second quadrant will occur at $-K_i$ (*x* axis) and at $1/V_{max}$ (*y* axis). Note that if you find where on the $1/v$ versus $[I]$ line that $1/V_{max}$ occurs, you can actually obtain $-K_i$ with a single-substrate concentration. However, measurements with more than one-substrate concentration will clearly be more reliable. Figure 8-6 also shows Dixon plots for noncompetitive and uncompetitive inhibition, which give characteristic recognizable patterns.

You might note, especially from Fig. 8-4, that for all types of inhibition, the direct (Michaelis-Menten) plots are very similar, and visually, you may have difficulty in identifying them. Transforming the data into one of the linear forms can help you recognize which type of inhibition is most consistent with your data. However, as mentioned above, because these transformations weight the data unevenly, it is better to use nonlinear least-squares procedures on the untransformed data and then replot the fitted data on linear transformed plots to help you identify the type of inhibition. When dealing with mixed inhibition (referring to Figs. 8-2 and 8-3), if the affinity of the inhibitor for E is greater than that for ES, it may be difficult to distinguish whether you have noncompetitive or competitive inhibition. When this difference in affinity is large, the inhibition will appear competitive. Similarly, if the affinity of the inhibitor for ES is greater than that for E, it may be difficult to distinguish between noncompetitive and uncompetitive inhibition. Again, when these differences are large, the inhibition will appear to be uncompetitive. Scatter in the data can make it difficult to distinguish various types of inhibition. The quality of data necessary to distinguish inhibition patterns depends upon the ratio of the affinities of the inhibitor for E and ES. Often you will need further corroborative experiments to reach a satisfactory conclusion.

We have only discussed reversible inhibition. Many inhibitors form irreversible complexes with the enzyme, which may be only partially active or even completely inactive. Many drugs fit the description of irreversible inhibitors. Consult the references below for a more complete treatment of enzyme inhibitors.

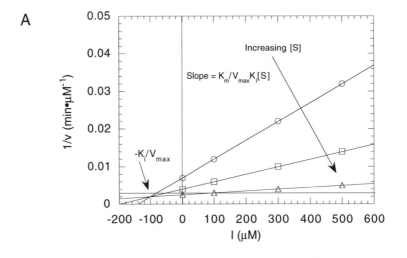

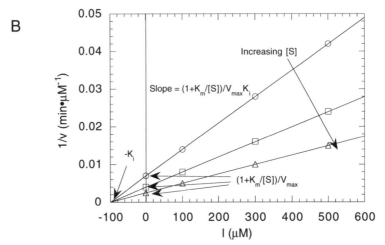

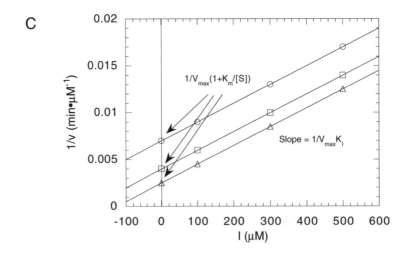

FIGURE 8-6 Dixon plots for various types of enzyme inhibition: *(A)* competitive, *(B)* noncompetitive, and *(C)* uncompetitive. Same conditions as in Fig. 8-4

REFERENCES

Cleland, W. W. 1979. Statistical analysis of enzyme kinetic data. *In* D. L. Purich, (ed.), p. 103. Methods Enzymol, vol. 63. Academic Press, New York.

Cornish-Bowden, A. 1995a. Fundamentals of enzyme kinetics, Portland Press, Ltd., London.

Cornish-Bowden, A. 1995b. Analysis of enzyme kinetic data, Oxford University Press. Oxford.

Garrett, R. H., and C. M. Grisham. 1995. Biochemistry. Saunders College Publishing. Fort Worth, Tex.

Kyte, J. 1995. Mechanism in protein chemistry. Garland Publishing, Inc. New York.

Purich, D. L., ed. 1979. Methods Enzymol, vol. 63. Academic Press, New York.

Segel, I. H. 1975. Enzyme kinetics, behavior and analysis of rapid equilibrium and steady-state enzyme systems. John Wiley & Sons, Inc., New York.

EXPERIMENT 8-1

DETERMINATION OF K_m AND V_{max} FOR ALKALINE PHOSPHATASE

MATERIALS

1. 0.4 mM PNPP in 0.2 M Tris-HCl (pH 8.0)
2. 0.2M Tris-HCl buffer (pH 8.0)
3. Stage 4 enzyme diluted with buffer to ~60 mU/0.2 mL

PROCEDURE

1. To five 13×100 mm tubes, pipette different volumes of 0.4 mM PNPP so that the amount of PNPP ranges between 30 and 600 nmol.

The final concentration of PNPP in 3 mL of reaction mixture should be between 0.01 and 0.2 mM.

2. Add 0.2 M Tris-HCl to make the volume in each tube 2.8 mL. Mix.

3. Adjust the spectrophotometer to read zero absorbance with the first tube, and then add 0.2 mL of enzyme to this same tube. Mix quickly and measure the increase in absorbance at 410 nm at 20-s intervals for at least 3 min.

4. Measure the enzyme activity in an identical manner in the four other tubes. Do not forget to zero the spectrophotometer with each tube before you add the enzyme.

CALCULATIONS

1. Determine the velocity of the enzymatic reaction v for each substrate concentration $[S]$ by plotting the A versus time, as you have done in previous experiments.

2. Plot v versus $[S]$ and present the graph.

3. Plot $1/v$ versus $1/[S]$ (Lineweaver-Burk plot) and calculate the values of K_m and V_{max} from the intercepts. Give units.

The Lineweaver-Burk plot has been criticized as not being very accurate for a number of reasons, especially because the experimental points are not equally spaced. A more precise line would be obtained by other plotting formats, e.g., by the Eadie-Hoffstee method (v versus $v/[S]$). You may try to plot your experimental data in this way and compare K_m and V_{max} with those obtained from the Lineweaver-Burk plots.

EXPERIMENT 8-2

PRODUCT INHIBITION OF ALKALINE PHOSPHATASE

Alkaline phosphatase is inhibited by inorganic phosphate, a product of the reaction. You will determine the type of kinetic inhibition and the inhibition constant K_i using graphical analysis.

MATERIALS

1. 0.5 mM PNPP solution
2. 1 mM sodium phosphate (P_i)
3. 0.2 M Tris-HCl buffer, pH 8.0
4. Stage 4 enzyme diluted with buffer so that activity is ~60 mU/0.2 mL
5. 10 N sodium hydroxide

PROCEDURE

1. Set up five 13×100 mm test tubes and pipette 0.3 mL of PNPP (0.5 mM) to each tube.
2. Pipette different volumes of 1 mM P_i to the tubes to cover a range from 10 to 600 nmoles (0.01 to 0.6 mL).
3. Add enough buffer (0.2 M Tris-HCl) to each tube to make the final volume 2.7 mL.
4. Add 0.2 mL of enzyme solution to each (strictly timing the addition), mix, and allow to stand at room temperature for 3 min.
5. Add 0.1 mL of 10 N NaOH and mix quickly.
6. Read the absorbances at 410 nm against a "no-enzyme" blank (i.e., 0.3 mL of PNPP + 2.6 mL of 0.2 M Tris-HCl + 0.1 mL of 10 N NaOH).
7. Set up another assay system as above but use 0.9 mL of 0.5 mM PNPP in each tube. All other conditions, such as P_i concentration, buffer to make 2.7 mL, enzyme (0.2 mL), and alkali (0.1 mL) should be the same.

 Final volumes in both these assays should be 3.0 mL.

CALCULATIONS

1. Calculate the velocity of the enzyme in terms of nanomoles per minute (milliunits).

2. Plot the reciprocal of velocity $1/v$ against the concentration of phosphate $[P_i]$. You should obtain two straight lines from the two different concentrations of PNPP used.

3. Find the point of intersection of these two lines. From this intersection, calculate the inhibitor constant K_i and determine the nature of inhibition (competitive or noncompetitive).

K_i **like** K_m **is another kinetic constant for interaction between the inhibitor and enzyme, and it is a measure of the affinity of the inhibitor for the enzyme.**

MATERIALS NEEDED FOR CHAPTER 8

✓ Enzyme. Most of the students will be able to use their own enzyme preparations (Chapter 7). Commercially obtained enzyme may be used. (Sigma P4069). Store commercially obtained enzyme at -20 °C (it is provided as a 50% glycerol solution).

✓ p-Nitrophenyl phosphate (PNPP) (Sigma N6260). To obtain a 0.5 mM solution (for experiment 8-2), dissolve 0.23 mg/mL in water. Each group will need 1.5 mL, and it is advisable to provide 5 mL/group. To obtain a 0.4 mM solution in 0.2 M Tris-HCl, pH 8.0 (for experiment 8-1), combine 4 volumes of 0.5 mM PNPP with 1 volume of 1 M Tris-HCl, pH 8.0. Each group will need less than 5 mL, so we provide 10 mL. We usually prepare within a week of the session and store at -20 °C.

✓ Sodium phosphate ($NaH_2PO_4 \cdot H_2O$, Sigma S9638). To obtain a 1 mM solution, dissolve 14 mg in 100 mL of water. Each group will need ~3 mL. Store at room temperature.

✓ Sodium hydroxide (NaOH, S0899). For a 10 N solution, dissolve 200 g in 500 mL of water. Store at room temperature.

✓ 0.2 M Tris-HCl, pH 8.0. We generally prepare a 1 M solution and dilute to 0.2 M as needed. To prepare the 1 M solution, dissolve 121.1 g Tris base (Sigma Trizma T1503) in 800 mL of water. Adjust pH to 8.0 with HCl, and bring the volume to 1 L. Each group should be provided with 50 mL of 0.2 M Tris-HCl. Store at room temperature.

9 Enzymatic Methods of Analysis

ENZYMATIC ANALYSIS OF SUBSTRATES

Enzymes are often used as analytical tools to identify and/or quantify specific chemicals in solutions. The advantages of using enzymes for this purpose are that enzymatic reactions are fast and very specific to the nature of the substrates to be used and that, even under very mild conditions, enzymes catalyze reactions that rarely proceed at significant rates without the enzyme present. In these assays, the reaction is usually allowed to go to completion, and the extent to which the reaction proceeds permits the determination of the specific compound. The concentrations of the enzyme and other reagents are chosen to make the reaction proceed to completion within a convenient time period. The specificity of enzymes can overcome the necessity of removing interfering materials from samples, a process that is often tedious but necessary in other types of analytical procedures. Numerous purified enzymes are becoming commercially available for analyzing specific chemicals.

Specificity can be of several types, and enzymes can be used as reagents that take advantage of one or more of these types. A few enzymes have absolute specificity, in which the enzyme will catalyze only one reaction. Some enzymes have group specificity, in which the enzyme acts only on molecules that have specific functional groups, e.g., phosphates, esters, sugars, and amino groups. Other enzymes have specificity for particular types of linkages, almost regardless of the actual groups attached, e.g., glycosidases and peptidases. Still other enzymes will catalyze only reactions that involve a particular stereoisomer.

ASSAYS FOR ENZYMATIC ACTIVITY

It is also possible to determine the amounts of particular enzymes in biological fluids by measuring activities with specific substrates. For example, you have already assayed β-galactosidase and alkaline phosphatase (AP) in crude extracts. For these determinations, the rate of conversion of a particular substrate to product is measured rather than the extent to which the reaction proceeds. A sufficient concentration of substrate is included to ensure that the enzyme is operating at the maximal rate ($[S] \gg K_m$ so that V_{max} is measured, if possible). Thus, as the substrate concentration decreases as the reaction progresses, the reaction velocity will not be appreciably affected. This condition is known as **zero-order kinetics**; there is "no" dependence on the concentration of substrate. Under such conditions, the rate observed will (hopefully) be proportional to the amount of enzyme present. Therefore, we generally try to adjust the conditions to achieve zero-order kinetics. In Experiment 9-4, you will use a rate assay that involves a coupled enzyme system to determine the amount of creatine phosphokinase (CPK) in a solution. For both types of enzymatic assays, extent and rate, it is the specificity and efficiency of enzymes that make them so useful.

Enzymatic analysis is a particularly useful aid in clinical diagnosis. Coupled with physical examination, enzymatic analysis can help make the proper diagnosis for several diseases that have similar physical symptoms. The normal level of many enzymes in serum is very low. When disease or injury occurs to a specific organ, such as the heart, liver, or pancreas, increased levels of organ-specific enzymes can also be found in serum. Thus, diseases to these organs may be correctly diagnosed by determining the enzyme levels. Some of the enzymes that frequently are clinically assayed include the following: glutamate oxaloacetate transaminase (GOT), CPK, lactate dehydrogenase (LDH), and AP.

One use of enzymatic tests is in the differential diagnosis of myocardial infarction (heart attack) and pulmonary embolism. Both disorders exhibit very similar physical symptoms and EKG patterns, but the diseases need to be treated differently. Myocardial infarction results in elevated levels of GOT, LDH, and CPK, whereas pulmonary embolism results in elevated levels of LDH but not elevated levels of GOT or CPK. It should be emphasized that such clinical tests usually are not sufficient for diagnoses. They must be used to complement thorough physical examination and other diagnostic procedures. For example, elevated CPK will be observed not only in myocardial infarction, but also in muscle and brain diseases or injuries.

PRACTICAL CONSIDERATIONS

When using enzymatic methods of analysis, it is important to be cautious on several counts. Many enzymes are inhibited by various substances. The researcher must be aware of such conditions and use proper controls. Metal chelators, such as citrate and other compounds capable of forming bidentate complexes with metals, are often severe inhibitors of metalloproteins such as kinases. If you are measuring a substance such as alcohol in a fluid (Experiment 9-3), it can be useful to spike the reaction with a known amount of alcohol to ascertain whether the test is reliable. This would serve as an **internal standard**. Enzymatic activity is also influenced by pH. This can derive from pK_a values on the substrate and/or the enzyme. Therefore, the pH chosen and the selection of an appropriate buffer are important.

Enzymatic reactions, similar to most chemical reactions, are sensitive to temperature, with the rate typically being affected by twofold or more by a change of 10° C. In addition, if the temperature becomes too high, the enzyme may become denatured and loss of activity will occur. Thus, if one is carrying out rate assays to determine levels of enzymes in solutions, it is critical to control the temperature of the assay mixture. Using the Spectronic 20® spectrophotometers in our laboratory does not permit tight control over the temperature, and therefore, temperature variation constitutes a source of uncertainty and error. If you are carrying out slow reactions that require several minutes for their assay, it might be prudent to incubate the reaction tube in a constant temperature water bath between readings.

When possible it is wise to use **continuous assays** rather than **fixed-time point assays**. The latter are quite convenient, since a whole series of tubes can be set up and the reaction can be initiated by the addition of one of the critical reagents. Then in sequence, the reactions in the several tubes can be quenched at regular intervals. However, if lags or bursts of product formation occur, they will not be detected by this procedure, and the results will be inaccurate. Therefore, if you intend to use a fixed-time point assay, it is important to demonstrate the linearity of the assay by checking it with a continuous assay, with a series of enzyme concentrations, or with a series of fixed-time points over the range you intend to use.

COUPLED ASSAYS

Frequently (but not always), spectrophotometric methods are used to monitor enzymatic reactions. These can be quite convenient if the substrate or the product contains a distinct **chromophoric** (colored) group. If neither of these is colored, the reaction can often be coupled to another reaction (enzymatic or nonenzymatic) that will produce a colored compound. For example, if you want to measure "A" with the reaction A→ B and both A and B are colorless, a second reaction, in which B is converted to a colored product C (or a second substance that drives B to C changes color), could be coupled to the first reaction as

$$A \longrightarrow B \longrightarrow C$$

Colorless   Colorless   Colored

(9-1)

The amount of C formed will be equivalent to the initial amount of A present if the reactions proceed to completion. If the equilibrium of the reaction to be measured is not favorable, the reaction can sometimes be coupled to a second reaction to drive it to completion

$$A \rightleftharpoons B \longrightarrow C$$

Reaction 1   Reaction 2

(9-2)

In equation (9-2), where A is being determined, the formation of B from A (reaction 1) may not be favored because the equilibrium lies toward the formation of A. If it is possible to couple this reaction to another reaction (reaction 2) that is essentially irreversible, all of A can be quantitatively converted to C so that determination of C allows us to quantify A. Experiment 9-3, the determination of ethanol, illustrates this technique.

Many compounds are spectrophotometrically measured by coupled enzymatic reactions. Generally, the enzyme(s) and all other required chemicals, except the one to be measured, are added in relatively high concentrations to ensure a fast and complete reaction. The compound to be measured is present in a limited quantity so that the amount of final product formed will be directly proportional to that compound. Therefore, you can record the initial absorbance on the spectrophotometer (which may be set to zero), allow the reaction to proceed to completion or to equilibrium, and record the final reading. The rate of the reaction is not important as long as it occurs in a reasonable period of time. The end point gives the quantification.

In the first three experiments described in this chapter, you will measure reactions involving only one enzyme. It is often useful to couple other enzymes and substrates to reactions so that (1) a convenient chromophoric substrate or product can be observed, (2) reactions that are thermodynamically unfavorable can be driven fully in the desired direction by removing products, (3) product inhibition of the first enzyme in the sequence can be minimized, or (4) unstable products or substrates that might interfere with the reaction can be removed before they accumulate.

Coupled assays can also be used to measure the amount of an enzyme for which direct measurement of the substrate or product is difficult. As in any assay of an enzyme, the rate of catalysis is measured. When product cannot be directly measured, it is often possible to couple the reaction to one or more enzymes that use the product as substrate to yield a substance that can be conveniently measured. For example, in equation 9-2, if we are interested in determining the enzyme that catalyzes reaction 1 and neither A nor B can be easily measured, we could couple this to an enzyme that catalyzes reaction 2 to form the product C, which can be easily measured. For the coupled enzyme assay to be valid, the rate of appearance of the measured product C must be proportional to the concentration of the enzyme catalyzing reaction 1. Thus, the coupling system catalyzing reaction 2 must be able to keep up with reaction 1. In practice, you should try to adjust the coupling enzyme system to have the capacity to turn over products and substrates at least 10-fold faster than the enzyme system being measured. In Experiment 9-4, you will use these principles to develop a coupled assay for CPK.

EXPERIMENTS WITH PYRIDINE NUCLEOTIDE-REQUIRING ENZYMES

Pyridine nucleotide-requiring reactions are widely used in enzymatic methods of analysis. These coenzymes (NAD, NADH, NADP, and NADPH) are ideal because they are used stoichiometrically in a large number of oxidation-reduction reactions. The general reaction is shown in Scheme 9-1. In biological reactions, transfers of the hydride at position 4 are always stereospecific, with some enzymes transferring specifically the H_R and others transferring specifically the H_S hydride.

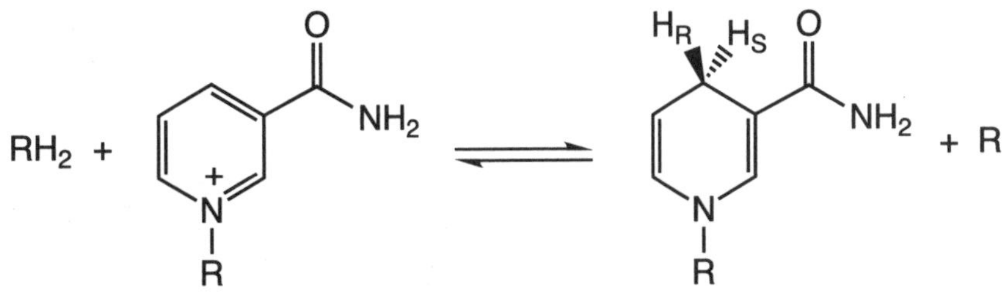

Scheme 9-1

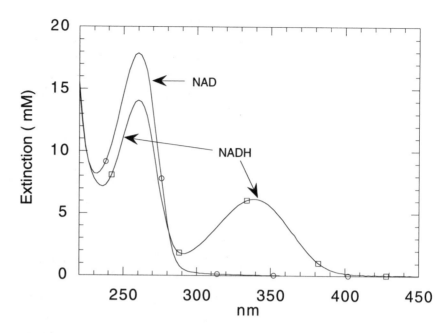

FIGURE 9-1   Spectra of NAD and NADH forms of pyridine nucleotides.

Only the reduced forms (NADH and NADPH) absorb light at 340 nm. The absorbance spectra of these compounds (Fig. 9-1) show that both the oxidized and the reduced pyridine nucleotides have absorbance maxima at 260 nm because of the adenine group, which is part of the R group in Scheme 9-1 ($\varepsilon = 1.78 \times 10^4$ M^{-1} cm^{-1} for the oxidized form and $\varepsilon = 1.44 \times 10^4$ M^{-1} cm^{-1} for the reduced form). However, at 340 nm, only reduced pyridine nucleotides (pyridine ring in the quinoid form) absorb strongly ($\varepsilon = 6.22 \times 10^3$ M^{-1} cm^{-1}). High-performance spectrophotometers can detect concentrations as low as 1 µM (an absorbance of 0.0062 in a 1-cm cuvette). Reduced pyridine nucleotides also fluoresce (see Chapter 2), and the fluorescence excitation spectrum corresponds to the absorbance spectrum, while the fluorescence emission spectrum has a broad maximum at ~400 nm. By using fluorescence, it is often possible to detect concentrations as low as 1–10 nM, which is an increase in sensitivity over that with absorbance spectroscopy of nearly 10^3-fold. Therefore, reactions that generate or utilize reduced pyridine nucleotides can be used to measure substrates very sensitively. For example, in equation 9-3, the concentration of AH_2 can be determined by measuring the formation of NADH from NAD (an increase in absorbance at 340 nm), since 1 mol of NADH is formed for each mol of AH_2 oxidized.

$$AH_2 + NAD \quad \rightleftharpoons \quad A + NADH + H^+ \qquad (9\text{-}3)$$

Alternatively, A could be determined by measuring the formation of NAD from NADH (a decrease in absorbance at 340 nm). Note that this reaction is reversible, and the experimental conditions must be adjusted so that either the forward or reverse reaction is favored, depending on the substrate to be assayed. For example, since H^+ is one of the products of the reaction, if the generation of NADH from NAD is required (estimation of AH_2), the reaction can be run at high pH (low $[H^+]$). One might also use a higher concentration of NAD to drive the reaction to the right. If A is to be estimated, the reaction can be run at low pH (of course, it must be within the pH range in which the enzyme is active). Using a high concentration of NADH to drive the reaction to the left is not usually practical since it would give a high background absorbance, making determination of small changes in absorbance inaccurate.

Because the absorbance coefficient of the reduced pyridine nucleotides at 340 nm is high (a solution of 5×10^{-5} M will have an absorbance of 0.311; be sure that you know how to calculate this), reactions with pyridine nucleotides can be used to estimate very small amounts of substrates as indicated above. This will be demonstrated in Experiment 9-3. Even with the simple spectrophotometers described in this book, it is possible to measure as little as 0.05 µmol of substrate accurately. As discussed above, by using a fluorimeter (reduced pyridine nucleotides are fluorescent), the sensitivity can be increased approximately another 10^3-fold. Lowry, whose methods for protein assay we use, devised a recycling technique with pyridine nucleotide-dependent reactions by which femtomole (10^{-15} mol) amounts of substrates could be measured (Lowry and Passonneau, 1963). Many reactions can be coupled to dehydrogenases to take advantage of this convenient and sensitive measurement technique. Experiment 9-4 illustrates this. The GOT reaction shown below shows no color changes in the visible portion of the spectrum; however, the reaction can be coupled to malate dehydrogenase (MDH) as shown, so that NADH is stoichiometrically used with the conversions in the GOT reaction.

$$\alpha\text{-ketoglutarate + aspartate} \underset{}{\overset{\text{GOT}}{\rightleftharpoons}} \text{glutamate + oxaloacetate}$$

$$\text{oxaloacetate + NADH} \underset{}{\overset{\text{MDH}}{\rightleftharpoons}} \text{malate + NAD}$$

To illustrate enzymatic methods of analysis, you will perform four short experiments that use pyridine nucleotides: in Experiment 9-1, the molar absorbance of NAD(P)H at 340 nm will be determined in your spectrophotometer; in Experiment 9-2, the equilibrium constant of the reaction catalyzed by alcohol dehydrogenase will be measured, and some thermodynamic parameters of the reaction will be calculated; and, in Experiment 9-3, the amount of ethanol in an unknown sample will be measured with alcohol dehydrogenase. To minimize variation, the same spectrophotometer should be used for all three of these experiments. In Experiment 9-4, you will develop a coupled assay system for the measurement of CPK.

REFERENCES

Bucher, T., and G. Pfleiderer. 1955. Pyruvate kinase from muscle. Methods Enzymol. **1:**435.

Cahn, R. C., N. O. Kaplan, L. Levine, and E. Zwillingk. 1962. Nature and development of lactic dehydrogenases. Science **136:**962.

Kayne, F. J. 1973. Pyruvate kinase, p. 353. *In* P. D. Boyer (ed.), The enzymes, vol. VIIIA, 3rd ed. Academic Press Inc., New York.

Lowry, O., and J. V. Passonneau. 1963. Measurement of pyridine nucleotides by enzymatic cycling. Methods Enzymol. **6:**792.

Rudolph, F. B., B. W. Bangher, and R. S. Beissner. 1979. Techniques in coupled enzyme assays. Methods Enzymol. **63:**22.

Tanzer, M. L., and C. Gilvarg. 1959. Creatine and creatine kinase measurement. J. Biol. Chem. **234:**3201.

DETERMINATION OF THE EXTINCTION COEFFICIENT OF NAD(P)H

The oxidation of glucose-6-phosphate by NAD(P) catalyzed by glucose-6-phosphate dehydrogenase (G-6-PDH) is shown in Scheme 9-2. G-6-PDH obtained from bakers yeast uses only NADP(H) efficiently, whereas the G-6-PDH from *Leuconostoc mesenteroides* uses either NADP(H) or NAD(H). In your laboratory exercises, you will probably use the latter to avoid making up another expensive solution. Although the equilibrium of the reaction does not favor full production of the 6-phospho-glucuronolactone, the nonenzymatic hydrolysis of the latter to 6-phosphogluconic acid is nearly irreversible and drives the enzymatic reaction to completion. Thus, the conversion of NAD(P) is essentially complete, permitting you to equate the final concentration of NAD(P)H to the initial concentration of NAD(P). This experiment will also be used to calibrate your spectrophotometer. Because of the wide bandwidth of these simple spectrophotometers (20 mm) and, more importantly, the variation in diameter and optical quality of your cuvette test tubes, the theoretical value for molar absorbance of NAD(P)H may not be obtained. Nevertheless, the value you obtain will be valid for your experimental conditions because you will have determined the absorbance coefficient under these experimental conditions.

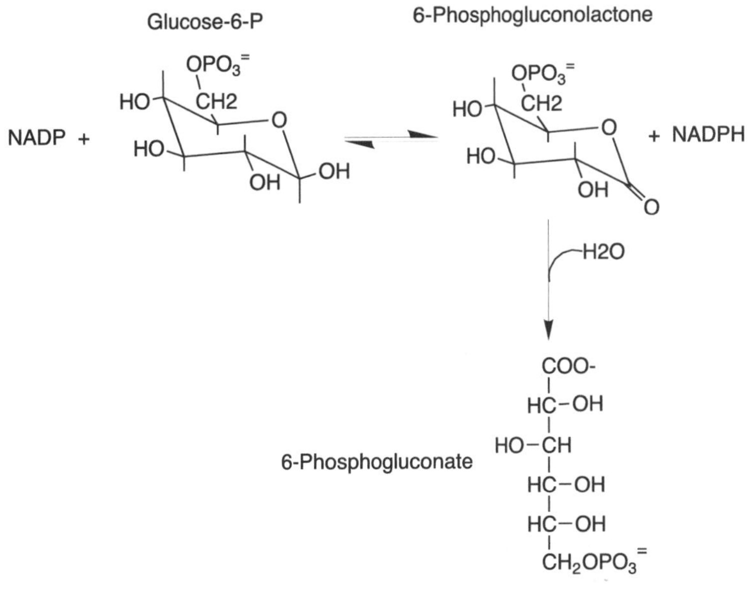

Scheme 9-2

EXPERIMENT 9-1

PROCEDURE

1. Add to the spectrophotometer cell (13-mm outside diameter test tube) the following:

   1.0 mL of G-6-PDH (0.5 units in 0.5 M Tris-HCl buffer, pH 8.0)
   0.1 mL of $MgCl_2$ (0.1 M)
   0.1 mL of glucose-6-P (0.1 M) (i.e., 10^{-5} moles)
   2.8 mL H_2O

2. Mix well, and place the tube into the spectrophotometer. Adjust the 100% knob so that the meter reads 0 absorbance. (If your instrument does not give 100% T at 340 nm, use 345 or 350 nm.)

3. Add 0.02 mL of NAD(P) (2.5 mM); mix and measure absorbance at 340 nm until you see no further increase in absorbance. Note the final absorbance.

4. Add an additional 0.03 mL NAD(P) to the reaction mixture, and measure the absorbance until a constant value is attained.

5. Repeat step 4.

6. Plot the concentration of NAD(P) added (which should equal the NAD(P)H measured) in millimolar (in the reaction mixture) versus the absorbance. Is the plot linear? Does it pass through zero?

Calculate the millimolar absorptivity (mM^{-1} cm^{-1}) of NAD(P)H. Assume that the path length of the tube is 1 cm. Note that the round test tubes you are using as cuvettes have ~1-cm optical paths. If you use very similar tubes for all of your measurements, your calibrations should be valid.

QUESTIONS

1. What do you think is the greatest source of error in this experiment? Explain.

2. How would you modify this procedure to measure D-glucose-6-phosphate?

3. How would you modify this procedure to measure D-glucose? Hint: in this case, you have to use another enzyme and cofactors to convert D-glucose to D-glucose-6-phosphate. (There are several answers to this question.)

4. Why was NAD(P) added in several aliquots?

EQUILIBRIA OF ENZYME-CATALYZED REACTIONS

In this experiment, you will examine the equilibrium of the biochemical reaction catalyzed by the enzyme, alcohol dehydrogenase (ADH) (equation 9-4), which converts acetaldehyde to ethanol in fermenting yeast. It is very important for certain industries that this reaction be driven to the left. In fact, all of human history would be different if this reaction were always driven to the right.

Although enzyme-catalyzed reactions are rarely in equilibrium in vivo (i.e., they are usually essentially in steady state), the study of the equilibria of such reactions in vitro has contributed to a semiquantitative understanding of the energy relationships among numerous reactions occurring in living systems. Indeed, the use of enzymes has facilitated the study of the equilibria of a number of chemical reactions formerly intractable to study because of the long times required to establish equilibria.

Enzyme systems can facilitate the establishment of equilibrium in a timely manner, with great specificity, and with quantitative conversions without side products, thus largely eliminating the establishment of multiple equilibria, which would be difficult to dissect and quantify. It is this specificity that controls and establishes particular metabolic pathways in biology rather than allowing true thermodynamic equilibria to occur. Thus, although thermodynamics determines what direction a reaction will proceed, it is the kinetic properties of enzymes that direct what metabolic pathways will be important.

A balanced equation for the ADH-catalyzed reaction is

$$CH_3CH_2OH + NAD \rightleftharpoons CH_3CHO + NADH + H^+ \quad (9\text{-}4)$$

Note the stoichiometric relationships among the components of the reaction. You will determine an equilibrium constant for this reaction and carry out calculations of several thermodynamic constants based on K_{eq}.

$$K_{eq} = \frac{[NADH][CH_3CHO][H^+]}{[NAD][CH_3CH_2OH]} \quad (9\text{-}5)$$

PROCEDURE

1. Add to the spectrophotometer tube the following solutions:
    0.5 mL of 0.17 M ethanol
    0.1 mL of 1 M Tris, pH 8.0
    2.9 mL H_2O
    0.1 mL of 3.3 mM NAD

2. Mix well, and place the tube in the spectrophotometer cell compartment. Adjust the absorbance to zero, and initiate readings.

3. Record readings for 1 min at 15-s intervals, and note any change in absorbance. Add 0.4 mL of ADH (120 units), mix well, and quickly put tube back into the spectrophotometer.

4. Read the absorbance at 15-s intervals for the first 2 min and then at 1-min intervals until no further increase in absorbance is seen.

5. Add 0.1 mL of enzyme solution (30 units); read immediately after adding enzyme and at 1-min intervals until no further change occurs.

6. Repeat steps 1–5 two more times with different quantities of ethanol and/or NAD.

7. From the absorbance at equilibrium and the extinction coefficient previously determined in Experiment 9-1 (if applicable, assume that the molar absorbance of NADPH at 340 nm is the same as for NADH), calculate the concentrations of NADH formed at equilibrium for each of the experiments above in mol/liter (M).

8. From this, calculate the concentrations of NAD at equilibrium (initial NAD concentration minus NADH concentration at equilibrium). The acetaldehyde concentrations will be equal to the NADH concentrations.

9. Calculate the concentrations of ethanol at equilibrium (it will be equal to the initial ethanol concentration minus the acetaldehyde at equilibrium). The $[H^+]$ is 10^{-8} M.

10. From these concentration values, calculate the equilibrium constants for the alcohol dehydrogenase using equation 9-5.

QUESTIONS

5. What was the effect of the second addition of enzyme? Is this what you would have expected?

6. Will the addition of more enzyme shift the reaction further to the right?

7. Do you expect the individual determinations of K_{eq} to be the same?

8. Did you find the same answer for the individual determinations of K_{eq}? Was the variance within your estimated experimental uncertainty?

EXPERIMENT 9-3

DETERMINATION OF ETHANOL

The ADH reaction that you have just utilized is often used to determine ethanol in blood serum or in foodstuffs, where it may be present in very low concentrations. However, as you may have noticed, the equilibrium of the reaction is more toward the reduction of acetaldehyde by NADH than toward the oxidation of ethanol. (The enzyme could be called acetaldehyde reductase rather than alcohol dehydrogenase.) However, a number of tricks are used to drive the reaction toward the oxidation of ethanol. The first trick is the use of NAD in excess of ethanol. A second trick is the use of high pH. As discussed above, because H^+ is a product of the reaction, at low $[H^+]$ (high pH), the reaction will be driven from left to right (formation of NADH). The third trick (and most important) is the addition of semicarbazide to the reaction mixture, which by reacting with the product, acetaldehyde, prevents the reaction from going backward, i.e., right to left. Semicarbazide reacts with acetaldehyde rapidly and quantitatively to form a Schiff's base (the semicarbazone) in the following manner:

$$CH_3CHO \; + \; H_2N-\underset{H}{N}-\overset{\overset{\displaystyle O}{\|}}{C}-NH_2 \; \longrightarrow \; CH_3CH = N-\underset{H}{N}-\overset{\overset{\displaystyle O}{\|}}{C}-NH_2 \; + \; H_2O$$

Therefore, the equilibrium of the ADH reaction (9-4) is displaced to the right by trapping the acetaldehyde. Note that this is an example of equation 9-2 above. The amount of NADH formed under such conditions is stoichiometric with the amount of ethanol in the original solution. The overall reaction is depicted below.

$$CH_3CH_2OH + NAD \rightleftharpoons NADH + H^+ + CH_3CHO \xrightarrow{\text{Semicarbazide}} \text{Semicarbazone}$$

PROCEDURE

1. Add to the spectrophotometer tube the following:
    3.5 mL of buffer (sodium pyrophosphate, 75 mM; glycine, 22 mM; semicarbazide, 75 mM, pH 8.7)
    0.2 mL of NAD (16 mM)
    0.1 mL of alcohol dehydrogenase (120 units)

2. Mix well, put the tube into the Spectronic 20 cell compartment, and zero the absorbance at 340 nm.

3. Add 0.2 mL of an ethanol-containing solution of unknown concentration (note which unknown was used) and mix.

4. Record a reading every few minutes until there is no further increase in absorbance. From the increase in absorbance (final reading) and the extinction coefficient determined in Experiment 9-1, calculate the concentration of ethanol present in the original unknown solution in millimolar and also in milligrams per milliliter.

NOTE✔ **This method is generally used to determine blood alcohol concentration, which is a well-used measure of the degree of intoxication of an individual. Ethanol levels in abstaining subjects are reported to be <0.01% (<10 mg percent or <100 mg/L). Impairment of driving ability starts at a level of 0.04% (40 mg percent). At a level of 0.08%, there is a definite deterioration of driving ability. At a level of 0.1% (100 mg percent) or above, a person is considered to be too intoxicated to drive.**

QUESTIONS

9. How much blood serum sample that contains 0.05% ethanol would you use in the above experiment to obtain a reading of 0.3 absorbance? (It is much smaller than you think)

10. How would you modify the experimental protocol as used in Experiment 9-3 to determine enzymatically NAD? acetaldehyde?

EXPERIMENT 9-4

COUPLED ENZYME ASSAYS: DEVELOPMENT OF A LINEAR CPK ASSAY

CATALYZED REACTIONS

In this experiment, you will develop appropriate conditions for assaying the enzyme CPK, which is found in muscle and in brain tissues. To enable the detection of CPK activity, you will couple this reaction (9-6) to the PK (9-7) and LDH (9-8) reactions. CPK is widely distributed and accounts for 10–20% of the cytoplasmic protein of muscle. Its function is to keep ATP levels high by regenerating ATP on demand from phosphocreatine via reaction 9-6 (right to left) shown below. During periods of rest, phosphocreatine is regenerated by the reverse reaction (left to right). CPK is a "high-energy" storage compound ($\Delta G^{0\prime} = -10.3$ kcal/mol). CPK is the only enzyme that catalyzes this interconversion, and therefore, it is important in maintaining the proper levels of ATP during intense physical activity.

CPK is often assayed to determine if serum levels have been elevated as a result of muscle damage, e.g., when it is suspected that a heart attack has occurred. CPK can be found in three isoforms: as an MM dimer in muscle, as a BB dimer in the brain, and as a BM dimer in various tissues. For this experiment, we use CPK from rabbit muscle. It is known that rabbit muscle CPK is inhibited by a number of compounds, including sulfhydryl reagents, metal chelating agents, adenosine, chloride, and others. ADP acts strongly as a competitive inhibitor; thus, the coupled assay we use, which removes ADP as it regenerates ATP, alleviates this problem. Phosphocreatine is a competitive inhibitor with respect to creatine. CPK is activated by divalent cations such as Mg^{2+}, Ca^{2+}, and Mn^{2+}, which is why metal chelating agents are inhibitory. These metals, which are critical to the enzymatic efficiency, are thought to act as Lewis acids or electrophilic catalysts in the overall reactions. The pH optimum for the kinase activity of CPK is 9.0, while the reverse reaction is optimal at pH 7.5. Since we want to measure the kinase activity, we use glycine buffer at pH 9. Be sure that when mixing your reagents you have a final pH of 9.

The reaction catalyzed is as follows:
1. Creatine Phosphokinase (CPK)

$$ATP + creatine \leftrightarrow ADP + phosphocreatine \qquad (9\text{-}6)$$

We couple this reaction to two other enzyme-catalyzed reactions:
2. Pyruvate Kinase (PK)

$$ADP + PEP \leftrightarrow pyruvate + ATP \qquad (9\text{-}7)$$

3. Lactate Dehydrogenase (LDH)

$$H^+ + pyruvate + NADH \leftrightarrow lactate + NAD \qquad (9\text{-}8)$$

The overall assay that you will develop can be easily adapted for other important biological assays. Note that the PK-LDH system uses ADP, which could derive from any of a wide variety of sources. For example, if you wanted to develop a continuous assay for a signal transduction process that involved the phosphorylation of a protein and one of the products was ADP, you could simply substitute your kinase reaction for the CPK reaction. The PK-LDH system can also be used to assay for ATPase by the same means. Hexokinase catalyzes the phosphorylation of glucose to form 6-phosphoglucose and ADP. The ADP can therefore be monitored with the PK-LDH system. Try to think of other assays for which this system would be valuable.

EXPERIMENT 9-4

GOAL OF THE EXPERIMENT

It will be your task to design experiments to optimize the conditions for determining the amount of CPK in a sample. You want the reaction velocity determined to be proportional to the quantity of CPK added. In an optimal assay for an enzyme (which is not always possible), the assay should be insensitive to small variations in the concentration of substrate. Thus, small changes (less than twofold) in [ATP] or [creatine] should not appreciably affect the overall rate. For assays, we therefore try to choose concentrations of substrates $>>K_m$. As mentioned, this is not always possible because of such factors as expense, solubility, or availability of substrates.

As discussed above, for a coupled assay to function properly, the coupling enzymes should be capable of catalyzing the reaction at a rate that does not limit the rate of the reaction being measured (i.e., CPK). For example, if you are trying to measure CPK that is converting ATP into ADP at 0.001 mM/min, the consumption of ADP to regenerate ATP, as well as the consumption of pyruvate and NADH to generate NAD and lactate, must keep up with this 0.001 mM/min. A 10-fold excess capacity of the coupling enzyme system is the usual goal.

TESTS TO DESIGN IN DEVELOPING THE ASSAY

For your assay development, you should first check that LDH is operating at a sufficient rate. Taking the example above, ask whether the LDH will use NADH at ≥ 0.01 mM/min ($0.124\ A$ min^{-1} cm^{-1} path).

If this criterion is met, check the coupled system with the PK system added to it. You will put together a mixture with ADP, PEP, PK, NADH, and LDH. The pyruvate generated by the PK reaction will feed the LDH reaction.

Ask whether this combined (coupled) system will sustain 0.01 mM/min (i.e., 10-fold safety factor). If not, try to figure out what is limiting the system and adjust your conditions.

Now you are ready for the complete system. Use your mixture as above, except now you leave out ADP because this will be generated from the CPK reaction. Add to your mixture ATP and creatine (and the appropriate buffer).

Check the background rate, and then add the CPK. What do you think could cause a background rate? Now measure the rate of the assay as a function of [CPK].

Design experiments to answer the following questions:

1. Is the overall rate proportional to the amount of CPK added?

2. Do small variations in [ATP] and/or [creatine] affect the rate?

3. What is the limiting rate at which you would not trust the assay without further adjustment of conditions?

PROTOCOL SUGGESTIONS

The following are suggestions for the concentrations of reagents in the spectrophotometric measuring tube. These are based on some literature suggestions from the references above. For all of these reagents, especially the enzymes, test what is necessary. Enzymes are expensive, and you want to use as little as possible and still obtain reliable results.

We use glycine buffer, 0.05 M, pH 9.0, for all experiments. However, all of the reagents are made up in 50 mM Tris, pH 7.5. Be sure that your glycine buffer compensates for the Tris effects on the pH.

Mg^{2+} is required for activity for nearly all kinases. The suggested protocols have Mg^{2+} included. You should design experiments that check to ensure that optimal amounts of Mg^{2+} are added. Thus, you want maximal activity without causing any inhibition.

PRECAUTIONS

1. Do not put creatine solutions on ice because at this concentration they will precipitate.

2. Take one 5-mL, one 0.2-mL, and eight 0.1-mL pipettes and label them appropriately. You will use them for several additions of the same solutions. Avoid contamination by being organized.

3. For assays of reaction 3 alone, or 2 + 3, dilute your enzyme stock to the suggested range. Otherwise, the reaction will be too fast to record reliable data, and you will run out of enzyme. Extra enzyme is not available.

4. For the CPK assay, when you add everything except CPK, wait until you obtain a stable reading at 340 nm before adding CPK.

EXPERIMENT 9-4

REACTION 3 ALONE

Pyruvate, 1 mM
NADH, 0.05–0.1 mM
LDH, 0.005–0.05 units/mL

REACTION 2 AND 3

ADP, 1 mM
PEP, 1 mM
NADH, 0.01–0.1 mM
LDH, 0.1–1.0 units/mL
PK, 0.005–0.05 units/mL
0.3– 6 mM $MgCl_2$

REACTION 1, 2, AND 3

NADH, 0.05–0.1 NADH
PEP, 1 mM
ATP, 1 mM
Creatine, 1 mM
LDH, 0.1–1.0 units/mL
PK, 0.1–1.0 units/mL
CPK, various amounts, 0.001–0.05 units/mL
$MgCl_2$, 6 mM (you might want to test this)

QUESTIONS

11.   This coupled reaction series removes ADP (a competitive inhibitor) by regenerating ATP (to keep ATP constant). You can notice, however, that phosphocreatine is not removed by the coupling reactions. How would you determine if this is a problem?

12.   Upon planning and completing this experiment, you will have developed an assay that can determine the amount of CPK in a solution. How would you modify this assay to study the steady-state enzyme kinetics of CPK? Remember that steady-state kinetics studies are carried out to determine how the reaction velocity varies as a function of the substrate concentration, not as a function of the

enzyme concentration. The goal is to determine V_{max}, K_m, and possibly various inhibition constants and patterns. Describe in your laboratory report how you would change the conditions to carry out steady-state kinetics studies.

MATERIALS FOR ENZYMATIC METHODS OF ANALYSIS

 Use glass distilled water for making up these reagents.

EXPERIMENTS 9-1 TO 9-3: DAY 1

1. Glucose-6-phosphate dehydrogenase: Prepare a solution containing 0.5 units/mL in 0.5 M Tris-HCl, pH 8.0. Each group of students will receive 1 mL. This can be stored in refrigerator for up to 2 days (Sigma G 8404 or the Sigma G 6378).

2. Alcohol dehydrogenase: Prepare a solution containing 300 units/mL (~1 mg/mL) in distilled water. Store in the refrigerator for as many as 2 days. Each group receives 1 mL. Do not freeze (Sigma A 7011, 30,000 units).

3. NADP (2.5 mM): Prepare solutions containing 2.5 µM NADP/mL H_2O (~2.1 mg/mL; check the MW and purity on the label). Check the concentration by measuring the absorbance at 260 nm (1.8×10^4 M^{-1} cm^{-1}). You will need to dilute the stock 1:100. Each group receives 0.5 mL. Store at 4 °C. (Sigma N 5755, 50 mg). Note that if you use the *Leuconostoc mesenteroides* G-6 PDH (Sigma G 8404), you can use NAD (below).

4. NAD (3.3 mM): Prepare a solution containing 3.3 µmol/mL (~2.5 mg/mL; check the MW on the label) in distilled H_2O. Check the concentration by measuring absorbance at 260 nm (1.8×10^4 M^{-1} cm^{-1}). Each group receives 0.5 mL. Store in refrigerator for up to 5 days (Sigma N7004, NAD, 500 mg).

5. Glucose-6-phosphate (0.1 M): Prepare a solution containing 0.1 mmol/mL in distilled H_2O. Each group receives 0.2 mL. Store in refrigerator or freeze (Sigma G 7250 Glc-6-P, Na_2, 1 g).

EXPERIMENT 9-4

6. Tris-HCl (1 M, pH 8.0): Dissolve 1.21 g of TRIZMA in 5 mL of H_2O. Add 2.9 mL of HCl, and bring the volume to 10 mL with H_2O. Check pH, and store in refrigerator. Each group receives 0.2 mL.

7. Buffer for ethanol determination: Dissolve 20 g of sodium pyrophosphate ($Na_4P_2O_7 \cdot 10\ H_2O$), 5.0 g of semicarbazide hydrochloride (Sigma S 4125, 100 g) and 1.0 g of glycine (Sigma 100 g) in 500 mL of distilled H_2O. Adjust pH to 8.7 with ~4 mL of 5 M NaOH and bring the volume to 600 mL. Can be stored in refrigerator for up to 2 weeks.

8. NAD (16 mM): Dissolve 144 mg of NAD in 12 mL of distilled H_2O. Stable in refrigerator for up to 1 week.

EXPERIMENT 9-4: DAY 2

1. Pyruvic acid (0.1 M) (Sigma P2256): Dissolve 0.396 g into 36 mL of H_2O. Keep refrigerated; stable for at least 5 days.

2. NADH (10 mM) (Sigma N4505): Dissolve 0.262 g in 36 mL of 10 mM Tris base (no pH adjustment). Refrigerate and avoid light; stable for at least 1 month.

3. ADP (0.1 M) (Sigma, A2754, 1 g): Dissolve 0.149 g into 36 mL of 50 mM Tris, pH 7.5. Adjust the pH to 7.5. Keep frozen; stable for at least 5 days.

4. PEP (0.1 M) (Sigma, P3637, 1 g): Dissolve 0.684 into 36 mL of H_2O. Keep frozen; stable for at least 5 days.

5. ATP (0.1 M) (Sigma, A7699, 1 g): Dissolve 1.984 g into 36 mL of 50 mM Tris, pH 7.5. Adjust the pH to 7.5. Keep frozen; stable for at least 5 days.

6. Creatine (0.1 M) (Sigma C0780): Dissolve 0.472 g into 36 mL of 0.1 M glycine, pH 9.0. Keep at room temperature; it will precipitate at ice temperature.

7. Glycine buffer (1 M, pH 9.0) (Sigma G7403, 1 kg): Dissolve 75.82 g in 990 mL of H_2O. Adjust the pH to 9.0 with NaOH (10 M).

8. $MgCl_2$ (0.3 M) (Sigma M9272, 500 g): Dissolve 2.195 g into 36 mL of H_2O.

9. Pyruvate kinase (0.03 U/µL) (Sigma P1506): Dilute 6.2 mL of $(NH_4)_2SO_4$ suspension into 46 mL of 0.05 M Tris-HCl, pH 7.5. (There should be ~0.03 U/µL at pH 9 in the glycine buffer.) Keep refrigerated; after 5 days it should retain ~86% of its activity.

10. Creatine phosphokinase (0.002 U/μL) (Sigma C3755): Dissolve 40 mg of lyophilized enzyme into 10 mL of cold 0.05 M Tris-HCl, pH 7.5. Refrigerate; stable for at least 5 days.

11. Lactate dehydrogenase (0.1 U/μL) (Sigma L-2500): Dilute 1.3 mL of $(NH_4)_2SO_4$ suspension into 11.7 mL of 0.05 M Tris-HCl, pH 7.5. Refrigerate; retains < 75% of its activity after 5 days; better to make up on day of experiment. Note that pH 9 is not optimum. Therefore, a unit at pH 7.5 has considerably less enzyme than one unit at pH 9.

APPENDIX 9-1. SAMPLE CALCULATIONS

Some of the calculations for this experiment may be unfamiliar to you. Therefore, we have included some sample calculations. More detailed discussions can be found in any textbook of physical chemistry. Try and understand the reasoning and the basic procedures as you use your data in equivalent calculations. If you have any trouble, contact one of your instructors.

EXPERIMENT 9-1

The initial incubation mixture contained 0.1 mL of 3.3 mM NAD in a total volume of 4.0 mL. Assume that all of the NAD was converted to NADH, the path length was 1 cm, and you obtained an absorbance change of 0.492. The final concentration of NADH is

$$(0.1 \text{ mL})(3.3 \text{ mM NAD})/4 \text{ mL} = 0.0825 \text{ mM}$$

$$\varepsilon = A/(c \cdot l) = 0.492/0.082 = 6 \text{ mM}^{-1}\text{cm}^{-1}$$

EXPERIMENT 9-2

Calculation of Thermodynamic Constants for Alcohol Dehydrogenase Reaction

The protocol was as follows:

0.5 mL ethanol (0.17 M)
0.1 mL buffer (pH 8)
0.1 mL NAD (3.3 mM)
0.4 mL ADH
2.9 mL water

Suppose that you obtained an absorbance value of 0.198 at equilibrium. To calculate K_{eq}, ΔG^0, $\Delta G^{0\prime}$, and $\Delta E^{0\prime}$ for this reaction, you will need to know the concentrations of all products and substrates at equilibrium. You can use the extinction coefficient determined in experiment 9-1 to calculate the concentration of products and substrates. Assume that the reaction (equation 9-4 repeated below) was carried out at 25 $^\circ$C.

$$NAD + CH_3CH_2OH \times NADH + H^+ + CH_3CHO$$

Calculation of K_{eq}

$$K_{eq} = \frac{[NADH][CH3CHO][H^+]}{[NAD][CH_3CH_2OH]} \quad (9\text{-}9)$$

The principles of mass balance require that $[NADH] = [CH_3CHO]$ (at equilibrium). $[NADH] = 0.198/(6 \text{ mM}^{-1} \text{ cm}^{-1}) = 0.033$ mM in 1 cm, which is your path length. Therefore, the concentration of both NADH and CH_3CHO is 3.3×10^{-5} M.

$$[NAD]_{initial} = \frac{(0.1 \text{ mL})(3.3 \text{ mM})}{4.0 \text{ mL}} = 0.083 \text{ mM or } 8.3 \times 10^{-5} \text{ M}$$

$$[NAD]_{eq} = [NAD]_{initial} - [NADH]_{eq}$$

$$(8.3 - 3.3) \times 10^{-5} \text{ M} = 5.0 \times 10^{-5} \text{ M}$$

Ethanol was diluted eightfold so that the initial concentration was $[Ethanol] = (0.17 \text{ M})/8 = 0.021 \text{ M} = 2.1 \times 10^{-2}$ M

$$[Ethanol]_{eq} = [Ethanol]_{initial} - [Acetaldehyde]_{eq}$$

$$= 2.1 \times 10^{-2} \text{ M} - 3.3 \times 10^{-5} \text{ M}$$

$$= 2.1 \times 10^{-2} \text{ M}$$

And, since our reaction was at pH 8,

$$[H^+] = 10^{-8} \text{ M}$$

$$K_{eq} = \frac{(3.3 \times 10^{-5} \text{ M})(3.3 \times 10^{-5} \text{ M})(10^{-8} \text{ M})}{(2.1 \times 10^{-2} \text{ M})(5.0 \times 10^{-5} \text{ M})} = 1.04 \times 10^{-11} \text{M}$$

Calculation of ΔG^0 and $\Delta G^{0'}$

$$\Delta G^{0'} = -RT \ln K_{eq} \quad (9\text{-}10)$$

ΔG^0 is the change in free energy for the conversion of one mol of each substrate into product when all reagents are held at 1 M, $P = 1$ atm, and $T = 25$ °C (298 K). These are called **standard conditions**. R is the gas constant = 1.987 cal/K•mol.

$$\Delta G^0 = -(1.987 \text{ cal/K} \cdot \text{mol}) \times (298) \text{ K} \times \ln(1.04 \times 10^{-11})$$

$$= 15,000 \text{ cal/mol}$$

ΔG^0 is defined for standard conditions, which means that $[H^+] = 1$ M. Since most biological materials are denatured at pH 0, actual measurements of K_{eq} to calculate ΔG^0 are not possible. Therefore, biochemists have adopted a modified standard state in which all reactants and products except H^+ are at 1 M. The H^+ concentration is taken at pH 7. The relationship between ΔG^0 and the modified standard state free energy change, designated $\Delta G^{0'}$, can be calculated as illustrated below. For example, a reaction that yields H^+ as a product (such as the ADH reaction):

$$R \rightleftharpoons P + H^+$$

$$\Delta G^{0'} = \Delta G^0 + RT \ln \frac{[P][H^+]}{[R]} = \Delta G^0 + RT \ln \frac{[P]}{[R]} + RT \ln[H^+]$$

where $RT = 592$ cal/mol

At standard conditions, when $[P] = [R] = 1$ M this simplifies to

$$\Delta G^{0'} = \Delta G^0 + RT \ln [H^+] \tag{9-11}$$

For this example,

$$\Delta G^{0'} = \Delta G^0 + (592 \text{ cal/mol}) \times 2.3 \times \log[H^+]$$

$$= \Delta G^0 + (1362 \text{ cal/mol}) \times \log[H^+]$$

$$= \Delta G^0 - (1362 \text{ cal/mol}) \times \log(1/[H^+])$$

$$= \Delta G^0 - (1362 \text{ cal/mol})(\text{pH})$$

$$= 15,000 \text{ cal/mol} - 10,900 \text{ cal/mol}$$

$$= 4,100 \text{ cal/mol}$$

If H^+ is a reactant rather than a product, this equation becomes:

$$\Delta G^{0'} = \Delta G^0 + (1362)(\text{pH}) \tag{9-12}$$

Calculation of Redox Potentials ($\Delta E^{0'}$)

Redox potentials can be measured in electrochemical cells. In such a cell, there are two redox-active systems electrically linked by electrodes and called **reaction half-cells**; one becomes reduced, and the other becomes oxidized. Electrons flow from the species being oxidized to the species being reduced, and the tendency for electrons to flow gives rise to an electromotive force (emf) measured with units of volts (V). To make it possible to compare how likely any particular substance might reduce another, a standard convention is used. In chemistry, we reference redox-active components to the H^+/H_2 half-reaction cell. This reference half-reaction is defined as having a reduction potential $E^0 = 0.000$ V at standard conditions (1 M, $P = 1$ atm, and $T = 298$ K) . If the hydrogen system can reduce a substance under these conditions, the redox potential E^0 of that half-reaction system has a positive potential; however, if that system can reduce H^+ to H_2, it has a negative redox potential. Thus, by convention, the E^0 values are given as reduction potentials. As for $\Delta G^{0'}$, biochemists also define $\Delta E^{0'}$ as the potential measured at pH 7. Tables of $E^{0'}$ values for redox half-reactions can be found in most biochemistry and physical chemistry textbooks, as well as in many chemistry handbooks.

The $\Delta E^{0'}$ for a system is a thermodynamic parameter indicating the likelihood that a reaction will proceed. It is related to $\Delta G^{0'}$ by the following relationship:

$$\Delta E^{0'} = -\frac{\Delta G^{0'}}{nF} \tag{9-13}$$

where F is the Faraday constant, 23,063 cal/(mol•V), and n is the number of electrons transferred in the reaction

Thus, you do not need data from electrochemical measurements to determine redox potentials. This is particularly useful because electrochemical measurements require special instruments and are often difficult to perform. For our example,

$$\Delta E^{o'} = -\frac{\Delta G^{o'}}{nF} = \frac{-4100 \text{ cal / mol}}{(2)(23,063 \text{ cal / mol} \bullet \text{V})} = -0.089 \text{ V}$$

$\Delta E^{0'}$ comprises the $E^{0'}$ values for the two half-reactions:
$\Delta E^{0'} = [E^{0'}$ of the half reaction containing the oxidizing reagent] – [$E^{0'}$of the half reaction containing the reducing agent]

If you know the $E^{0'}$ value for one of the half-reactions and the $\Delta E^{0'}$ for the overall reaction (as determined from the equilibrium constant), you can calculate the $E^{0'}$ for the other half-reaction. $E^{0'}$ for the NAD/NADH half-reaction (shown below) is known:

$$\text{NAD} + 2\,H^+ + 2\,e^- \rightleftharpoons \text{NADH} + H^+ \tag{9-14}$$

$$E^{0'} = -0.320 \text{ V}$$

Thus, we can determine the $E^{0'}$ for the acetaldehyde/ethanol half-reaction: From our example for reaction 9-4, as written

$$\Delta E^{0'} = -0.089 \text{ V}$$

$$= (-0.320 \text{ V}) - (E^{0'} \text{ for the acetaldehyde/ethanol half-reaction})$$

Thus, from our measurements, we calculate the $E^{0'}$ for the acetaldehyde/ethanol half-reaction as

$$E^{0'} = -0.221 \text{ V}$$

Points to consider:

1. The fact that a reaction has a very high K_{eq} and therefore a high negative ΔG^0 does not necessarily mean that it will proceed at a high rate. These calculations only tell us in which direction the reaction would go under standard-state conditions and what is the position of the equilibrium when it is attained. They do not tell us anything about the rate of the reaction. Note from your experiments that without the enzyme, no detectable reaction occurs. Experiment 9-2 illustrates how you can readily determine $E^{0'}$ for the acetaldehyde/ethanol half-reaction using the enzyme ADH, which speeds up the reaction to make it possible to reach equilibrium in a reasonable time.

2. The $\Delta E^{0'}$ must be positive to obtain a negative $\Delta G^{0'}$; i.e., a positive $\Delta E^{0'}$ indicates a spontaneous reaction.

Useful Relationships:

At pH 7 and standard-state conditions,

$$\Delta G^{0'} = -nF\Delta E^{0'}$$

$$\Delta G^{0'} = -RT \ln K_{eq}$$

$$\Delta E^{0'} = RT/nF \ln K_{eq}$$

$$\Delta E^{0'} = 0.06 \text{ V/n} \log K_{eq} \text{ at } 30 \text{ }^{\circ}\text{C}$$

$$\Delta E^{0'} = 0.059 \text{ V/n} \log K_{eq} \text{ at } 25 \text{ }^{\circ}\text{C}$$

The reduction potential of a half-reaction in which the oxidized and the reduced forms of the compound are present at nonstandard concentrations may be calculated from the **Nernst equation**:

$$E' = E^{o'} + \frac{RT}{nF} \ln \frac{[\text{Oxidized Form}]}{[\text{Reduced Form}]} \qquad (9\text{-}15)$$

This important equation enables you to calculate the potential for a reaction to occur regardless of the concentrations of reactants present.

10

Ligand Binding

LIGAND BINDING IS THE KEY TO MOST BIOLOGICAL PROCESSES

The binding of ligands to enzymes and protein receptors is the basis of all enzymatic specificity and of molecular recognition in biology. **Signal transduction** occurs because of specific ligand interactions, and DNA is replicated with fidelity because of numerous molecular recognition processes. The specific transport of small molecules across membranes is another example of this process. Compounds that are not able to diffuse through lipid membranes can be transported across membranes by proteins known as **permeases**. The binding of a specific small molecule to the extracellular surface of its permease is the first step in the translocation process. An example of a permease is the LacY protein of *Escherichia coli*, which is required for the efficient uptake and utilization of the disaccharide lactose.

Allosteric enzymes are enzymes that are regulated by ligands binding to regulatory sites on the surface of the enzyme. These ligands, known as allosteric (Greek *allos,* other space) effectors, may be either activators or inhibitors of an enzyme. Binding of the allosteric effector at a site distinct from the catalytic site causes a conformational change that affects the catalytic site. Often, regulatory sites are located on enzymes far from the catalytic site. Indeed, in hetero-oligomeric proteins, allosteric regulatory sites may even be found on different subunits than those containing the catalytic site. A classical example of such an enzyme is aspartate transcarbamylase of *E. coli*. The binding of small-molecule effectors sometimes directly controls the activity of **transcriptional activators** and **repressors**. The cyclic AMP (cAMP)-receptor protein of *E. coli*, known as CRP or CAP, provides an example of this phenomenon. This protein is a transcriptional activator and a repressor of many genes, but it is able to bind DNA only when it is complexed with its ligand, cAMP.

More complicated **signal transduction** systems often contain receptor proteins that span the cell membrane. The binding of a ligand to the extracellular domain of such receptors results in conformational changes that are propagated through the membrane-spanning segment to change the structure of the intracellular domain of the receptor. The latter may then activate or inhibit the activity of particular cytosolic proteins. An example of this class of proteins is provided by the chemotaxis receptors of bacteria. When "looking" for an attractant (the ligand for the receptor), the bacteria tumble and continually change directions. The receptor is not yet bound to its ligand, and these chemotaxis protein receptors activate the signaling pathway that leads to cells changing their direction. However, when a ligand is "found" and binds to the receptors, the signaling pathway to promote tumbling is shut down. Thus, bacterial cells continue to swim (chemotax) toward these ligands (energy sources that are known as attractants) when the receptors are full, but they tumble and "search" when they are empty.

The interaction of **antibodies** with **antigens** is another example of biologically important interactions between small-molecule ligands and protein molecules. Such interactions are highly specific and often characterized by dissociation constants $<10^{-9}$ M.

In each of the cases noted above, the interaction of ligands with the receptor is through a combination of hydrophobic, electrostatic, and hydrogen bonds. Covalent bonds are rarely encountered.

In the examples provided above, the receptors are proteins and the ligands are small molecules, such as amino acids, disaccharides, or cyclic nucleotides. However, ligands may also be DNA sequences, proteins, or macromolecular complexes. Antibodies are receptors that frequently bind ligands that are also proteins. Receptors may also be nucleic acids or macromolecular complexes containing protein and nucleic acids. Ligands can be any of the following: (1) small organic compounds such as metabolic intermediates, (2) metal ions or protons, (3) nonmetal inorganic ions such as phosphate, pyrophosphate, NH_4^+, Cl^-, Br^-, (4) peptides and oligonucleotides, or (5) other proteins, carbohydrates, lipids, and nucleic acids.

The examples described above illustrate the importance of the **specificity** of receptor-ligand interactions in biological systems. While many permeases can recognize equally well several compounds in a class of related molecules, other receptors are remarkably specific, similar to the specific interactions of enzymes with their substrates. The great specificity of these interactions is the result of millions of years of evolution and natural selection.

Thus, the interaction of ligands with receptors is of profound biological importance, and in this chapter, we develop equations describing these interactions and discuss experimental methods for analyzing them. There are two main methods to study receptor-ligand interactions. The most common method is to study the degree of saturation of the receptor at equilibrium with various concentrations of ligand. The second method is to study the kinetics of the association and dissociation of the ligand and receptor as a function of concentration. Both methods provide insights into the binding process.

ANALYSIS OF LIGAND BINDING AT EQUILIBRIUM

A ligand's association with and dissociation from its receptor are equivalent (and treated similarly) to a proton's association with and dissociation from its conjugate base. Thus, many of the equations that follow are essentially identical to those already presented in Chapter 1, with the terms renamed. For example, in acid-base chemistry L in equation 10-1 would be H^+ and R would be A^-. Let us start with a simple system where a single ligand L binds to a single site on the receptor R. At equilibrium, the binding of L to R will be dependent on the concentrations of L and R.

$$R + L \underset{k_{-1}}{\overset{k_1}{\rightleftharpoons}} LR \qquad (10\text{-}1)$$

$$K_d = \frac{[L][R]}{[LR]} = \frac{k_{-1}}{k_1} \qquad [LR] = \frac{[L][R]}{K_d} \qquad (10\text{-}2)$$

$[L]$ and $[R]$ are the concentrations of **free L** and **free R**. K_d is defined as the equilibrium dissociation constant and has units of molar (M). In this chapter, we refer to $[L]$ as the concentration of free ligand, not the total concentration of L that was added. In biochemistry, we usually talk about dissociation constants as defined above. In this case, the smaller the value of K_d, the tighter the binding. However, chemists often talk about the K_f, the association or formation constant. $K_f = 1/K_d$ and thus has the units of M^{-1}. A larger value for K_f implies tighter binding. For all of the discussion below you could, of course, use K_f values just as well as K_d values. K_f (or $1/K_d$) is a direct measure of the strength of the binding interaction. From equation 9-6, it is seen that $\Delta G^0 = -RT \ln (K_f) = RT \ln (K_d)$. At 25 °C, RT $= 0.592$ kcal/mol. Thus, we can calculate the free energy change on binding a ligand to a receptor from this relationship. Note that the more negative the ΔG^0, the stronger the binding interaction. The following examples may help you to better appreciate how binding constants and ΔG^0 are related. If for a particular LR the $K_d = 10^{-4}$ M (5.4 kcal/mol), and a solution is made up by mixing R and L so that the total added is 10^{-4} M of each, only 38% of L and R will be in the complex. (Remember that some of the added L is used to form the complex, so that free L $< 10^{-4}$ M.) If the K_d were 10^{-7} M (9.5 kcal/mol), ~97% of L and R would be in the complex. By checking these calculations yourself with the above relationships, you will become more familiar with the manipulations required for this laboratory session. Tables of K_d values are often tabulated as logarithms of the formation constants because they are proportional to the free energies. Thus, $\log K_f = -\log K_d = pK_d$. This is also directly related to acid/base nomenclature, where we talk about pH and pK_a values.

DETERMINATION OF DISSOCIATION OR BINDING CONSTANTS

Dissociation or binding constants are determined by varying the concentrations of L and R and by observing how the concentration of LR changes. By measuring the concentrations of L, R, and LR, you can directly calculate the K_d using equation 10-2. Of course, it is a prerequisite that there be a suitable method of distinguishing L, R, and LR. In many treatments of ligand binding, you will deal with bound and unbound fractions rather than the actual concentrations of R and LR. This can simplify the equations and facilitate comparisons of binding reactions at different concentrations. The fraction of bound receptor sites can be defined as α, where $[R_t]$ is the total concentration of all of the receptor sites, bound or unbound.

$$\alpha = \frac{[LR]}{[R_t]} = \frac{[LR]}{[R] + [LR]} \tag{10-3}$$

Since

$$[LR] = \frac{[L][R]}{K_d}$$

(from equation 10-2),

$$\alpha = \frac{\dfrac{[L][R]}{K_d}}{[R] + \dfrac{[L][R]}{K_d}}$$

Multiplying by K_d and factoring out $[R]$:

$$\alpha = \frac{[L]}{K_d + [L]} = \frac{[LR]}{[R_t]} \tag{10-4}$$

Rewriting in terms of LR, equation 10-4 becomes

$$[LR] = \frac{[R_t][L]}{K_d + [L]} \tag{10-5}$$

Since all of the receptor molecules are either bound or unbound, $(1 - \alpha) =$ fraction unbound $= [R]/[R_t]$. Substituting the expression for α from equation 10-4 and solving for $[R]$, we obtain

$$[R] = \frac{[R_t]K_d}{K_d + [L]} \tag{10-6}$$

From these equations (which are often called binding isotherms because they are measured at constant temperatures), you can determine K_d from measurements of $[LR]$ and free $[L]$ (equation 10-5) or from measurements of free $[R]$ and free $[L]$ (equation 10-6).

COMMON PLOTS FOR THE ANALYSIS OF LIGAND BINDING EXPERIMENTS

An important point of the previous discussion is that at any concentration of the receptor, the fractions of both bound and unbound receptors are dependent on the concentration of the free ligand and the dissociation constant. This is an intrinsic property of the system. A graph of $[LR]$ (or α) versus $[L]$ will be a rectangular hyperbola with a limiting value of $[R_t]$ (or 1) (Fig. 10-1A). The concentration of L at which $[LR] = [R_t]/2$ (or $\alpha = 0.5$) is the K_d. **Linear** and **semilogarithmic** plots are frequently used to analyze the data (Figs. 10-1A and 10-1B). A disadvantage to both of these plots is that it is difficult to obtain graphically a value for $[LR] = [R_t]$, the value at full saturation. Equation 10-5, which describes a rectangular hyperbola, may be used to fit binding data that measure $[LR]$ as a function of $[L]$.

Direct fitting of the untransformed data to the equation by nonlinear regression is the preferred method of analysis (see below).

It should be clear that these equations parallel closely those used for enzyme kinetics in Chapter 8. Recall the Lineweaver-Burk transformation of the rectangular hyperbolic Michaelis-Menten kinetic equation where we plot $1/v$ versus $1/[S]$. We can carry out the same transformation on equation 10-6.

$$\frac{1}{[LR]} = \frac{K_d + [L]}{[R_t][L]} \tag{10-7}$$

$$\frac{1}{[LR]} = \frac{K_d}{[R_t]}\left(\frac{1}{[L]}\right) + \frac{1}{[R_t]} \tag{10-8}$$

A plot of $1/[LR]$ versus $1/[L]$ (a **double reciprocal** plot) will be linear with a y intercept of $1/[R_t]$ and a slope of $K_d/[R_t]$ (Fig. 10-1C). An advantage to this plot is that it extrapolates to $1/[R_t]$, enabling the estimation of full saturation.

Another transformation that is frequently used is the **Scatchard plot.** Equation 10-5 can be rearranged as shown below.

$$[LR]K_d = [L]([R_t] - [LR])$$

$$\frac{[LR]}{[L]} = \frac{[R_t]}{K_d} - \frac{[LR]}{K_d} \tag{10-9}$$

If $[LR]/[L]$ is plotted versus $[LR]$, the result will be a straight line with a slope of $-1/K_d$, a y intercept of $[R_t]/K_d$, and an x intercept of $[LR]$ at full saturation (i.e., $[LR] = [R_t]$, Fig. 10-2). This plot has the advantage that both K_d and $[LR] = [R_t]$ are determined.

DIGRESSION ON REGRESSION

Before nonlinear regression programs were available, a standard method of analyzing a rectangular hyperbola was to transform the data with an equation that would permit the researcher to plot the data as a straight line. The researcher could draw a line through the data and determine K_d and $[R_t]$. Alternatively, linear regression (a method of least squares) could be used to determine the constants. Linear regression by the method of least squares fits a straight line to the data that minimizes the sum of the squares of the deviations of that line from the data. It assumes that the random errors of the mea-

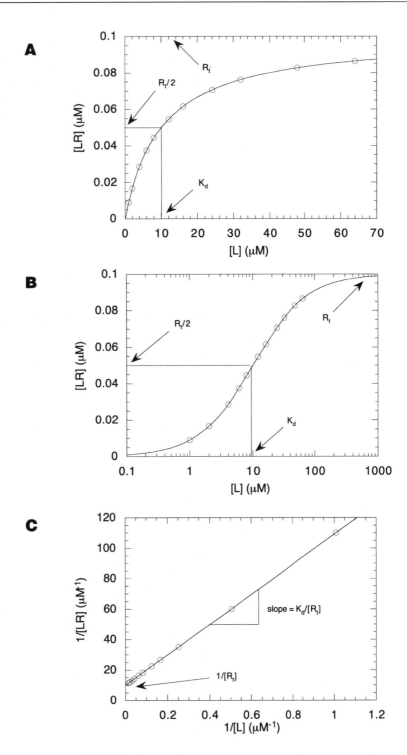

FIGURE 10-1   Plots commonly used for analysis of the binding of a ligand to a receptor: $[R] = 10^{-7}$ M and $K_d = 10^{-5}$ M. (A) Linear binding plot, (B) log-linear plot (often called a semilogarithmic plot), and (C) double reciprocal plot. These are plots of "perfect data" for a simple bimolecular binding system.

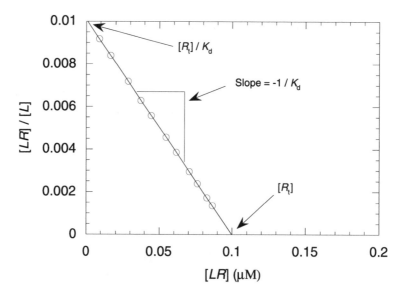

FIGURE 10-2 Scatchard plot of "perfect" ligand binding data from Fig. 10-1.

surements are distributed evenly with a Gaussian distribution around the average. It also assumes that there is only one variable (usually the y value) that is subject to error and that the errors between measurements are not correlated. For binding data, linearized plots, such as the reciprocal and the Scatchard plots (Figs. 10-1 and 10-2), are frequently used. However, it is important to realize that such transformations amplify and distort the experimental error with a bias; in particular, the transformation gives undue weight to data where very little binding has occurred. Furthermore, in the Scatchard plot, since the y axis is [bound/free] and the x axis is [bound], this linearization uses the independent variable to calculate the dependent variable, an operation that violates the assumptions of linear regression. An excellent depiction of this problem is given by H. Motulsky, who provides a general guide for analyzing binding data. This guide is available at http://www.graphpad.com/www/radiolig/radiolig1.html. For this reason, it is best to fit the binding data directly to equation 10-5. Nonlinear regression programs can be used to fit functions such as equation 10-5 directly without transforming and distorting the data. Nonlinear regression uses the same principle of minimizing the sum of the squares of the experimental deviations from the fit, but iteratively tries many solutions to arrive at the "best" values.

Several approaches can be used to find the best fit (see Press et al., 1992). After fitting the data with nonlinear least-squares procedures, it is often useful to replot the data in linearized forms, such as the Scatchard plot, along with the transformed fit. Such plots can be visually useful in deciding how well the data set fits classical simple binding. You might also note that a similar problem of fitting data is discussed in Chapter 8, Enzyme Kinetics.

EFFECTS OF THE CONCENTRATION OF L AND R

Figure 10-1 is a plot of [LR] versus [L]. When [L] >> K_d, [LR] approaches [R_t], although it does so more gradually than you might expect. For example, when [L] = 4K_d, [LR] is 0.8[R_t]; when [L] is 9K_d, [LR] = 0.9[R_t]; and when [L] = 99K_d, [LR] = 0.99[R_t]. These relationships between [LR], [L], and K_d hold true for all receptor-ligand interactions that occur by this simple bimolecular model, even though the K_d values for different receptor-ligand pairs are not the same. The logarithmic binding curve (binding isotherm) such as shown in Fig. 10-1B is simply shifted to higher or lower K_d values in the different systems, but the shape of the curve remains the same. This is analogous to titration curves of weak acids where the shapes remain the same for different acids, but the curves are shifted by the relative pK_a values (Chapter 1).

To obtain reliable results when determining binding parameters, it is important to measure [LR] when [L] is near K_d. The graphs and equations shown above yield the most accurate values when [R_t] << [L_t]; thus, [L_t] ≈ [L] because depletion of the free ligand concentration due to binding to the receptor is insignificant. This happens when [R_t] << K_d. In many experiments, this assumption is not true; it may be necessary (e.g., for sensitivity) to have [R_t] ≈ K_d to facilitate measurement of [LR]. In such cases, titration of the receptor with ligand causes significant ligand depletion. Indeed, in some of the methods described below, the extent of depletion of the ligand resulting from binding to the receptor is the measured variable. Procedures for dealing with ligand depletion are presented by Hulme and Birdsall (1992), Swillens (1995), Goldstein and Barrett (1987), and Gains (1979) and take into account the fact that [L] = [L_t] – [LR] and [R] = [R_t] – [LR]. For equilibrium binding experiments when ligand depletion is significant, the experimental data may be fit to equation 10-10, which is derived from equation 10-5 with the quadratic equation.

$$[LR] = \frac{(K_d + [R_t] + [L_t]) - \sqrt{(K_d + [R_t] + [L_t])^2 - 4[R_t][L_t]}}{2} \quad (10\text{-}10)$$

In more extreme cases where [R_t] >> K_d, R essentially directly titrates L, so that [L] is nearly zero during the titration until [LR] = [R_t]; then [L] ≈ [L_t] – [R_t]. In such a titration, very few useful data are obtained for determining binding constants, since there is never a significant amount of free L and free R at the same time. Thus, calculations with equation 10-5 become very imprecise. However, in the situation of [R_t] >> K_d, [LR] is directly proportional to [L_t] until [L_t] = [R_t], after which, of course, no further LR can form. Thus, with high but unknown concentrations of receptor, it is possible to determine [R_t] directly from titrations with L.

EFFECTS OF TWO SITES OR COOPERATIVITY

What if the shape of your binding data does not match the binding isotherm for a simple bimolecular interaction? Then there are several possibilities to consider. The first trivial possibility is that there is too much scatter in the primary data to make a good fit to anything; however, this can be resolved by performing more or better experiments. The second possibility is that there may be two or more classes of binding sites, each with a distinct R_t and K_d. Equation 10-5 can be adapted as shown in equation 10-11 for a system with two classes of nonequivalent sites:

$$[LR] = \frac{[R_{t1}][L]}{K_{d1} + [L]} + \frac{[R_{t2}][L]}{K_{d2} + [L]} \qquad (10\text{-}11)$$

Graphical methods will usually give complex plots that are difficult to evaluate unless K_{d1} and K_{d2} differ by ≥ 50-fold. Nonlinear regression techniques can fit such data if they are of sufficiently high quality. In the special case where the sites are identical ($K_{d1} = K_{d2}$) and do not interact (for example, a protein that is a dimer or a tetramer of identical subunits that behave independently), binding will be equivalent to that for simple, single-site binding, but $[R]$ will be that of the monomeric unit. From equation 10-11, it can be deduced that for two identical sites $R_t = R_{t1} + R_{t2}$; therefore, equation 10-11 reduces to equation 10-5 with the adjusted R_t.

Another factor that may alter the shape of the binding isotherm and thus indicate that a simple bimolecular mechanism does not apply, is positive or negative **cooperativity** in the binding. **Negative cooperativity** means that when one binding site is occupied, the occupancy of additional sites is disfavored. In contrast, **positive cooperativity** means that when one site is occupied, the occupancy of additional sites is favored. Such relationships can occur if a receptor has more than one site for the ligand, and binding of the ligand causes a conformational change in the receptor. The binding of oxygen to hemoglobin is a classic case of positive cooperativity and is discussed in most biochemistry textbooks.

PRINCIPLES OF COOPERATIVITY

The following briefly develops the principles of cooperativity. You should consult Cantor and Schimmel (1980) for a much more rigorous discussion. Consider the following binding reaction:

$$R + nL \; \rightleftharpoons \; L_nR \qquad (10\text{-}12)$$

If the ligand binds with infinite cooperativity in the presence of a substoichiometric amount of ligand, some of the receptors will be fully bound while others will contain no ligand at all. This happens when, on binding the first ligand, the $n-1$ other ligands for a particular receptor molecule are bound much more tightly. For this situation,

$$K = \frac{[R][L]^n}{[L_n R]} \qquad (10\text{-}13)$$

The fractional saturation is defined by Y_L, which is equivalent to α in equation 10-3. Traditionally Y_L is used for equations describing cooperativity.

$$Y_L = \frac{[L_n R]}{[R]+[L_n R]}$$

Substituting for $[L_n R]$ from equation 10-13,

$$Y_L = \frac{\dfrac{[R][L]^n}{K}}{[R]+\dfrac{[R][L]^n}{K}}$$

This is rearranged to give the **Hill equation**:

$$Y_L = \frac{[L]^n}{K+[L]^n} \qquad (10\text{-}14)$$

Nonlinear regression methods (Press et al., 1992) can be used to fit your data to equation 10-14 to yield K and n. Equation 10-14 was derived with the assumption that the system was infinitely cooperative, a situation that never occurs. When real data are analyzed, the value determined for n is always less than or equal to the actual number of sites (n). The degree of cooperativity is indicated by how close the measured Hill coefficient approaches n. It might be noted that if there is only one binding site, no cooperativity will be observed. Equation 10-14 can be manipulated as shown below to yield a form used for the **Hill plot**, another common way of visualizing and analyzing cooperative binding data.

$$\frac{Y_L}{1-Y_L} = \frac{[L]^n}{K}$$

$$\log\left(\frac{Y_L}{1-Y_L}\right) = n\log[L] - \log K \qquad (10\text{-}15)$$

An often used method of analyzing binding data for cooperativity uses the Hill plot: log $[Y/(1-Y)]$ versus log $[L]$ (Fig. 10-3). In a simple bimolecular mechanism, the slope of this line n, known as the Hill coefficient, will be 1 (Fig. 10-3A); a slope of >1 indicates positive cooperativity. A Hill plot for hemoglobin is shown plotted in Fig. 10-3B. There are three distinct regions in this plot. The first is indicative of simple ligand binding where there is only enough ligand to partially populate the first site, and a slope of 1 is observed. This part of the plot defines K_d for the binding of the first ligand. If this line is extrapolated to log $[Y_L/(1 - Y_L)] = 0$, K_{d1} can be calculated. As this site becomes populated, because of the positive cooperativity, subsequent sites bind ligand more tightly, and the slope increases. When there is only one site remaining, normal simple binding ensues, and a slope of 1 returns. Similarly, this part of the curve can be extrapolated to log $(Y_L/1-Y_L) = 0$ to determine K_{d2} (see Fig 10-3). Note that $K_{d2} < K_{d1}$ as expected as a result of the cooperativity. A slope of <1 in a Hill plot may indicate negative cooperativity, or possibly multiple classes of sites. The distinction between these latter two possibilities cannot be made from binding studies alone; additional data must be acquired to resolve this issue (see Cantor and Schimmel, 1980).

TWO LIGANDS AND ONE SITE

When two ligands that compete for the same receptor are present, the situation becomes slightly more complex.

$$R + L_1 \rightleftharpoons L_1R \qquad\qquad R + L_2 \rightleftharpoons L_2R$$

$$[L_1R] = \frac{[R][L_1]}{K_{L_1}} \qquad [L_2R] = \frac{[R][L_2]}{K_{L_2}} \qquad (10\text{-}16)$$

With two ligands

$$[R_t] = [R] + [L_1R] + [L_2R] \qquad (10\text{-}17)$$

The fraction bound as L_1R is calculated from equations 10-16 and 10-17 and is shown in equation 10-18.

$$f = \frac{[L_1R]}{[R_t]} = \frac{\dfrac{[R][L_1]}{K_{L_1}}}{[R] + \dfrac{[R][L_1]}{K_{L_1}} + \dfrac{[R][L_2]}{K_{L_2}}} \qquad (10\text{-}18)$$

Multiply numerator and denominator by K_{L1}, factor out $[R]$, and regroup:

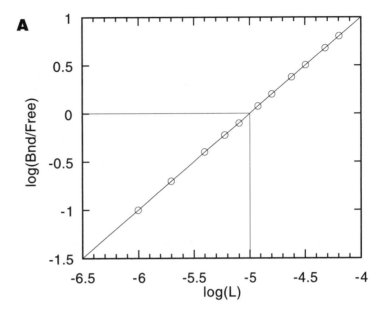

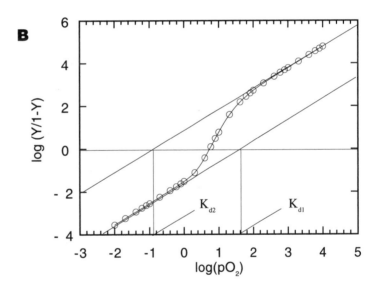

FIGURE 10-3 *(A)* Hill plot of the data in Figure 10–1 and *(B)* Hill plot of the data for the binding of oxygen to hemoglobin.

$$f = \frac{[L_1]}{K_{L_1} + [L_1] + \dfrac{K_{L_1}[L_2]}{K_{L_2}}} = \frac{[L_1]}{[L_1] + K_{L_1}\left(1 + \dfrac{[L_2]}{K_{L_2}}\right)} \qquad (10\text{-}19)$$

Since $[L_1 R] = f[R_t]$

$$[L_1 R] = \frac{[L_1][R_t]}{[L_1] + K_{L_1}\left(1 + \dfrac{[L_2]}{K_{L_2}}\right)} \qquad (10\text{-}20)$$

This is a normal binding isotherm like equation 10-5 with an apparent binding constant:

$$K_{L_1, app} = K_{L_1}\left(1 + \frac{[L_2]}{K_{L_2}}\right) = K_{L_1} + \frac{K_{L_1}}{K_{L_2}}[L_2] \qquad (10\text{-}21)$$

Thus, a plot K_{L1app} versus $[L_2]$ yields a straight line with K_{L1} as an intercept, and a slope of K_{L1}/K_{L2}. By analogy,

$$[L_2 R] = \frac{[L_2][R_t]}{[L_2] + K_{L_2}\left(1 + \dfrac{[L_1]}{K_{L_1}}\right)} \qquad (10\text{-}22)$$

so that K_{L_2} can be obtained similarly.

COMPETITION STUDIES

Competition binding studies are often used to determine the effectiveness of an unknown ligand to compete with a known ligand for a particular receptor site. This could be an antagonist competing with an agonist on a drug receptor or a synthetic compound competing with a natural hormone for its receptor. Such determinations frequently depend on being able to measure one of the complexes (L_1R) when the other (L_2R) is not easily measured. The most common method involves measuring a well-characterized complex containing labeled L_1 (to be measured by radioactivity, fluorescence, etc.) in the presence of various concentrations of an unknown L_2, which is unlabeled (see below for more discussion of radioactive ligands). Customarily, R is at a low concentration and one ligand L_1, which can be measured by radioactivity or some other parameter, is present at a fixed concentration that is sufficient to cause >90% complete formation of L_1R. L_1R is measured at various concentrations of the competing ligand L_2 (Fig. 10-4). At very low concentrations of L_2, no competition is observed, and the amount of L_1 binding is the same as in the absence of L_2. As L_2 increases, it competes for the receptor sites, and diminished binding of L_1 is observed. Eventually, the concentration of L_2 is high enough to block completely the binding of L_1 to the receptor. At any particular concentration of L_1, the concentration of competitor L_2 that diminishes the specific binding of L_1 to

50% of the original level is referred to as the **IC_{50}** or inhibition concentration for 50% inhibition for L_2.

The IC_{50} is not the K_d (usually called the K_i since L_2 is an inhibitor) for L_2; it is a function of the following parameters: K_d (or K_i) for L_2, K_d for L_1, and $[L_1]$ (see equation 10-23). The IC_{50} depends on the experimental conditions. Thus, if $[L_1]$ is high, the IC_{50} will appear higher than if $[L_1]$ is low. If L_1 binds to the receptor very tightly, the IC_{50} for L_2 will appear to be very high, whereas if the affinity of the receptor for L_1 is low, the IC_{50} for L_2 will appear to be low. In contrast to IC_{50}, K_i (K_d) for the binding of L_2 to R is an intrinsic property of the interactions between the ligand and receptor, and this constant does not depend on the nature of the competing ligand. From equation 10-22, we can extract an apparent binding constant for L_2 as shown in equation 10-23. This can be rearranged according to Cheng and Prusoff (1973) to yield equation 10-24, which is useful for calculating the K_i for a competing ligand L_2 from competition data such as in Fig. 10-4.

$$K_{L_2,app} = K_{L_2}\left(1+\frac{[L_1]}{K_{L_1}}\right) = IC_{50} \tag{10-23}$$

$$K_{L_2} = K_i = \frac{IC_{50}}{1+\dfrac{[L_1]}{K_{L_1}}} \tag{10-24}$$

Calculations of $K_i = K_{L_2}$ from Fig. 10-4 according to equation 10-24 are shown in Table 10-1. The values for IC_{50} were obtained by reading the point on the x axis where the observed fraction of L_1R has decreased to one half of its original value. It can be seen that the most reliable determinations are made (even with perfect data) when $[L_1]$ is considerably greater than K_{L1}. The reason K_i is not well-determined when $[L_1]$ is not considerably greater than K_{L1} is that at the start of the experiment only a fraction of L_1R is preformed. It should be noted that in this case (Fig. 10-4), K_i is 10-fold larger (weaker binding) than K_{L1}. When $K_{L2} = K_{L1} = 10^{-5}$ M, it is necessary to make L_1 as high as 10^{-2} M and to titrate L_2 to concentrations higher than 10^{-1} M to obtain a curve that gives truly reliable data.

ANALYSIS OF THE KINETICS OF LIGAND BINDING

Equation 10-1 (above and repeated below) shows the simple binding of L to R as a bimolecular reaction:

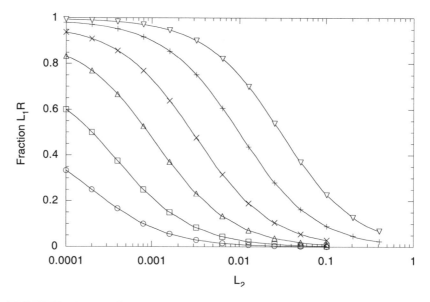

FIGURE 10-4 Competitive ligand binding. Measurement of the decrease in $[L_1R]$ as $[L_2]$ is increased. $K_{L_1} = 10^{-5}$ M; $K_{L_2} = 10^{-4}$ M; $[L_1]$: ($\circ$), 10^{-5} M; ($\square$), 3×10^{-5} M; ($\triangle$), 10^{-4} M; ($\times$), 3×10^{-4} M; ($+$), 10^{-3} M, and ($\triangledown$), 3×10^{-3} M.

$$R + L \underset{k_{-1}}{\overset{k_1}{\rightleftharpoons}} LR$$

Here k_1 is the association rate constant and k_{-1} is the dissociation rate constant. Association is a second-order reaction and is dependent on $[L]$, $[R]$, and k_1 (rate $= k_1[L][R]$). Dissociation is a first-order process, and its rate $= k_{-1}[LR]$. The association rate constant has units of $M^{-1}s^{-1}$, while the dissociation rate constant has units of s^{-1}. At equilibrium, the rates of forma-

TABLE 10-1		
Calculations of K_{L2} from IC_{50} Observed with Different $[L_1]$		
[L1] (M)	**IC50 (M)**	**K_{L2} (M)**
1.0×10^{-5}	4.2×10^{-4}	2.2×10^{-4}
3.0×10^{-5}	6.0×10^{-4}	1.5×10^{-4}
1.0×10^{-4}	1.3×10^{-3}	1.2×10^{-4}
3.0×10^{-4}	3.3×10^{-3}	1.1×10^{-4}
1.0×10^{-3}	1.0×10^{-2}	1.0×10^{-4}
3.0×10^{-3}	3.0×10^{-2}	1.0×10^{-4}

tion and of breakdown of the LR complex are equal, so that $k_1[L][R] = k_{-1}[LR]$; therefore, there is no net change in concentrations as a function of time. Equation 10-2 (repeated below) shows how $k_{-1}/k_1 = K_d$.

$$K_d = \frac{[L][R]}{[LR]} = \frac{k_{-1}}{k_1}$$

When L and R are first mixed together, the system is not at equilibrium, and the rate of change of the concentration of the LR complex will equal the difference between the association and dissociation rates as described by

$$\frac{d[LR]}{dt} = k_1[R][L] - k_{-1}[LR] \qquad (10\text{-}25)$$

The rate of formation of LR is greatest when L and R are first mixed together, and this rate gradually decreases until $d[LR]/dt = 0$ when the system attains equilibrium. This higher rate at $t = 0$ occurs because initially there is no LR, so $k_{-1}[LR] = 0$ and $d[LR]/dt = k_1[R][L]$. If we define $[LR_{eq}]$ as the concentration of LR when the system has attained equilibrium, then at any given time,

$$[LR] = [LR_{eq}]\left(1 - \exp\left(-\left(k_1[L] + k_{-1}\right)t\right)\right) \qquad (10\text{-}26)$$

Thus, LR increases exponentially to its equilibrium value. This is shown in Fig. 10-5A. Note that the effective rate constant is $(k_1[L] + k_{-1})$, the sum of the pseudo first-order forward $(k_1[L])$ and the reverse rate constants. The net reaction finishes sooner than if it were an irreversible reaction, since changes in LR occur only until equilibrium is attained. In Fig. 10-5, the effective rate constant k_{eff} for reaching equilibrium is $(10^4$ $M^{-1}\ s^{-1})(10^{-3}\ M) + 10\ s^{-1} = 20\ s^{-1}$. The time required for reaching one half of the total amplitude change can be calculated from the following relationship: $t_{1/2} = \ln(2)/k_{eff} = 0.693/20\ s^{-1} = 0.0347s$. This relationship holds for any first-order reaction. The data in Fig. 10-5 are typical for many biological reactions. You might note that the rates are considerably greater than you can measure in a simple spectrophotometer. Measurements of such rates require rapid-reaction devices such as stopped-flow spectrophotometers.

Suppose that you were to mix and incubate L and R (as in Fig. 10-5A) for a sufficient length of time to permit the formation of an equilibrium mixture of LR, L, and R. The kinetics of the dissociation reaction can be studied if further binding of L to R is prevented. This can be accomplished in several ways. You can add a large concentration of a competitor for L, so that once LR dissociates the free R is immediately bound by the competitor, not allowing it to rebind L. You can dilute the system greatly, so that once LR dissociates, it has very little chance to re-form. Alternatively, you can add a reagent that quickly reacts irreversibly with L, as soon as it dissociates, so that it cannot re-form any LR complex. In any of these cases, the concentration of LR present at any given time will be

$$[LR] = [LR_0] \exp(-k_{-1}t) \qquad (10\text{-}27)$$

where $[LR_0]$ is the concentration of the complex at equilibrium. That is, LR_0 disappears via an exponential decay at a rate governed by k_{-1}. This is shown in Fig. 10-5B. Using the same k_{-1} ($10\ \text{s}^{-1}$) as in Fig. 10-5A, the reaction half time can be calculated as 0.0693 s. If you can measure k_1 and k_{-1} by suitable techniques, you should be able to obtain K_d. If this K_d is different than that determined by the concentrations of L, R, and LR, you can probably assume that the binding is more complex than that shown in equation 10-1.

The kinetic analyses presented above pertain only to the simplest cases in which a single type of binding site is present and binding occurs under conditions where $[L] \gg [LR]$ such that $[L]-[LR] \approx [L]$. However, in nature, far more complex systems can be encountered. For example, there may be multiple types of sites present, and binding of L to these sites either may be independent or may demonstrate positive or negative cooperativity. Furthermore, in many experiments, depletion of ligand cannot be ignored. Cantor and Schimmel (1980), Hulme and Birdsall (1992), Swillens (1995), Goldstein and Barrett (1987), Gains (1979), and Cheng and Prusoff (1973) present a good discussion of such systems.

METHODS USED TO STUDY RECEPTOR-LIGAND INTERACTIONS

BASIC STRATEGY

The ability to analyze quantitatively ligand-binding processes depends on being able to distinguish between bound and free receptors. When possible, it is best if the LR complex can be distinguished from the free L or R without separating the compounds. Absorbance, fluorescence, nuclear magnetic resonance (NMR), or other physical measurements can be particularly good for measuring LR complexes in purified systems. With such means of analysis, receptors R can be titrated with L, and changes can be measured to obtain values for fitting by equations such as equation 10-10. This is discussed further below.

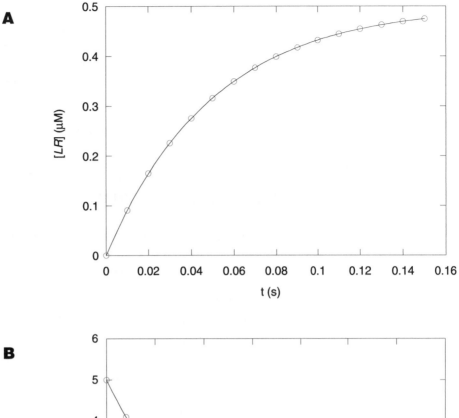

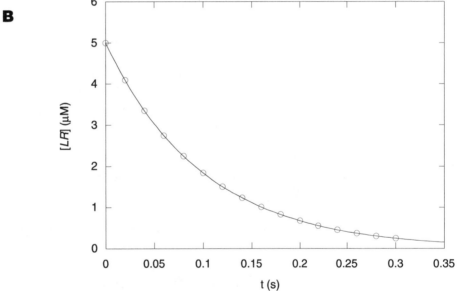

FIGURE 10-5 *(A)* Binding of L to R: $k_1 = 10^4$ M^{-1} s^{-1}; $k_{-1} = 10$ s^{-1}; $[R]$ = 10^{-6} M; $[L]$ = 10^{-3} M. *(B)* Dissociation of L from LR: $k_{-1} = 10$ s^{-1}; $[LR_0]$ = 5×10^{-6} M (initial concentration).

In Experiment 10-1, you will carry out binding experiments with absorbance spectrophotometry. When no such convenient method for distinguishing L, LR, or R is available, other approaches must be used.

Usually, in ligand-binding experiments in which a direct measurement of the binding is not possible, the following basic steps are followed. First, the receptor (which may be purified or not) is mixed with a series of known concentrations of the ligand. Second, the receptor and the bound ligand are physically separated from the free ligand. Finally, the free ligand, the bound ligand, or both are measured. The main experimental problems associated with these experiments are (1) verifying the completeness and suitability of the separation method and (2) dealing with the nonspecific binding of the ligand to nonreceptor sites, such as to the walls of the vessels, to the filters used in the separation, or to other macromolecules present in the receptor preparation.

USE OF RADIOLABELED LIGANDS

Many experiments in which ligand binding is studied make use of radiolabeled ligands to facilitate highly sensitive measurements of ligand concentration. Many radioactively labeled compounds are commercially available and may be obtained with very high specific activity. If a ligand is not commercially available in radiolabeled form, it may be possible to synthesize it in vitro (either enzymatically or chemically) from a labeled compound that can be commercially obtained. If the compound is not available, it may be possible to grow cells in the presence of a labeled precursor and isolate the desired, labeled ligand. Finally, it may be possible to synthesize a labeled analog of the desired ligand. For example, it is possible to iodinate many compounds in vitro by using ^{125}I. Although such a labeled compound is not identical to the original ligand, in many cases, such alterations do not appear to change the binding properties of the ligand and can therefore be useful.

METHODS BASED ON THE SEPARATION OF BOUND AND FREE LIGAND

A variety of methods of separation are available including the following: ultrafiltration, ammonium sulfate precipitation of receptor and bound ligand, equilibrium dialysis, and gel filtration.

In **ultrafiltration** methods a special filter with uniform small pores is used. The filter retains the large receptor-ligand complexes, but free ligand is permitted to pass through. Devices are available that allow convenient collection of the filtrate. Before filtration, a sample is taken from the mixture and used for measurement of total ligand (bound + free). A sample of the mixture is then filtered, and material passing through the filter is used for the measurement of free ligand. The amount of bound ligand is then the difference in the ligand concentration of these two samples. Ultrafiltration methods are often used when a large number of measurements must be made, because the technique is simple and easy to perform repetitively.

However, ultrafiltration methods, as well as other methods that depend on calculating the fraction bound as (bound + free) − (free), are limited to binding interactions with low a K_d relative to the concentration of receptor that can be used. For example, if the highest receptor concentration that can be attained is 1 μM, it would not be possible to measure reliably the binding of a ligand with a K_d = 20 μM. At a free ligand concentration of 20 μM, 0.5 μM of the ligand would be bound by the receptor. Thus, (bound + free) would equal 20.5 μM, and free would equal 20 μM. This difference, amounting to ~2.5% of the total ligand concentration, is probably not distinguishable from the uncertainty of the measurement. On the other hand, if the K_d in our example were 2 μM, then at 2 M free ligand, 0.5 μM would be bound. The experimental method would have to be able to detect reliably the difference between 2.5 M ligand (bound + free) and 2.0 M ligand (free). This difference of 20% in the total ligand concentration can usually be measured in ultrafiltration experiments, which usually have uncertainties of 5–10% or less.

Another variation of the ultrafiltration method measures the bound fraction retained by the filter, and free-ligand concentration is calculated as (total – bound). This method is useful in experiments for which a very low concentration of receptor is available; in these experiments, ligand depletion is usually negligible. However, in many experiments of this type, non-specific binding becomes a major problem and must be accounted for. This will be described below.

Ammonium sulfate precipitation of LR provides another method of separation; in this case, the bound ligand fraction is usually measured. Before using the method, it is necessary to demonstrate that precipitation occurs with retention of the bound ligand. Also, the method is a bit cumbersome, so it is rarely used.

Equilibrium dialysis is a method that is useful for measuring the binding of small ligands (<1000 MW) to large receptors. Although it can be somewhat cumbersome, it is, nevertheless, a useful method. In this method, a small membrane similar to the cellulose material for dialysis (Chapter 7) separates two chambers. One chamber is loaded with a solution containing the receptor, while the other chamber is loaded with a solution lacking the receptor but containing the ligand. The membrane is chosen so that the ligand may pass through, but the receptor is retained. Ligand diffuses through the two chambers until it reaches equilibrium, at which point the free ligand concentration in both chambers is identical. After equilibrium has been attained, the ligand concentration in the chamber lacking the receptor is the free ligand concentration, and the ligand concentration from the chamber containing receptor is the (bound + free ligand). The difference in these values is the bound ligand concentration. Since the depletion of the free ligand is what is measured in these experiments, equilibrium dialysis is useful only for those ligands with K_d low relative to the receptor concentration.

Gel filtration can also be used to separate the LR complex from L if the k_{-1} is very small, such as for some protein complexes or for antibody complexes. An important adaptation of the gel-filtration method is the method of **Hummel and Dryer**. In this method, a gel-filtration column is equilibrated with buffer containing the ligand at a given concentration [L]. The gel matrix is chosen such that the receptor molecules will be fully excluded. The receptor is mixed with the ligand at the same concentration [L] that is present in the column, and this sample is applied to the column and eluted with buffer containing [L]. Thus, in the applied sample, the total ligand consists of both free L and LR, so that the remaining free [L] is less than [L] originally added. Fractions are collected as the receptor percolates through the column and is eluted. Fractions eluting before the receptor contain ligand at the original (free) concentration chosen. The LR complex moves through the column faster than the free L so that when the receptor elutes, it will be in a zone containing [L], bringing with it bound ligand. Thus, when receptor elutes, the concentration of ligand in those fractions will be increased by [LR]. The fractions eluting just behind the receptor will contain less than the original free ligand concentration, since some of this ligand was bound by the receptor. Thus, in a graph of ligand concentration versus fraction number, a peak of ligand coincident with the receptor is observed, followed by a trough in the ligand concentration of equal value. Quantification of the peak (or trough) by integration reveals the bound fraction of ligand. Since the receptor concentration in each fraction can be easily determined, direct estimation of the K_d is possible.

ESTIMATION OF NONSPECIFIC BINDING

Ligands bind specifically to their receptors, but they also can bind nonspecifically to vessels used in the experiments and to other macromolecules present in receptor preparations. Since the number of such nonspecific sites may be very large in comparison to the number of receptor sites in the experiment, the contribution of nonspecific binding to total binding may be significant (Fig. 10-6). For example, if you are measuring the binding of a hydrophobic ligand to a cell surface receptor, you may find that the ligand binds nonspecifically to both the membrane and other membrane proteins. It is necessary to determine the level of nonspecific binding for most experimental procedures. One way to do this is to repeat the binding measurements in the presence of a very high concentration of competitor that specifically binds the receptor. For example, if a radiolabeled ligand is used, a large concentration ($\sim100\ K_d$) of the same unlabeled molecule could be used. The unlabeled species will completely occupy the receptor sites, and thus, any binding of the radiolabeled ligand observed must be due to nonspecific binding of other sites. Note that the nonspecific binding component will be highest whenever the ligand concentration is highest, as shown in Fig. 10-6. Figure 10-6 shows an example that can be analyzed graphically, because it is a simple function of a hyperbola plus a straight line. In other cases the nonspecific binding may be nonlinear and more difficult to analyze. As is obvious from Fig. 10-6, failure to measure and sub-

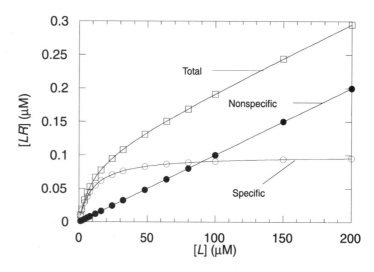

FIGURE 10-6 Specific and nonspecific binding of ligand to
R. $K_d = 10^{-5}$ M; $[R] = 10^{-7}$ M; nonspecific binding is $0.001[L]$.

tract the nonspecific binding can lead to large errors in the estimation of K_d
and R_t. In general, specific binding will saturate at a much lower concentration of L than will the nonspecific binding (Fig.10-6). Hulme and Birdsall (1992) discuss these problems in more detail.

METHODS USED TO DETECT PROTEIN-PROTEIN INTERACTIONS

In the preceding examples, the receptor and ligand were of vastly different
size. For example, the equilibrium dialysis method requires that a membrane be used that completely retains the receptor but permits the ligand to
freely pass through. However, when both the ligand and the receptor are
proteins, methods that distinguish L and R on the basis of large differences
in size, will not be useful. For these cases, there are methods available, but
they are generally less quantitative than the methods discussed in the previous section. These methods are thoroughly reviewed in Phizicky and
Fields (1995). One commonly used method is the detection of complexes
of **co-immune precipitation**. For this method, a sample containing
receptor and ligand proteins is mixed with an antibody highly specific for
one of these proteins. The antibody precipitates its antigen protein from the
solution. When the antigenic protein is bound to its partner, both proteins
will co-precipitate. The specificity can be demonstrated by incubating the
antibody with a similar mixture of proteins that lacks the antigenic protein.
In this control experiment, the antibody should precipitate no proteins.
This type of evidence is sometimes used to show that two proteins interact
in solution, but it should be used with caution.

Affinity chromatography (see Chapter 4) is another way to study protein-protein interactions. For this method, one protein is immobilized on a particulate support, such as Sephadex beads, and poured into a column. The proteins being tested as ligands are then passed through the column. If a test protein binds to the immobilized protein, it will be retained on the column, or its passage through the column will be retarded. If the test protein does not bind the immobilized protein, it will pass directly through the column.

Cross-linking is another way that protein-protein interactions have been studied. This method is designed to trap the bound fraction as an irreversible covalent complex. Many different types of cross-linking agents may be used; these must show high reactivity under the conditions used to trap completely the bound complex. In a typical experiment, the mixture of receptor protein and ligand protein is treated with the cross-linking agent, and the protein mixture is subjected to a fractionation technique such as SDS-gel electrophoresis. Since cross-linking involves covalent bond formation, the cross-linked species will migrate more slowly than the unlinked free protein species. By quantification of the amount of a test protein in the unlinked and cross-linked bands, the degree of binding may be estimated.

METHODS WITHOUT PHYSICAL SEPARATION OF BOUND FROM FREE LIGAND

For many receptor-ligand systems the binding of a ligand changes some measurable property of L and/or R. The most convenient and precise methods for measurement, use spectrophotometric, fluorimetric, or other techniques that do not require physical separation of bound and free species. In some cases, the binding of a colored ligand results in alteration of the extinction coefficient or a shift in the adsorption spectrum (or both). Thus, by difference spectroscopy, you can directly measure ligand binding without physical separation of the bound and free species. A very elegant example of the method is shown in Jones and Waley (1979). Similarly, the fluorescence or some other property of a ligand or receptor may be altered upon binding the receptor, thereby permitting direct measurement of binding without physical separation of bound and free forms.

In our laboratory exercise, we will study the binding of the anionic dye, 2-(4-hydroxybenzeneazo) benzoic acid (sometimes called 4´-hydroxy-azobenzene-2-carboxylic acid (HABA)), an analog of biotin, to the biotin-binding protein avidin. The structures of biotin and HABA are shown in Figs. 10-7 and 10-8. HABA absorbs light with an absorption maximum of 348 nm. When HABA is bound by avidin, its absorption maximum is shifted to 500 nm, and its extinction coefficient is greatly increased. At 500 nm, the unbound and bound forms of HABA have extinction coefficients of 600 M^{-1} cm^{-1} and 34,500 M^{-1} cm^{-1}, respectively. Thus, the binding of HABA to avidin can be easily measured spectrophotometrically.

REFERENCES

Cantor, C. R., and P. R. Schimmel. 1980. Biophysical chemistry, part 3: the behavior of biological macromolecules. W. H. Freeman & Co., New York.

Cheng, Y., and W. H. Prusoff. 1973. Relationship between the inhibition constant K_i and the concentration of inhibitor which causes 50 percent inhibition (I_{50}) of an enzymatic reaction. Biochem. Pharmacol. **22:**3099.

Dryer, R. L., and G. F. Lata. 1989. Experimental biochemistry. Oxford University Press, Oxford.

Gains, N. 1979. The determination of binding parameters when the total and free substrate concentrations are not approximately equal. Biochem. J. **179:**697.

Goldstein, A., and R. W. Barrett. 1987. Ligand dissociation constants from competition binding assays: errors associated with ligand depletion. Mol. Pharmacol. **31:**607.

Hulme, E.C. 1992. Receptor-ligand interactions, a practical approach. Oxford University Press, Oxford.

Hulme, E. C., and N. J. M. Birdsall. 1992. Strategy and tactics in receptor-binding studies, p. 64. *In* E. C. Hulme (ed.), Receptor-ligand interactions, a practical approach, chapter 4. Oxford University Press, Oxford.

Jones, R. B., and S. G. Waley. 1979. Spectrophotometric studies on the interaction between triose phosphate isomerase and inhibitors. Biochem. J. **179:**623.

Phizicky, E. M., and S. Fields. 1995. Protein-protein interactions: methods for detection and analysis. Microbiol. Rev. **59:**94.

Press, W. H., S. A. Teukolsky, W. T. Vetterling, and B. P. Flannery. 1992. Numerical recipes in C, the art of scientific computing, 2nd ed. Cambridge University Press, Cambridge.

Swillens, S. 1995. Interpretation of binding curves obtained with high receptor concentrations: practical aids for computer analysis. Mol. Pharmacol. **47:**1197.

EXPERIMENT 10-1

THE BINDING OF AVIDIN AND BSA TO BIOTIN
BACKGROUND

There are three main carriers of one-carbon groups in cells, S-adenosyl-methionine, tetrahydrofolate, and biotin. Biotin carries one-carbon groups in their most oxidized form, CO_2 (Fig. 10-7). This vitamin plays a key role in many carboxylation reactions, such as the pyruvate carboxylase-catalyzed carboxylation of pyruvate to form oxaloacetate (Fig. 10-7). Biotin is not produced by humans and thus is an essential nutrient (a vitamin) of the diet. However, biotin is abundant in many foods, so dietary deficiency is rare.

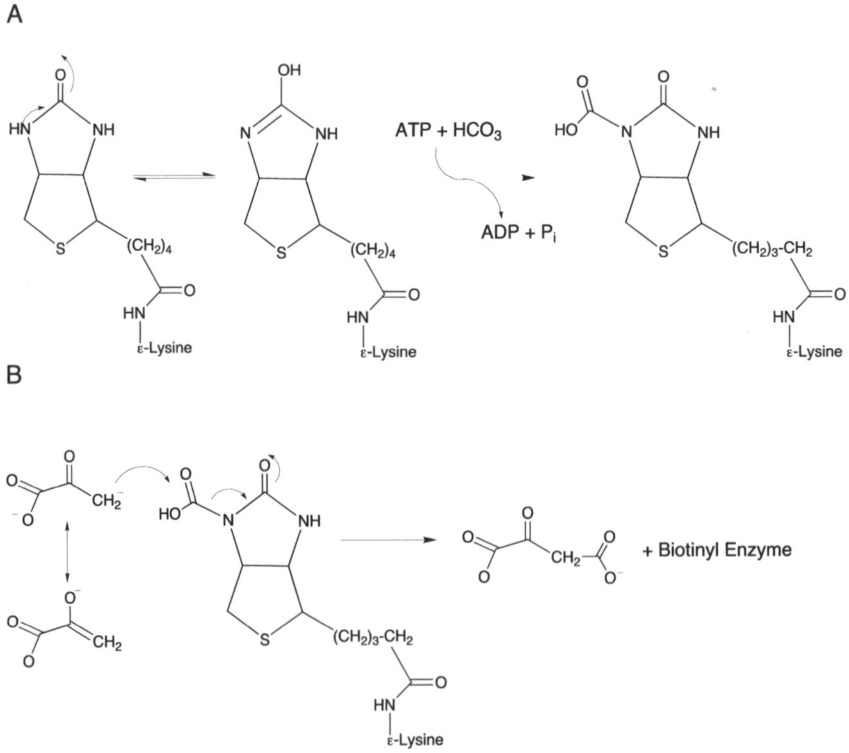

FIGURE 10-7   Structures and reactions of biotin. *(A)* Charging of biotin with concomitant hydrolysis of ATP. A carbonylphosphate intermediate is likely to be formed in the activation the CO_2. This attacks the biotinyl enzyme to give the carboxybiotin. *(B)* Carboxylation of pyruvate enolate by carboxybiotin.

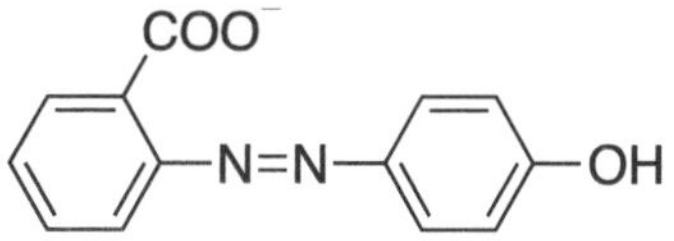

FIGURE 10-8 Structure of HABA, whose chemical name is 2-(4′-hydroxybenzeneazo) benzoic acid.

Deficiency in biotin can be caused by excessive consumption of uncooked eggs, a condition known as "egg-white injury." Egg whites contain avidin, a protein that binds biotin very tightly; therefore, intake of large quantities of uncooked eggs may result in excessive avidin accumulation, which can bind sufficient biotin obtained from other foods to produce an apparent dietary deficiency. Biotin was discovered by the realization that the damage to rats (loss of hair, etc.) caused by eating large amounts of raw egg whites could be prevented by inclusion of extracts of liver. The active ingredient was found to be biotin. Heating inactivates avidin, so egg-white injury does not result from eating cooked eggs.

Avidin has been purified from egg white by a combination of ammonium sulfate fractionation, cation-exchange chromatography, and gel filtration. Each molecule of avidin is a tetramer (68 kDa) containing identical subunits of ~17 kDa. There are four binding sites for biotin in each tetramer. The binding of biotin to avidin is one of the tightest biological interactions known ($K_d \approx 10^{-15}$ M). The activity of avidin is defined by units that correspond to the quantity of avidin binding 1 μg of biotin. Because of the exceptionally strong interaction between avidin and biotin, avidin is a potent inhibitor of biotin-containing enzymes. This inhibition is diagnostic for biotin-containing enzymes.

In this experiment, you will study the binding of biotin and the biotin analog HABA to avidin.

OBJECTIVES

No procedure is given for this experiment. You are expected to use the information in this chapter, as well as the skills developed in the previous experiments, to design protocols that will give answers to questions listed in the objectives.

1. Carry out a saturation binding study of the binding of HABA to avidin using a fixed concentration of purified avidin. Plan your experiment carefully, since avidin is somewhat expensive. (Hint: add increasing concentrations of HABA to a tube containing avidin and measure the complex formation spectrophotometrically after each addition) Deter-

mine the R_t and K_d for HABA. Be sure to account for nonspecific binding. You can estimate the quantity of avidin required since the change in absorbance coefficient of HABA is 33,900 M^{-1} cm^{-1}. Check your calculations with your instructor.

2. Determine whether biotin binds to the same sites as HABA. To do this, you need to conduct a binding competition study. (Hint: add increasing concentrations of biotin to a tube containing the HABA-avidin complex and measure the binding of HABA to avidin after each addition of biotin). Be sure to account for nonspecific binding.

3. Determine the K_d of biotin for avidin by first determining the IC_{50} and then calculating the K_d using the Cheng-Prusoff relationship (equation 10-24).

4. Determine whether measurable concentrations of avidin are present in egg whites. Eggs will be provided. Carefully break an egg without breaking the yolk, and pour off the egg white into a beaker. If any of your colleagues are good cooks, they may be able to show you the technique for separating egg whites. The egg white will be very viscous. If a sonifier is available, brief sonication of the sample reduces its viscosity without altering its HABA binding properties. If a sonifier is not available, carefully remove the least viscous 0.5 mL and put it into a separate tube. Use this as the egg-white sample. Use this sample to measure the binding of HABA and the competition of this binding by biotin. Determine how much avidin activity is present in the egg whites. If you have time, determine the protein content of the pure avidin and of the egg white, and calculate the fraction of egg white protein that is avidin. Otherwise, freeze a sample of the egg white and determine the protein concentration at a later time to perform this calculation.

5. Determine whether purified bovine serum albumin binds HABA. If it does bind, determine whether biotin competes for the same site on BSA.

REAGENTS NEEDED FOR CHAPTER 10

✓ HABA (4-hydroxyazobenzene-2-carboxylic acid): Sigma H5126, 25 g

✓ Avidin: Sigma A9275, 25 mg

✓ Biotin: Sigma B4501, 1 g

✓ Eggs: We have tried only white eggs to date and barely detected avidin. It would be interesting to try brown eggs or other kinds of eggs.

11

Recombinant DNA Techniques

INTRODUCTION

Recombinant DNA technology involves the biochemical manipulation of genes by using enzymes that normally participate in DNA or RNA metabolism. These enzymes are isolated from a variety of organisms and can be commercially obtained. Genes are manipulated for several reasons, including the following: to facilitate the study of gene expression and physiological regulation, to identify the product of a gene and/or bring about the overexpression of the gene product, to study the structure/function relationships in proteins, or to identify cellular components that interact with particular nucleic acid sequences or protein domains.

The genetic information of all living organisms is contained within the sequence of purines and pyrimidines in nucleic acids. For most organisms, the genetic information is contained in deoxyribonucleic acid (**DNA**). In certain viruses, the genetic information is contained in ribonucleic acid (**RNA**). The complete set of the nucleic acids containing the genes for an organism is referred to as the **genome**.

The term **chromosome** is used to describe particular individual DNA or RNA molecules that are part of the genome. The genome of the bacterium *Escherichia coli* consists of a single, double-stranded DNA molecule; thus this organism is said to contain a single chromosome. Eukaryotic organisms and some bacteria contain genomes that consist of more than one molecule of DNA, and each of these DNA molecules is referred to as a chromosome. If an organism contains only one copy of each chromosome, it is referred to as **haploid**; if there are two copies of each chromosome, the organism is referred to as **diploid**; and if there are four copies of each chromosome, the organism is referred to as **tetraploid**. In some cancer cells, the number of copies of each chromosome is variable, and these cells are said to be **polyploid**.

The maintenance, storage, repair, replication, and expression of the genome, be it RNA or DNA, require a large number of different enzymes and other proteins as well as nucleic acids. The study of such protein–nucleic acid and nucleic acid–nucleic acid interactions is important for developing an understanding of the physiological processes central to life. Furthermore, understanding these central physiological processes may permit the identification of the cause of various disease states and facilitate the development of clinical strategies to cure or ameliorate the disease state.

In the exercises in this chapter, we examine the activity of several enzymes that interact with and alter DNA. Then we construct recombinant (chimeric) DNA molecules in vitro, introduce the recombinant DNA molecules into bacterial cells, and purify the recombinant DNA molecules from the cells and characterize them.

OVERVIEW OF THE STRATEGY OF RECOMBINANT DNA TECHNIQUES

The DNA genomes of most free-living organisms are quite large, consisting of from 4×10^6 base pairs in the simplest bacteria to roughly 1000 times this amount in eukaryotes. Since an individual gene constitutes only a very tiny fraction of the genome, the study of a gene often starts with manipulations designed to purify the gene or to monitor the expression or some other property of the desired gene.

Genes are inherited, of course, owing to the fact that the DNA comprising them is replicated and partitioned into dividing cells. Replication of DNA involves the action of several enzymes and can occur for various reasons. For example, DNA that is damaged by noxious environmental agents is typically excised from the genome, degraded, and replaced with undamaged DNA by enzymes responsible for DNA repair. Alternatively, the normal replication of undamaged DNA before cell division requires a separate set of enzymes responsible for this process. The replication apparatus is assembled at DNA sequences known as origins of replication (**ori** sequences in bacteria, autonomously replicating sequences (**ARS**) in eukaryotes). In order to be replicated and stably maintained in an organism, each chromosome must contain at least one origin of replication. In some circumstances, chromosomes will require sequences necessary to ensure partition to each of the daughter cells at cell division. This is especially necessary in eukaryotes, where the partition function is dependent on the **centromere**. A piece of DNA that is able to function like a chromosome in vivo, i.e., capable of being replicated and partitioned stably over many generations, is called a **replicon**.

To study genes efficiently, it is often desirable to construct a small DNA molecule consisting of the gene of interest, an origin of replication, and, if necessary, sequences necessary for partitioning during cell division. That is, the target DNA sequences are linked with DNA sequences constituting a replicon. Many different replicons have been used for such DNA manipulations, and these have different properties, such as the capacity to function as a replicon in various host organisms, the capacity to confer a selectable phenotype on the host cells that contain them, the capacity to bring about the high-level expression of the passenger gene, etc. Replicons that are designed to "carry" passenger DNA sequences of interest and facilitate their manipulation and study are referred to as **vectors**. In the laboratory exercises described in this chapter, you will use vectors that are replicons in the bacterium *E. coli*. These vectors will exist in the cell as small double-stranded circular DNA molecules that remain separate from the bacterial chromosome. This type of vector is called a **plasmid**.

The first steps in a recombinant DNA study of a gene of interest are usually to isolate the gene of interest, insert the gene into a plasmid vector to generate a composite molecule with both vector sequences and the gene of interest, and introduce the composite molecule into living cells so as to propagate it. To accomplish this, DNA sequences from both the genome and the vector are spliced together to form a composite molecule that is a replicon containing the gene of interest. These composite molecules are called **recombinant DNA**. Once successfully introduced into living cells, the recombinant DNA molecule will be replicated many times (since it is a replicon), and it is possible to grow large quantities of the cells and recover many copies of the recombinant DNA molecule from the cells.

CUTTING AND SPLICING DNA

In order to construct recombinant DNA molecules, we take advantage of enzymes that are involved in DNA metabolism. For example, double-stranded DNA may be specifically cleaved at particular sequences by the action of **restriction endonucleases**. These enzymes are normally involved in the protection of various cells from invading DNA such as the DNA of viruses. The process of DNA **restriction** was discovered by the microbiologist S. Luria (who with Max Delbruck won the Nobel Prize in 1960 for their earlier work on the nature of mutations) and M. L. Human (1952) many years before the first restriction endonucleases were isolated. Luria observed that bacteriophage (viruses that infect bacteria) that had been grown on one race of *E. coli* could then be grown with high efficiency on that same race of *E. coli*, but not on distantly related races of *E. coli*. However, any phage that survived after infecting a distantly related race (usually a very small fraction of the total), could be easily propagated on that race, but not on the original *E. coli* race. Thus, growth of the phage was partially "restricted" to the race that served as the most recent host (reviewed by Luria (1970)).

Some phage restriction processes are due to enzymes known as Class I restriction endonucleases. These restriction endonucleases recognize particular DNA sequences and cleave both strands of the DNA double helix at a nearby site. Cells prevent their restriction endonucleases from cleaving their own DNA by methylating the DNA sequences that are recognized by the restriction endonucleases. Usually, the restriction endonucleases can cleave only DNA that is completely unmethylated (unmethylated on both strands). Upon replication of the (methylated) DNA of the host, hemimethylated DNA (modified on only one strand of the duplex DNA) is produced, and this hemimethylated DNA is not cleaved by restriction endonucleases; instead, it is methylated by specific methylase enzymes unique to that race of bacteria. To survive, a cell cannot contain a restriction endonuclease unless it also contains the methylase that modifies the specific sequence recognized by the restriction endonuclease.

DNA restriction operates as follows. When phage infect a new race of *E. coli* and inject their DNA into the *E. coli* cytoplasm, the phage DNA is usually degraded by the host cell restriction enzymes. This is because the phage DNA is not methylated at the sites recognized by the host cell restriction endonuclease. However, occasionally but very rarely, the host methylation enzyme will modify the phage genome before it is cleaved, protecting it from the restriction endonuclease. This phage DNA will be capable of causing a productive infection that results in the destruction of the host cell and the release of many phage (virus) particles. Each of these progeny phage will contain a phage genome that is modified appropriately for the *E. coli* race on which they last grew. Different *E. coli* races (such as *E. coli* B, *E. coli* C, *E. coli* K, etc.) contain different restriction and methylation enzymes that specifically recognize different DNA sequences. Thus, the phage are restricted to growth on the race that served as the last host, since their DNA will only be resistant to the restriction endonuclease of that host.

There are two different classes of restriction endonucleases. Class I enzymes each recognize a particular DNA sequence and cleave the DNA at sites adjacent to the recognized sequence. Class II enzymes recognize a specific DNA sequence and cleave the DNA within the recognized sequence. Most of the class II enzymes recognize either a four-base-pair sequence or a six-base-pair sequence. A four-base sequence does not allow much specificity because, statistically, all four base sequences will occur every 256 base pairs (if the DNA is 50% GC base pairs and 50% AT base pairs). However, if a six-base recognition sequence is used, considerably more specificity will be realized, since a particular six-base sequence will occur statistically every 4,096 base pairs. Cleavage within the recognized sequence may occur on both strands in the center of the recognized sequence, generating a blunt cleavage terminus, or may occur at staggered positions on each strand, generating termini with one strand or the other having an "overhang." Such overhangs are called "sticky ends." Examples of the types of cleavage termini resulting from the action of restriction endonucleases are shown in Fig. 11-1. Many Class II restriction endonucleases are currently available commercially, and the catalogues of vendors that supply these enzymes often contain a listing of all the currently known

Blunt or flush termini

```
5'-pApCpGpApA-OH          pTpTpCpGpT-3'
   | | | | |               | | | | |
3'-TpGpCpTpTp           HO-ApApGpCpAp-5'
```

3'-Extensions

```
5'-pApCpGpApA-OH          pTpTpCpGpT-3'
   | | |                   | | | | |
3'-TpGpCp               HO-TpTpApApGpCpAp-5'
```

5'-Extensions

```
5'pApCpG-OH              pApApTpTpCpGpT-3'
  | | |                       | | |
3'-TpGpCpTpTpApAp       HO-GpCpAp-5'
```

FIGURE 11-1 Termini generated by restriction endonuclease cleavage. The figure is modified from Sambrook et al. (1989).

enzymes. A list of the Class II restriction endonucleases that we will use in our studies and the sequences they recognize and cleave is presented in Table 11-1.

Since cleavage by restriction endonucleases depends on the DNA sequence, a given DNA molecule will have a characteristic pattern of cleavage when treated with several different restriction endonucleases. This property allows us to distinguish or identify different DNA molecules of the same length, but with different sequences, without sequencing them. The process of determining the locations of restriction endonuclease cleavage sites within a piece of DNA is known as **restriction mapping**. Experiment 11-1 involves the restriction mapping of a plasmid DNA molecule or phage genome. The fragments of DNA that result after cleavage with restriction endonucleases are commonly referred to as **restriction fragments**.

Restriction fragments with compatible termini may be spliced together by DNA ligase, which normally provides this activity in vivo. In particular, we use the enzyme encoded by bacteriophage T4 since this enzyme is easily purified, relatively stable, commercially available, and fairly inexpensive. In *E. coli* cells infected with phage, the T4-DNA ligase plays an important role in linking together short pieces of phage DNA during DNA replication. The reaction catalyzed by this enzyme, known as ligation, is shown in Fig. 11-2.

Various other enzymes are available and are frequently used in the construction of recombinant DNA molecules. For example, by allowing DNA polymerase I to extend the recessed end (fill in the missing nucleotides of a sticky end), the conversion of DNA termini with 5´ overhangs to molecules that are blunt-ended can be achieved. Since all known DNA polymerases add onto the 3´ end of the DNA chain, the DNA polymerase activity of DNA polymerase cannot extend the recessed 5´ ends found on molecules with 3´ extensions. However, many DNA polymerases also contain a 3´ → 5´ exonuclease activity that permits one to digest away the 3´ overhang on molecules with such termini. For more information on the types of tools available for DNA manipulations, the reader is urged to consult Sambrook et al. (1989) or Ausubel et al. (1987).

Table 11-1 Restriction Endonucleases Used in Our Studies		
ENZYME	**MICROORGANISM**	**SEQUENCE, CLEAVAGE PATTERN**
*Bam*HI	*Bacillus amyloliquefacens* H	G GATCC
*Eco*RI	*Escherichia coli* RY13	G AATTC
*Hind*III	*Haemophilus influenzae* Rd	A AGCTT
*Pst*I	*Providencia stuartii* 164	CTGCA G

Ligation of cohesive DNA termini or nicks

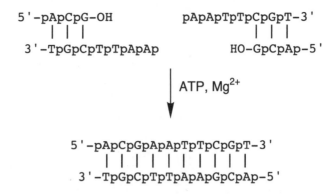

Ligation of blunt ends

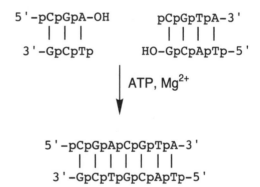

FIGURE 11-2 Reactions catalyzed by T4 DNA ligase.

GEL ELECTROPHORESIS OF DNA

Gel electrophoresis is frequently used to monitor the progress of DNA cleavage and ligation reactions. The principles of electrophoresis are discussed in Chapter 5. At alkaline pH, DNA and RNA strands are negatively charged and therefore migrate in an electric field. For small DNA fragments, **polyacrylamide** gel electrophoresis (PAGE) is the method of choice, and the technique used is quite similar to the simple protein gel electrophoresis method that you are familiar with, except that a stacking gel is not used.

For larger DNA molecules, electrophoresis in **agarose** gels is the method of choice. Agarose is a complex polysaccharide extracted from algae; when heated, the polysaccharide molecules dissolve in aqueous solutions and fully extend. When cooled, the polysaccharide strands fold and twist up with each other, forming a meshwork. This meshwork (or gel) contains pores that are of various sizes, depending on the concentration of agarose in the heated solution. The higher the concentration of agarose used, the smaller the pores will be in the solidified gel. The fractionation ranges for gels consisting of various agarose or acrylamide concentrations are listed in Table 11-2.

All nucleic acids are of very similar chemical composition and have a high charge-to-mass ratio at physiological pH. Therefore, the mobility of various DNA and RNA molecules in agarose and acrylamide gels depends primarily on the frictional coefficient, which is related to the length and conformation of the nucleic acid. Most linear double-stranded DNA fragments will be separated by electrophoresis on the basis of length (molecular mass). In double-stranded DNA, the two strands coil around each other once per turn of the helix (every 10 base pairs). Supercoiling results from unwinding (or overwinding) the double helix. In cells, most of the DNA is supercoiled. Circular DNA molecules and supercoiled DNA circles migrage anomolously when compared to linear duplex DNA (Fig. 11-3). Supercoiled DNA, which has a more compact structure, has higher mobility than linear DNA of the same size. Circular DNA may have a higher or lower mobility than linear DNA, depending on the electrophoresis conditions.

TABLE 11-2 **Fractionation Ranges for Polyacrylamide and Agarose Gels**	
AGAROSE IN GEL (%)	**FRACTIONATION RANGE FOR LINEAR DNA (kb)**
0.3	60–5
0.6	20–1
0.7	10–0.8
0.9	7–0.5
1.2	6–0.4
1.5	3–0.1
ACRYLAMIDE IN GEL (%)	**FRACTIONATION RANGE FOR LINEAR DNA (kb)**
3.5	100–1000
5.0	80–500
8.0	60–400
12.0	0–200
20.0	10–100

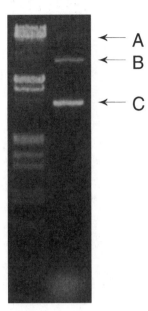

← A
← B
← C

FIGURE 11-3 Comparison of the mobility of *(A)* nicked circular *(B)* linear and *(C)* supercoiled plasmid DNA. The direction of electrophoresis was top to bottom. The sample in the left lane was a mixture of different linear DNA fragments, while the sample in the right lane was a mixture of the nicked circular, linear, and supercoiled forms of a bacterial plasmid. Note that the nicked circular form has lower mobility than the linear form of the plasmid, while the supercoiled form has the greatest mobility.

Single-stranded molecules present a more complex phenomenon, since these can have considerable secondary structure, such as loops. To avoid the effects of secondary structure in electrophoresis, single-stranded nucleic acid molecules are separated in the presence of strong denaturants. As with protein SDS-PAGE, the length of nucleic acid fragments that have been separated on gels can be deduced, as long as the gel contains appropriate "marker" fragments of known molecular mass to enable the generation of a standard curve. The migration distance of fragments in the gel will be inversely proportional to the log of the molecular mass or, for a more convenient approximation, the log of the number of base pairs, analogous to SDS electrophoresis of proteins (Fig. 5-6). As part of Experiment 11-1, you will determine the size of DNA fragments generated by restriction endonuclease digestion of DNA.

Various methods are available for the detection of DNA fragments within agarose or polyacrylamide gels. The simplest way is the method of "UV shadowing." When UV light strikes a glass plate or a piece of paper, fluorescence from impurities appears. Because nucleic acids absorb UV light, passing UV light through a gel will result in shadows appearing on such surfaces where the gel contains nucleic acid bands. One can simply see and photograph the shadow pattern with no further manipulations. Unfortunately, the gel must contain a considerable amount of nucleic acid in the band for the shadow to be visible. A more common method to detect nucleic acid bands within gels is to stain them with compounds that bind nucleic acids. Dyes, such as ethidium bromide, that intercalate between the bases of the nucleic acids are prominent among these. Ethidium bromide is highly florescent when exposed to UV light only when it intercalates within nucleic acids. Finally, the most sensitive detection of nucleic acids in gels is made possible by labeling the nucleic acid, either in vivo or in vitro, with radioactive nucleotides. For example, the incorporation of nucleotides that contain radioactive ^{32}P at the position of the α-phosphate into DNA will result in DNA containing ^{32}P. The position of the radioactivity in the gel can then be easily detected by several means. The most common of these is to expose x-ray film to the gel, a process known as **autoradiography**.

In addition to analytical applications, agarose gel electrophoresis is often used to purify DNA restriction fragments. After the gel has been stained, the bands of interest are excised from the gel with a razor blade, and the DNA is recovered from the excised band by soaking in a high-salt solution, by electrophoresis into a dialysis bag (**electroelution**), or by enzymatically digesting away the agarose.

INTRODUCING DNA INTO CELLS

The introduction of recombinant DNA molecules into host organisms, a process known as **transformation**, may be achieved by various procedures, and the method used will depend on the type of host cells to be used. Bacterial cells such as *E. coli* represent the simplest case. These cells may be treated with agents that partially disrupt the cell membrane, rendering it permeable to added DNA molecules. For example, as part of Experiment 11-3, we treat *E. coli* cells with $CaCl_2$, which partially disrupts the membrane, allowing DNA uptake. Cells that are able to take up added DNA are said to be **competent**. Some strains of bacteria, such as *Hemophilis influenza* and *Bacillus subtilis* (but not *E. coli*) are naturally able to take up DNA under certain conditions (natural competence). (The natural competence of certain bacterial cells and the transformation of these cells with pure DNA are historically important because they permitted the demonstration that DNA constitutes the genetic material of bacterial cells. These elegant experiments by Avery et al. (1944) are very instructive, and you may find them worth reading.

Another way to introduce DNA into cells is to treat the cells with a short burst of electric current. During this process, known as **electroporation**, the cell membrane is transiently disrupted and then re-forms. DNA can get inside the cell during the very brief period when the cell membrane is disrupted by the electric current. Electroporation has been successfully used with many different types of cells, including eukaryotic cells.

An additional method of introducing DNA into bacterial cells is to "package" the DNA in vitro into phage particles (**in vitro packaging**). This can be accomplished by using extracts of phage-infected cells that contain all of the phage components necessary to assemble intact phage. The phage packaging enzymes recognize sites on phage DNA known as *cos* sites. In order to be packaged the DNA must have two *cos* sites separated by 36–50 kb of intervening DNA. Plasmid and phage vectors bearing *cos* sites are available. After splicing the target DNA into these vectors, the recombinant DNA can be packaged in phage particles. Infection of bacteria with these phage particles containing the recombinant DNA molecules will then result in the very efficient injection of the recombinant DNA molecules into the *E. coli* cytoplasm, as if the recombinant DNA molecule were a phage genome. The plasmid vectors bearing *cos* sites are known as **cosmids**. Cosmids are often used to construct recombinant DNA libraries consisting of clones of every gene in a genome. This is discussed in a later section.

IDENTIFYING TRANSFORMED CELLS

After inserting DNA molecules into cells, it is necessary to identify those cells that have inherited the desired recombinant DNA molecules. This is usually necessary because the vector replicon itself, although lacking the desired DNA fragment, is capable of existing independently in cells. Moreover, because the transformation of cells with DNA is quite inefficient, most cells put through the transformation procedure do not acquire a DNA molecule at all. In order to select the cells that contain either the vector or a recombinant molecule containing that vector, most vectors are constructed to contain genes that confer a selectable phenotype on the cells that inherit them. For example, the vectors commonly used in bacteria contain genes that confer resistance to antibiotics such as ampicillin or tetracycline. Cells that do not inherit a vector in the transformation experiment will be sensitive to such antibiotics and thus will be killed if exposed to them, while cells containing the vector will be drug resistant. Thus, a common approach is to select for cells that inherit the drug-resistance vector.

There are many mechanisms by which drug resistance occurs. For example, the resistance gene(s) may encode an enzyme that degrades the drug. This is true for ampicillin resistance (Ap), which we use in Experiment 11-3. In other cases, such as tetracycline resistance (tet), the resistance genes encode proteins that promote the export of the drug from the cell. When the drug targets are macromolecular components, such as RNA polymerase or ribosomes, the drug-resistance genes encode resistant polymerase or ribosomal components.

It is frequently more difficult to select for recombinant molecules that contain the desired gene of interest than it is to select for the vector markers (that is, to distinguish between cells that inherited only the vector from those that inherited the desired recombinant DNA molecule). Sometimes, the gene of interest will confer a selectable phenotype on the host cells (for example, the ability to grow on a new substrate), in which case the problem is simplified. However, this is usually not the case.

Various techniques have been developed to aid in the identification of cells containing the desired recombinant DNA molecule. If one has the antibody to the specifically expressed protein, low levels of expression of the target gene can occasionally be detected immunologically. Often, nucleic acid hybridization techniques are used to identify cells containing the appropriate DNA sequences. However, the existence of the desired recombinant DNA molecule can be certain only after its purification from the cells and characterization in vitro.

Certain vectors contain two different genes for antibiotic resistance (see Figs. 11-4 and 11-5). For example, the plasmid pBR322 (Fig. 11-5) contains genes for ampicillin resistance and tetracycline resistance. Each of

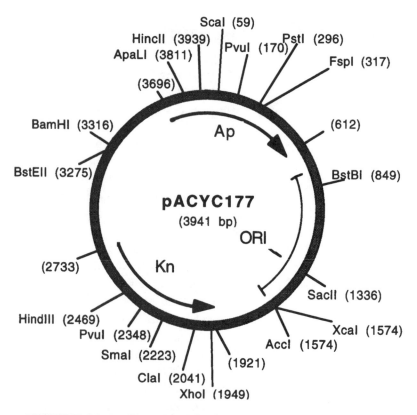

FIGURE 11-4 Plasmid pACYC177 is a low copy number *E. coli* cloning vector (Chang and Cohen, 1978). It carries the origin of replication from plasmid p15A so that it can coexist with vectors like pBR322 and pUC19 that carry the ColE1 origin. This feature of compatibility makes it useful for cloning experiments that require the presence of more than one recombinant plasmid per cell. pACYC177 carries the kanamycin-resistance gene (Kn) from Tn903 and the β-lactamase gene (Ap) from Tn3. Several of the important sites for cloning are indicated on the map. The origin of replication is also indicated (ORI). The plasmid can be maintained in *E. coli* in media containing ampicillin (50–100 μg/mL) and kanamycin (35–50 μg/mL). To obtain large amounts of plasmid DNA, it is necessary to amplify the plasmid pACYC177 with chloramphenicol because of the low copy number of pACYC177.

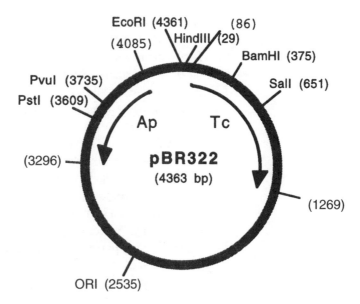

FIGURE 11-5 Plasmid pBR322 is an *E. coli* plasmid cloning vector (Bolivar et al., 1977). pBR322 was constructed in vitro with the tetracycline resistance gene (Tc) from pSC101, the origin of DNA replication (ORI) and rop gene from the ColE1 derivative, pMB1, and the ampicillin resistance gene (Ap) from transposon Tn3. pBR322 can be amplified with chloramphenicol.

these genes contains restriction endonuclease sites that are useful for the insertion of DNA restriction fragments; for example, the ampicillin resistance gene contains a site for the enzyme *Pst*I, while the tetracycline resistance gene contains a site for the enzyme *Bam*HI (Fig. 11-5). Thus, if a target DNA fragment is cloned into the *Pst*I site of pBR322, the resulting recombinant plasmid will confer tetracycline resistance but will not confer ampicillin resistance, because the ampicillin resistance gene will have been disrupted. Likewise, if a target DNA fragment is cloned into the *Bam*HI site of pBR322, the resulting recombinant plasmid will confer resistance to ampicillin but not to tetracycline, because the tetracycline resistance gene will have been disrupted.

The inactivation of genes by insertion of DNA fragments in vitro is known as **insertional inactivation**. Insertional inactivation may be used to distinguish between transformants that contain an unaltered pBR322 vector molecule and those that contain a recombinant plasmid, since the former will be resistant to both ampicillin and tetracycline, while the latter will be resistant to only one of these antibiotics.

A variation of the insertional inactivation method described above and known as α-complementation takes advantage of the unique ability of the *E. coli* enzyme β-galactosidase to be reconstituted from two separate polypeptides in vivo. This enzyme is a tetramer consisting of four identical subunits, encoded by the *lacZ* gene. Small deletions at the N-terminal end of the *lacZ* gene result in the production of a truncated polypeptide that forms the normal tetramer but is inactive. One such deletion is the often-used *lacZ15* allele. Surprisingly, the inactive *lacZ15* product is rendered highly active by addition in trans of the missing N-terminal polypeptide, known as the α–fragment (trans implies that the complementation is from a different DNA element). The process by which the α–fragment restores activity to the *lacZ15* polypeptide is known as α–complementation.

α–Complementation is very useful for distinguishing between transformants that contain a vector and transformants that contain the desired recombinant molecules. Typically, a host bacterial strain containing the *lacZ15* mutation is used; this strain will be *lac⁻* since it lacks β-galactosidase activity. A vector such as pUC19 containing the α-polypeptide gene (the beginning part of the *lacZ* gene called *lacZα*) is used (Fig. 11-6). Several such vectors are available, and these have been engineered to contain several restriction sites within the α-polypeptide gene. Insertion of target DNA at any of these restriction sites renders the α-polypeptide inactive, owing to disruption of the gene. Transformation of the *lacZ15* cells with vector alone will result in *lac⁺* cells, owing to α-complementation. Transformation of these cells with recombinant molecules in which the α-polypeptide gene is disrupted will result in *lac⁻* cells.

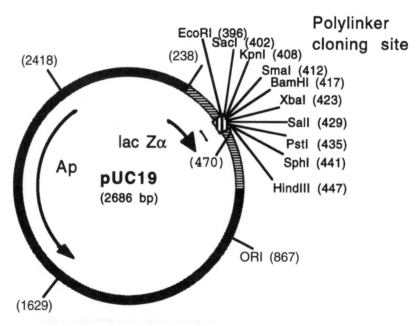

FIGURE 11-6 Plasmid pUC19 is a small, high copy number *E. coli* plasmid cloning vector. It is part of a series of related plasmids constructed by Yanisch-Perron et al. (1985), and it contains portions of pBR322 and M13mp19. It can be amplified with chloramphenicol, and it carries a 54-base-pair multiple-cloning-site polylinker.

Several methods for distinguishing *lac*⁺ and *lac*⁻ cells use bacteriological media, known as indicator media. The most commonly used method involves inclusion of 5-bromo-4-chloro-indoyl-β-D-galactoside (X-gal) in the media. This compound is colorless, unless cleaved by β-galactosidase, in which case a dark blue insoluble product is formed. Colonies that are *lac*⁺ are stained dark blue, while *lac*⁻ colonies appear white on this media. Thus, the desired *lac*⁻ colonies, containing recombinant plasmid molecules in which the α-polypeptide gene has been inactivated by insertion of the target DNA fragment, may be directly identified as white colonies. This process is referred to as **"blue/white screening."**

THERE ARE MANY USEFUL TOOLS FOR RECOMBINANT DNA EXPERIMENTS

Since the first recombinant DNA experiments (Cohen et al., 1973), a variety of tools have been (and are being) developed that make recombinant DNA techniques indispensable for the study of virtually all areas of biology. In this section we have summarized some of the developments of vector and host systems that make recombinant techniques practical. For more information, the cloning manuals by Sambrook et al. (1989) and Ausubel et al. (1987, with yearly updates) may be consulted. In addition, the catalogues and web sites of vendors supplying recombinant DNA materials such as restriction enzymes and cloning vectors are often a useful source of information on the availability of tools for recombinant DNA research. Several of the web sites are listed in Table 13-3.

Hosts. Host systems have been developed for the maintenance and expression of recombinant DNA in bacteria, yeast, insect, and mammalian cells. The host systems for bacterial experiments are the most highly developed, because historically the first recombinant experiments were done with bacteria. Recombinant DNA containing eukaryotic DNA fragments is routinely maintained and may also be characterized in bacteria. To illustrate some of the types of host manipulations possible, several examples of bacterial hosts will be noted below.

Host cells that are defective in their ability to restrict and modify DNA are used for most experiments. DNA that has been cloned in such a host is not modified by the methylases of the host restriction/modification systems, and therefore may be readily cleaved with restriction endonucleases. Also, the lack of modification makes the DNA vulnerable to most host restriction systems so that the possibility of recombinant DNA molecules becoming transferred to bacteria in the wild is minimized. Another host feature that is useful is the absence of genetic recombination capabilities, resulting from mutation of the *recA* gene. In such a host, recombination between host chromosomal sequences and plasmid sequences, or between different plasmids, does not occur, helping to stabilize the plasmid. Hosts bearing the *pcn⁻* mutation are defective in regulating the number of copies per cell of plasmids containing the ColE1 origin of replication. Use of *pcn⁻* mutant host cells results in a very high number of plasmid copies per cell. A high number of copies of plasmid per cell may or may not be useful, depending on the application.

We have already discussed the use of α-complementation to identify recombinant clones by blue/white screening. The host cells for these experiments must contain the *lacZ15* allele, encoding the truncated β-galactosidase enzyme, which is to be complemented in trans by the *lacZα* peptide.

Random mutagenesis experiments can be carried out using host cells that have been constructed to have very high mutation rates. You can mutate a gene that is cloned on a plasmid simply by transforming such cells with the recombinant plasmid and then re-isolating the plasmid DNA after several rounds of growth and dilution. In contrast, **site-directed mutagenesis**, which is for specifically altering plasmid genes, is carried out using hosts developed to permit various mutagenesis strategies. For example, one host used for these experiments lacks the ability to excise uracil from DNA, so that DNA produced in this strain contains uracil in place of thymine. To make mutations, one strand of this DNA is replicated *in vitro* with a mutagenic oligonucleotide primer and the normal deoxyribonucleotides. This results in a plasmid with a mutation on the strand that contains thymine, and the original sequence on the strand that contains uracil. Transformation of wild-type cells with this DNA results in the propagation of the mutant strand and the destruction of most of the unmutagenized strand, which contains uracil. Thus, recovery of the mutant DNA sequence is favored.

Recombinant DNA experiments generally pose no danger to laboratory workers or to society. However, for experiments in which the recombinant DNA molecules may be dangerous, such as clones from **pathogenic** viruses or bacteria, special **"compromised hosts"** have been developed. These hosts typically have a large number of mutations in genes involved in metabolism and therefore require supplementation with numerous nutrients for survival. The large number of nutritional requirements helps to ensure that these bacteria cannot survive outside the laboratory setting, where they are forced to compete with wild-type bacteria. Also, the mutation of the restriction/modification systems in these hosts helps ensure that the recombinant DNA molecules themselves will not be transferred to wild-type bacteria outside the laboratory. Even with these precautions, physical isolation in special laboratory facilities is required to prevent the escape from the laboratory of dangerous recombinant DNA that may have been transferred to "non-compromised" cells. Safety guidelines and requirements for recombinant DNA experiments have been established by Congress and are enforced by the National Institutes of Health and the Food and Drug Administration. Each institution performing recombinant DNA experiments, including industrial institutions, must conform to these regulations and have an Institutional Biosafety Committee to monitor compliance. The guidelines can be accessed from the web (http://www.nih.gov/od/orda/toc.htm) or obtained from your local committee.

Another group of host strains has been developed for the regulated **expression** of genes. For example, cells massively overexpressing the repressor of the *lac* operon are used in experiments in which the expression of target genes is controlled by the *lac* promoter. The presence of many copies of the *lac* repressor in these cells guarantees that the cloned DNA is not expressed at a high level until you add an inducer of the *lac* promoter system, such as IPTG or lactose. For genes whose expression must be maintained lower than the fairly inefficient repression that the *lac* system permits, a system using the *lac* system coupled with bacteriophage RNA polymerase is frequently used. In this system, the host cells contain the gene encoding a bacteriophage RNA polymerase that is under the control of the *lac* promoter, and the target gene is cloned downstream from the phage promoter sequence in a plasmid. Induction of the *lac* system results in the expression of the phage RNA polymerase, which then expresses the target gene on the plasmid. Variants of this system are widely used for regulated expression of target genes in *E. coli*.

Recombinant DNA may also be integrated into the host chromosome, where it replicates and functions like a typical host gene. For example, we have already used such a host and recombined DNA for expression of β-galactosidase (Chapter 4). The *lacZ* structural gene was cloned into a plasmid downstream from a strong regulated promoter (the *glnK* promoter of *E. coli)*. The plasmid vector used was specifically designed to permit the recombination of the fusion into the *trp* genes of the *E. coli* chromosome, resulting in a stable, single-copy, gene fusion. On each side of the *glnK-lacZ* fusion, the vector contained part of the chromosomal *trp* genes of *E. coli*. Recombination between plasmid and chromosomal *trp* sequences results in the insertion of the recombinant DNA into the chromosomal *trp* genes. This results in a tryptophan requirement (auxotrophy) for growth due to the insertional inactivation of the *trp* genes. The *glnK* promoter is expressed only in nitrogen-starved cells; to induce the expression of β-*galactosidase* one need only transfer the cells from a medium containing a good nitrogen source to one containing a poor nitrogen source.

Cloning Vectors. We have already noted some of the specialized vectors used for particular applications. For example, in Chapter 4 we discussed the use of recombinant DNA techniques for "tagging" proteins with protein or peptide sequences that facilitate purification by affinity chromatography. In Chapter 6, we noted the usefulness of genetic overexpression of proteins and the tagging of proteins with peptide sequences. A detailed list of the vectors available for recombinant DNA experiments, even those using just *E. coli* as the host, is beyond the scope of this book. An overview of the major types of vectors and some of their desirable features is presented below. More detailed information can be obtained from Sambrook et al. (1989) and from vendors (Table 13-3).

There are three basic types of cloning vectors used in *E. coli*: plasmids, **phage vectors** based on bacteriophage λ, and **cosmid** vectors, which replicate as plasmids but may be packaged in phage particles, as already mentioned. Plasmids are generally the most convenient vectors to use when the passenger DNA size is less than 5–10 kb. Larger plasmids tend to be unstable because any smaller molecules that result from infrequent spontaneous deletions have the advantage of replicating faster and thus come to dominate the plasmid population. For larger DNA target fragments, phage vectors are usually used. There are a wide variety of phage vectors with different features. For example, some cause the overexpression of cloned genes, facilitating the identification of clones on the basis of the expressed products. Some phage vectors may be stably maintained in *E. coli* as **lysogens**, while other phage vectors kill (**lyse**) the cells they infect. Even larger DNA fragments may be maintained in cosmid vectors, as noted earlier. The ability of phage and cosmids to maintain large recombinant DNA fragments makes these vectors ideal for constructing DNA libraries.

Useful features of plasmid vectors include the presence of unique restriction sites for the insertion of passenger fragments and drug resistance markers to permit selection. In some cases it is desirable to have multiple drug markers, permitting insertional inactivation, and/or the *lacZα* gene, permitting blue/white screening. In addition, as already noted, plasmids may be designed to bring about the expression of passenger genes, the fusion of passenger genes to "tags," or the excretion of the target gene product from the cell. Other plasmid vectors are designed to have low copy number, to permit the study of the effects of gene products at low concentrations or to study gene products that are toxic to the host at high concentration. Yet other plasmids are capable of self-transmission from one cell to another, by a process known as **conjugation** or bacterial mating. For this to occur, the plasmid must contain an extensive set of genes that encodes the mating pilus and the many proteins involved in DNA transfer. These plasmids are used when there is no better way to introduce DNA into certain strains of bacteria. The transmissible plasmid and passenger fragments may be introduced into *E. coli* by transformation or electroporation and then transferred to another strain of bacteria by mating.

Recombinant DNA Libraries. Collections of recombinant DNA clones that contain all of the DNA fragments from an organism are called libraries. Libraries are useful for isolating clones with various properties. For example, libraries may be screened to identify all clones that are homologous to a small oligonucleotide probe or that contain genes that complement a known mutation. Having a recombinant library on hand for the identification of new genes eliminates the effort required in isolating genomic DNA and ligating it into a vector each time a new gene needs to be cloned. Once a library is shown to be complete, it can be used for the cloning of many genes.

Libraries may be made in a number of ways. For example, **genomic libraries** are made by cleaving the genomic DNA with restriction endonucleases that cleave at 4 bp sequences. As already noted, these enzymes cleave an average of once every 256 bp. To make a library using these enzymes, the DNA is usually cleaved only partially with the enzyme, such that fragments of 20–50 kb are obtained. These are then cloned into a phage or cosmid vector to create the library. Another strategy is to shear the genomic DNA randomly, using sonication or by forcing the DNA through a narrow syringe. DNA fragments are then rendered blunt by using DNA polymerase and fractionated in an agarose gel to obtain fragments within the desired size range. These fragments are then ligated into a cosmid or phage vector. Because most genes are smaller than 20 kb, these methods assure that a few complete copies of every gene will be in the library.

In addition to the genomic libraries, gene libraries from eukaryotic cells are frequently made from the sequences found in mRNA. This is done to obtain the contiguous sequence of expressed amino acids for a protein, since in eukaryotes these sequences are usually separated by introns. The libraries made from mRNA sequences are called **cDNA libraries**, which stands for complementary DNA libraries. The first step in the construction of a cDNA library is to isolate the mRNA, which may be performed by chromatography on poly-U Sepharose (Chapter 4). The mRNA is then copied into DNA by the enzyme **reverse transcriptase**, which also makes the second DNA strand and degrades the RNA template strand. Then, the double-stranded DNA is cloned into an appropriate vector.

A wide variety of gene libraries from many organisms are available. In addition to commercial sources, many libraries are freely shared by researchers studying various organisms. As with hosts and cloning vectors, there are libraries available for almost every conceivable application. For example, there are **expression libraries** that cause the overexpression of passenger genes, and there are libraries in which the expression of passenger genes is prevented. There are libraries that may be maintained in the cell as phage lysogens or as plasmids, permitting the selection of cells with altered properties, and there are phage libraries that can only grow lytically on cells, forming plaques. Thus, you have a wide choice to consider for your research project.

APPLICATIONS OF RECOMBINANT DNA TECHNOLOGY

In this section, we discuss briefly some of the more common applications of recombinant DNA technology. The discussion that follows is intended to be illustrative. For further information, the reader is urged to consult Sambrook et al. (1989).

HYBRIDIZATION TECHNIQUES

Since complementary, single-stranded DNA strands will anneal to form duplex DNA, a DNA strand may be used to identify the complementary strand. Similarly, a single-stranded DNA strand will hybridize with a complementary RNA strand to form a DNA-RNA duplex. The techniques listed below all take advantage of the ability of complementary nucleic acid strands to hybridize.

Southern Blotting. This technique, developed by E. M. Southern, is designed to allow the researcher to identify DNA fragments that contain a particular sequence. For example, the technique is usually used to elucidate which genomic restriction fragments contain a particular DNA sequence. Such information is important when one needs to map the physical location of a gene, for example, when one wants to clone a gene from a large chromosome, or if one needs to know if the target cells contain only a single copy or more than one copy of a particular gene. The technique involves digesting the genomic DNA of the target organism with a restriction endonuclease and then separating these fragments on an agarose gel. The DNA fragments, separated on the basis of size, are then transferred to a membrane by capillary action; heating immobilizes the fragments on the membrane in a pattern that reflects their position within the agarose gel. The membrane is then allowed to interact with single-stranded DNA that contains sequences from the gene of interest (the probe). The probe DNA is usually made radioactive by the incorporation of labeled nucleotides during in vitro DNA replication. After incubation under conditions that favor the formation of duplex DNA, the excess probe is washed away, and the membrane is placed over an x-ray film and kept dark in a special, flat container. The radioactively labeled probe will expose the x-ray film, revealing the position of restriction fragments that contain sequences able to hybridize with the probe sequences. By manipulating the stringency of the hybridization conditions, you can selectively identify DNA sequences that are either a perfect match with the probe sequences or those that are related, but not perfectly complementary to the probe sequences.

Northern Blotting. This technique, which is analogous to the Southern blotting technique (and is called northern blotting in jest), is often used to monitor the level of expression of the gene of interest in cells grown under various conditions by determining the level of mRNA present. Typically, the mRNA pool from cells is extracted and separated on a polyacrylamide gel in the presence of a strong denaturant, such as urea. Under these conditions, single-stranded mRNA molecules are separated on the basis of their length (molecular mass). After electrophoresis, the mRNA molecules are transferred to a membrane by capillary action, and the membrane is heated, preserving the pattern obtained during electrophoresis. The membrane, containing the immobilized mRNA molecules, is then allowed to interact with an excess amount of probe DNA that contains sequences from the gene of interest. Conditions for the hybridization are usually chosen such that only perfectly complementary sequences will anneal. (This is called a high level of stringency.) After washing away the excess probe, the only probe attached to the membrane will be in duplexes with mRNA that is immobilized on the membrane, and you can deduce the amount of mRNA from the amount of probe that is retained in duplexes. Thus, because this technique permits the direct monitoring of the level of transcription of the target gene, it is very useful for investigating regulation of gene expression.

Primer Extension Analysis of Transcription Start Sites. Often in studies of gene expression it is desirable to identify the site of transcription initiation. Although the sequence of the DNA encoding the gene and the upstream region may be known, the site at which transcription is initiated may be unknown. This can be determined with the primer extension technique which makes use of reverse transcriptase, a special type of DNA polymerase that makes a complementary DNA strand to an RNA template. To identify the site of transcription initiation for the gene of interest, the mRNA pool is purified from the cell. The purified mRNA fom eukaryotes usually has a 50- to 250-residue poly A tail at the 3´ end. This makes it possible to use affinity chromatography with immobilized poly-U or poly-dT affinity matrices to selectively purify mRNA as described in Chapter 4. This mRNA pool is allowed to interact with a short, labeled DNA oligonucleotide (the primer) containing sequences from within the gene on the coding strand. The hybridization conditions are chosen so that only perfectly complementary sequences will anneal. Thus, this hybridization results in mRNA molecules containing a primer annealed within the gene coding sequence. The primer is then extended with reverse transcriptase, until the end of the mRNA molecule is reached, at which point the polymerase falls off. The extended primer DNA is then electrophoresed on a sequencing gel next to sequencing ladders of the gene formed by using the same primer and the Sanger dideoxy method of sequencing. By examining how long the primer has become after the end of the mRNA has been reached, you can deduce the site at which transcription must have initiated.

APPLICATIONS DESIGNED TO PERMIT THE STUDY OF GENE PRODUCTS

Overproduction of Gene Products So As to Facilitate Their Purification. Most cellular proteins constitute only a small fraction of the total protein content of cells. Consequently, it is often difficult to purify an individual protein away from all other proteins. This task is made simpler, of course, if you have available a tissue that is rich in the protein that is sought. Recombinant DNA technology enables you to create such an enriched source of a protein of interest from bacterial or insect cells (see Chapters 4 and 6). Typically, the gene of interest is ligated into a vector that contains strong transcriptional and translational initiation signals upstream from the gene of interest. In most cases, the expression vectors have been designed to make it possible to regulate the expression of the target gene. The strategy is usually to grow the cells containing the recombinant DNA well into log phase, so that a high density of cells is obtained. Then the cells are induced to give high levels of expression of the cloned gene during the last one or two cycles of cell division and the cells are harvested and the protein can be purified. This avoids trying to grow the cells in the presence of high concentrations of the expressed protein, which may be toxic. Thus, cells containing the recombinant molecules can be induced to bring about copious expression of the desired gene product by providing the appropriate induction signal.

Study of Gene Structure and Function. Usually, in the studies of protein structure and function, it is necessary to determine the nucleotide sequence of the gene encoding the protein. From the DNA sequence, the sequence of the amino acids in the protein may be deduced by conceptual translation of the DNA sequence. While there are several different ways one could go about determining the DNA sequence (see, e.g., Sambrook et al., 1989), the majority of these require that one have available a recombinant clone containing the DNA to be sequenced.

Site-Directed Mutagenesis is the process of selectively changing DNA. This is done by using a synthetic complementary oligonucleotide with one or more alterations in the sequence, so that when DNA polymerase extends this primer on the template, these changes are incorporated in the daughter DNA. After making the complement of this DNA (to obtain the proper coding sequence), the altered DNA is placed in a recombinant expression plasmid so that altered gene products can be produced. Thus, you can study the importance of particular amino acid side chains in processes such as protein folding, catalysis, ligand binding, etc., by altering the gene and observing the effects of these alterations on protein function in vivo or in vitro. The same method can also be used to study how changes in DNA sequences affect regulation of expression or other properties of the organism.

REFERENCES

Ausubel, F. M., R. Brent, R. E. Kingston, D. D. Moore, J. A. Smith, J. G. Saidman, and K. Struhl. 1987. Current Protocols in Molecular Biology. John Wiley & Sons, Inc., New York.

Avery, O. T., C. M. MacLeod, and M. McCarty. 1944. Studies on the chemical nature of the substance inducing transformation of pneumococcal types. J. Exp. Med. **79**:137.

Bolivar, F., R. L. Rodriguez, P. J. Greene, M. C. Betlach, H. L. Heyneker, and H. W. Boyer. 1977. Construction and characterization of new cloning vehicles. II. A multipurpose cloning system. Gene **2**:95.

Chang, A. C., and S. N. Cohen. 1978. Construction and characterization of amplifiable multicopy DNA cloning vehicles derived from the P15A cryptic miniplasmid. J. Bacteriol. **134**:1141.

Cohen, S. N., A. C. Y. Chang, H. W. Boyer, and R. B. Helling. 1973. Construction of biologically functional bacterial plasmids in vitro. Proc. Natl. Acad. Sci. U.S.A. **70**:3240.

Luria, S. E. 1970. The recognition of DNA in bacteria. Sci. Am. **222**:88.

Luria, S. E., and M. L. Human. 1952. A non-hereditary host-induced variation of bacterial viruses. J. Bacteriol. **64**:557.

Sambrook, J., E. F. Fritch, and T. Maniatis. 1989. Molecular cloning, a laboratory manual. 2nd ed. Cold Spring Harbor Laboratory, Cold Spring Harbor, N. Y.

Yanisch-Perron, C., J. Vieira, and J. Messing. 1985. Improved M13 phage cloning vectors and host strains: nucleotide sequences of the M13mp18 and pUC19 vectors. Gene **33**:103.

EXPERIMENT 11-1

CLEAVAGE OF DNA WITH RESTRICTION ENDONUCLEASES AND THE DETERMINATION OF THE SIZES OF THE RESULTING RESTRICTION FRAGMENTS

OBJECTIVES

In this experiment, you will cleave a bacterial plasmid vector with two different restriction endonucleases (*Bam*HI and *Pst*I), separate the fragments by agarose gel electrophoresis, and determine the size of the restriction fragments.

MATERIALS

1. Plasmid DNA unknowns: 50 ng/μL

2. Restriction endonucleases: 10 units of enzyme will be used for the cleavage reaction. A unit is defined as the amount of enzyme that completely digests 1 μg of bacteriophage DNA in 1 h at 37 °C in the optimal reaction buffer. The enzymes will be diluted to 2 U/μL at the beginning of the laboratory session.

3. Reaction buffers: the manufacturer of the enzymes provides a fivefold concentrated (5×) solution of the optimal reaction buffers. The optimal pH and ionic strength depends on the enzyme. In all cases, Mg^{2+} will be present.

4. Sterile H_2O: the H_2O is sterilized to minimize exonuclease contamination.

5. Molten agarose gel (0.7%) in 0.5× TBE (Tris-borate-EDTA) electrophoresis buffer, containing ethidium bromide at 0.5 μg/mL: this solution must be kept warm in a 60 °C water bath. One liter of 10× TBE buffer contains 108 g of Tris, 55 g of boric acid, and 7.44 g of disodium EDTA. The pH is adjusted to 8.2 with Tris or boric acid, as necessary.

6. Gel loading buffer (6×): this buffer contains a dye (bromophenol blue) to help visualize the samples, and 10% vol/vol glycerol to make the samples dense as an aid to loading the gels.

7. DNA size markers: these will usually be bacteriophage λ DNA cleaved with the restriction endonucleases *Eco*RI and/or *Hin*dIII. Your

instructor will inform you of the identity of the markers used in your experiment.

8. Polaroid-type 57 film

PROCEDURE

1. Digest 250 ng of plasmid DNA with 10 units of enzyme at 37 °C for 30 min. The final volume of each reaction should be 25 µL. Cut the plasmid with *Bam*HI, with *Pst*I, and with both enzymes simultaneously. When the digestion reaction is complete, add 5 µL of the 6× gel-loading buffer.

2. Prepare an agarose gel. The instructor will demonstrate how to cast the gel. Although the details will depend on the particular apparatus used, the object is to form a horizontal slab of agarose with wells for application of the sample. The wells are formed by molding the gel around a removable comb.

    The agarose contains 0.7% of agarose in 0.5× TBE buffer and 0.5 µg/mL of ethidium bromide. It is prepared by mixing the agarose dry powder with the buffer and boiling for a few minutes until a homogeneous solution is obtained. It is kept as liquid in a 60 °C water bath. The gelling temperature is ~40 °C.

    Ethidium bromide intercalates between bases in the DNA. This results in an enhancement of fluorescence of the dye (absorbs in the UV range and emits orange light). This property is used to detect the DNA in the gel.

The gel and the buffer that we use contain ethidium bromide. This compound is mutagenic so gloves should be worn when handling it!

3. After the gel solidifies, gently remove the comb from the gel (to avoid breakage of the wells).

4. Add buffer (0.5× TBE, 0.5 µg/mL of ethidium bromide) and load the gel (20 µL of each sample). Run the gel for at least 1 h at 100 V. Be sure to include the marker mixtures on the gel. Remember that DNA is an anion at this pH and will migrate in an electric field toward the anode (the positive electrode).

Use proper laboratory technique when using power supplies!

EXPERIMENT 11-1

5. Turn the power supply off before the dye migrates to the end of the gel. Then remove the gel.

6. Photograph the gel with a UV transilluminator.

Wear gloves when handling the gel and be sure to use eye protection when the UV light is on! Short wave UV is very harmful to the eyes. Even nonharmful exposure to UV may cause extreme pain. Do not expose your eyes to the UV light source.

7. Determine the sizes of the restriction fragments produced from the unknown plasmids. The maps of three commercially obtained plasmids are provided in this chapter. Can you determine which of these plasmids was your unknown?

DNA LIGATION

OBJECTIVES

In this experiment, you will digest bacteriophage λ DNA with the restriction endonuclease *Hin*dIII, and then recover the DNA fragments by phenol extraction and ethanol precipitation. Next, you will set up a ligation reaction and follow the time course of DNA ligation by subjecting samples to agarose gel electrophoresis.

MATERIALS

1. Bacteriophage λ DNA, 0.5 µg/µL
2. *Hin*dIII restriction endonuclease and 5× reaction buffer
3. Phenol/chloroform (1:1) saturated with TE buffer (TE buffer is 10 mM Tris-Cl, pH 8.0, 1 mM EDTA)
4. Chloroform ($CHCl_3$)
5. Ethanol (100%), ethanol (70% vol/vol)
6. 3 M NaOAc, pH 7.0
7. Ligation stop buffer (50 mM EDTA, pH 8.0)
8. DNA ligase and DNA ligase 5× reaction buffer
9. 6× gel loading buffer
10. Agarose gel (0.7% in 0.5× TBE with 50 µg/mL ethidium bromide)
11. 0.5× TBE electrophoresis buffer
12. Polaroid-type 57 film

PROCEDURE

1. Digest 2.0 µg of λ DNA with 10 units of *Hin*dIII restriction endonuclease for 30 min at 37 °C in a volume of 20 µL.
2. Save 0.5 µg of undigested λ DNA to serve as a negative control for the digestion reaction. When the digestion reaction is complete, remove an aliquot containing 0.5 µg of DNA and hold aside on ice. This will serve as the control for the ligation reaction.

EXPERIMENT 11-2

3. Add TE to the remaining portion of the cleavage reaction such that the final volume is 200 μL.

4. Extract the reaction mixture with 200 μL of phenol/chloroform. Separate phases by centrifugation in the microfuge for 3 min.

5. Remove the aqueous phase containing the DNA.

6. Extract the DNA with 200 μL of $CHCl_3$.

7. Separate the phases by centrifugation.

8. Recover the aqueous phase containing the DNA.

9. Add NaOAc to a final concentration of 0.3 M.

10. Add 2.5 volumes of (100%) ethanol.

11. Chill on ice for 10 min.

12. Recover the precipitated DNA by centrifugation (remove the supernate and hold aside).

13. Rinse the precipitate with 200 μL of 70% vol/vol ethanol.

14. Spin for 10 min, and remove the supernate.

15. Resuspend the DNA pellet in 15 μL of TE buffer.

Phenol and chloroform are caustic. They will cause painful burns if they come into contact with your skin. Wear gloves when using phenol or chloroform. Do not dispose of used phenol or chloroform; leave them at your bench, and the instructor will dispose of them properly.

The purpose of the phenol/chloroform extractions is to remove protein from the DNA. Extractions should be performed gently, using only enough agitation to cause mixing of the aqueous and organic phases. The solution should look milky white when properly mixed. When attempting to separate phases in the extraction steps, it is probably better to accept some losses than to carry forward the material that is at the interface of the organic and aqueous phases. DNA is precipitated in 70% ethanol. To help localize the pelleted DNA, it is probably best to orient the tube in the centrifuge so that the DNA pellet will be in a known part of the tube. This pellet is likely to be small and is easily lost (it does not stick very well to the tube), so be very careful not to lose your pellet when removing the ethanol.

16. Prepare an agarose gel while you are waiting during the above steps.

17. Ligate the fragments in a 25-μL reaction volume containing ligase buffer and ligase. Your instructor will inform you of the enzyme concentration and the correct number of units to use.

18. After 3 min, remove a 10-μL aliquot of the reaction mixture; stop the reaction with stop buffer (5 μL) and hold aside.

19. After 30 min, remove another 10 μL from the ligation reaction mixture, and stop the reaction as in step 18.

20. Add gel loading buffer as appropriate, subject the samples to agarose gel electrophoresis, and photograph the gel so as to have a record of the result.

Note the following: (1) Did the restriction digestion reaction go to completion? (2) Did DNA ligation occur within 3 min or within 30 min? (3) Was the recovery of the DNA fragments after extraction and precipitation complete or nearly so?

EXPERIMENT 11-3

DNA SUBCLONING AND *E. COLI* TRANSFORMATION

OBJECTIVES

In this experiment, you will ligate an *Eco*RI fragment into *Eco*RI-cleaved pUC19 vector. The ligation reaction will be used to transform *E. coli*, and α-complementation of *lacZ* will be noted on indicator plates. Putative clones will be inoculated into liquid medium and allowed to grow up for subsequent isolation of DNA. In Experiment 11-4, DNA will be isolated from these cultures, and the plasmid DNA samples will be analyzed to see if the desired clones were obtained.

MATERIALS

1. Purified *Eco*RI fragment (provided by the instructor)
2. pUC19 plasmid cleaved with *Eco*RI
3. Ligation buffer 5× and DNA ligase
4. Competent *E. coli* cells
5. 42 °C water bath
6. Sterile microtubes and tips
7. Luria broth (LB) liquid growth medium
8. Petri plates containing LB medium (solidified with agar) + ampicillin (100 μg/mL), isopropyl-thio-β-D-galactopyranoside (IPTG) (50 μg/mL), and X-gal (50 μg/mL)
9. Sterile glass beads for spreading the cells across the surface of the plates

PROCEDURE

1. Set up a ligation reaction containing the *Eco*RI cleaved pUC19 plasmid vector and the target *Eco*RI DNA fragment provided by the instructor. Use sterile tubes and tips for all manipulations, and try to avoid contamination of the samples. Allow the ligation reaction to proceed for 30 min at 37 °C. Be sure to set up control tubes lacking the target fragment ("vector alone") and lacking ligase ("no enzyme").

2. Transform 100 µL of the competent bacteria with each of your samples from the ligation reaction. Be sure to include all of the necessary controls in the transformation experiment. These should include controls for the ligation (no enzyme, vector alone), controls to test whether the competent cells were really competent (i.e., add a little uncleaved pUC19 plasmid), and controls to test whether the cells were really sensitive to the antibiotic before transformation and that the plates contain the antibiotic ("no DNA").

The transformation protocol is as follows:

a. Thaw the cells gently on ice.

b. Distribute 100 µL of cells to each tube that is needed.

c. Add the DNA samples (or buffer control) to the cells, and incubate on ice for 15 min.

d. Incubate in the 42 °C water bath for exactly 1 min.

e. Add 500 µL of LB liquid medium to each tube, and incubate at room temperature for 15 min.

f. Plate 100-µL aliquots of each sample on the LB + ampicillin + IPTG + X-gal plates. The instructor will show you the correct way to spread the cells on the plates.

3. Give the plates to the instructor. The plates will be incubated overnight at 37 °C and returned to you for the next laboratory period.

NOTE✔

LB medium is a "rich" or "complete" medium containing 10 g of Bactotryptone per liter, 5 g yeast extract per liter and 10 g of NaCl per liter. The addition of LB medium to the reaction mixture is designed to permit the cells to begin growing before being plated on the selective medium. Why do you think this is helpful?

DNA MINIPREPARATION AND ANALYSIS

OBJECTIVES

In this experiment, you will purify plasmid DNA by the alkaline lysis method and analyze the DNA by restriction digestion. The instructor will inoculate, into tubes of LB medium containing ampicillin, several white and blue colonies from your transformation experiment from Experiment 11-3, and these will serve as the samples to be analyzed.

MATERIALS

1. Solution 1: 50 mM glucose, 25 mM Tris-Cl, pH 8.0, 10 mM EDTA
2. Solution 2: 0.2 M NaOH, 1% wt/vol SDS
3. Solution 3: 3 M KOAc, pH 4.8
4. Ethanol, phenol/chloroform, chloroform
5. TE + RNaseA: This solution is TE buffer, pH 8.0, containing 10 mg/mL RNaseA.
6. 3 M NaOAc, pH 8.0

PROCEDURE

1. Put 1.4 mL of each bacterial culture to be analyzed into a microfuge tube, harvest the cells by centrifugation for 2 min, discard the supernate, and resuspend the cells in 100 μL of solution 1. Be certain that the cells are well dispersed.
2. Add 200 μL of solution 2, mix gently by inversion, and put on ice for 5 min. You should see that the cells are lysed by this treatment.
3. Add 150 μL of solution 3, mix by gentle inversion, and place on ice for 5 min. This treatment causes the precipitation of SDS, most proteins, and genomic DNA. Plasmid DNA, RNA, and some proteins remain in solution.
4. Clarify the suspension by centrifugation for 10 min in the microfuge, and carefully remove the supernate to a clean tube.
5. Extract successively with 200 μL of phenol/chloroform and 200 μL of chloroform, add NaOAc to 0.3 M, add 500 μL of ethanol, precipitate

the DNA on ice, collect the DNA by centrifugation, aspirate the ethanol and let the tube air dry, and resuspend in 50 μL of TE + RNase (see Experiment 11-2, step 1). These are the purified DNA samples.

6. Digest your DNA samples with *Eco*RI in a final volume of 25 μL. The instructor will tell you how many units of enzyme to use. After 1 h, remove your tubes and store them at –20 °C (the last step may be more convenient if an instructor does it.)

7. In the next laboratory session, prepare an agarose gel (0.7%) as before, subject your digestions and controls to electrophoresis as before, and record the results photographically. Did your ligation experiment result in the construction of the desired recombinant molecule?

QUESTION

1. The schematic gel shown below contains data from a hypothetical (idealized) Experiment 11-4. Determine the size in kb for all fragments seen. The sizes of the standards are shown to the left of the gel. The minipreparation DNA samples were cleaved with *Eco*RI and *Bam*HI. Explain the result for each of the four minipreparations, and explain why these samples gave the indicated colony color.

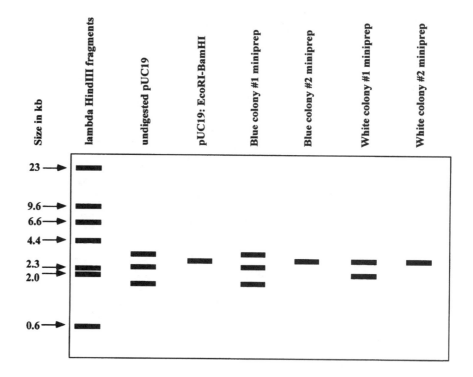

REAGENTS NEEDED FOR CHAPTER 11

✓ Plasmid DNA: pACYC177, available from New England Biolabs at no charge when ordering another item, or for the cost of shipping; pBR322, New England Biolabs, 303-3L, 250 µg; pUC19, New England Biolabs, 304-1L, 250 µg

✓ Restriction endonucleases: *Bam*HI, Sigma R0260; *Pst*I, Sigma R7002; *Hin*dIII, Sigma R1137

✓ Agarose: Sigma 99311, 100 g

✓ Tris base: Sigma T8524, 1 kg

✓ Boric acid: Sigma B7901, 1 kg

✓ EDTA: Sigma E5513, 50 g

✓ Bromophenol blue: Sigma B8026, 5 g

✓ Glycerol: Sigma G8773, 1 L

✓ Ethidium bromide: Sigma E8751, 5 g

✓ DNA size markers: we make our own standards by cleaving bacteriophage λ DNA with *Hin*dIII. λ DNA is from Gibco BRL, 25250-010, 500 µg. The same λ DNA is used for the ligation assay.

✓ Phenol: buffer saturated phenol, GibcoBRL 15513-039, 100 mL

✓ Chloroform

✓ Ethanol (100% and 70%)

✓ Sodium acetate: Sigma S7545, 250 g

✓ DNA ligase: Sigma D2886, 100 units

✓ Purified *Eco*RI fragment: virtually any *Eco*RI fragment will suffice, as long as it is not deleterious to the growth of *E. coli* when cloned into pUC19. If none are available locally, *Eco*RI fragments from λ DNA may be used. However, in this case, do not attempt to use the fragment containing the S gene.

✓ Competent *E. coli* cells: we produce our own locally, using the $CaCl_2$ method described by Sambrook et al. (1989). However, they may be obtained commercially: Gibco BRL, 18265-017, 2 mL.

✓ X-gal (5-bromo-4-chloro-3-indoyl-β-D-galactopyranoside): Sigma B9146, 100 mg

- ✓ Luria broth: Sigma L7275, 100 tablets
- ✓ Petri plates
- ✓ Luria broth agar: Sigma L7025, 100 tablets
- ✓ Ampicillin: Sigma A2804, 20 mg
- ✓ Sterile glass beads
- ✓ Glucose: Sigma G5400, 250 g
- ✓ Tris base: Sigma T8524, 1 kg
- ✓ EDTA: Sigma E5134, 100 g
- ✓ Sodium hydroxide: Sigma S0899, 500 g
- ✓ SDS: Sigma L3771, 100 g
- ✓ Potassium acetate: Sigma P1190, 500 g
- ✓ Acetic acid
- ✓ RNaseA: Sigma R6513, 10 mg
- ✓ Film: for photography of gels

12 Polymerase Chain Reaction (PCR) Technology

Introduction

Principle of the PCR Method

Applications of PCR

Experiments 12-1 to 12-5

Materials and Reagents Needed for Chapter 12

INTRODUCTION

The polymerase chain reaction (PCR) has revolutionized recombinant DNA technology by providing an extremely simple means by which specific DNA sequences can be amplified from highly complex DNA, such as that from a genome. This procedure was originally developed by Kerry Mullis, who won a Nobel Prize for this work and also received notoriety during the O. J. Simpson trial for his expertise about the PCR processes and their use in forensic DNA. Examples of uses of PCR include gene cloning and manipulation, gene mutagenesis, DNA sequencing, forensic DNA typing, and amplification of ancient DNA.

PRINCIPLE OF THE PCR METHOD

The principle of the PCR reaction is illustrated in Fig. 12-1. To amplify a sequence from DNA, two short oligonucleotide primers are annealed to denatured (strand-separated) DNA by using hybridization conditions ensuring that only primers perfectly complementary (or nearly so) with the desired sequence will anneal. The two primers are complementary to the two 3′ ends of the DNA segment to be amplified. The primers may contain additional noncomplementary sequences at their 5′ ends, such as sequences encoding sites for restriction endonuclease cleavage. The primers are then extended using DNA polymerase and the four deoxynucleotide triphosphates (dNTPs), generating two duplex DNA copies of the targeted region. After a period of time long enough to allow DNA replication of the desired region, the reaction is terminated, and the DNA strands are separated by heating the sample. This yields variable length fragments with an average length slightly longer than the distance between the two primers. These three steps (annealing, elongation, and thermal denaturation) constitute one "cycle" of PCR DNA replication. The process is repeated with new primers (already in the reaction mixture) annealing and elongating to produce another set of DNA fragments. Multiple cycles are performed, with each copy of the desired sequence being replicated each cycle. Note that in the second and subsequent cycles, any "additional" (noncomplementary) sequences that were added to the 5′ end of the primers will also be replicated, fusing these sequences to the targeted, amplified sequence (Fig. 12-1). Thus, after 30 cycles of amplification, the targeted sequence is, in principle, amplified 1,073,741,824 (2^{30} or 10^9) -fold. The investigator can have the target sequence flanked by restriction sites of choice so that the fragments can be cut to the same length after synthesis.

THREE STEPS OF PCR

The three steps of a PCR cycle (annealing, elongation, and denaturation) are usually accomplished by altering the reaction temperature. The annealing step is usually performed at a temperature that selects for good hybridization of only perfectly annealed DNA strands, such as 55 °C. (This temperature will depend on the length of the primer and the percentage of GC base pairs. Also, the annealing temperature may be varied to allow amplification from imperfectly matched primers.) The temperature for the extension phase is chosen to match the properties of the polymerase used. For the laboratory experiments in this chapter (Experiments 12-1 to 12-5), we use a thermostable DNA polymerase that works best at 70 °C. (This DNA polymerase is isolated from a bacterium that lives in thermal vents and hot springs at 90–100 °C, *Thermus aquaticus.*) Finally, the termination of each cycle and denaturing of the DNA strands are accomplished at temperatures of 95 °C.

PRIMERS IN PCR

The primers used in PCR amplifications are generally designed to anneal specifically to a known sequence. The part of the primer that anneals to the target DNA sequence is usually 16 or more nucleotides long. Since there are four possible nucleotides at any position in a 16-mer sequence, a par-

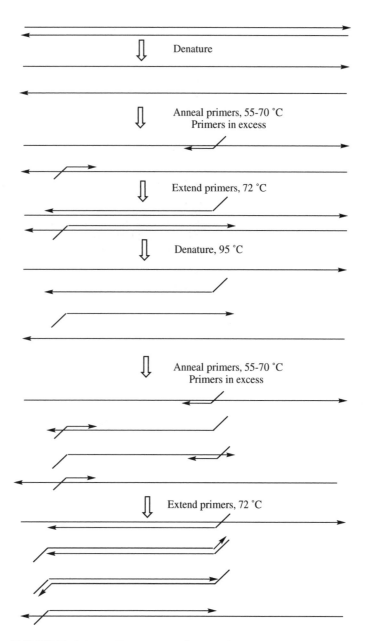

FIGURE 12-1 Principles of the PCR reaction. The lines correspond to DNA strands, with the 3′ end having an arrowhead. *(Top)* Genomic DNA is denatured and annealed to primers. These primers have additional sequences that are not complementary at their 5′ ends (the bent portions). Extension of the primers results in replication of the target sequence between the primers. After denaturation, annealing, and elongation steps, both the genomic and the copy of the target sequence made in the first cycle are present in solution. Note that the copy made in the first cycle results in the fusion and subsequent amplification of the noncomplementary sequence as well as the sequence of interest.

ticular 16-mer sequence will statistically appear in any genome only once in every 2.2×10^{10} base pairs (assuming 50% GC content). Thus, any 16-mer oligonucleotide is likely to be a unique sequence in a genome. The PCR primer can have additional useful sequences at the flanking ends, such as sites for restriction endonucleases, as mentioned above. Finally, a few extra G or C nucleotides are often added onto the 5´ termini, so that the amplified product has a few GC base pairs at each end. This increases the stability of the amplification product. To avoid artifacts, the primers used in an amplification should not easily anneal to each other and should not contain internal inverted repeats that can cause the primer to anneal to itself. To help in the design of PCR primers, there are several computer programs available. These programs not only help avoid problems like those mentioned above, but they can calculate the optimum annealing temperature for annealing any particular primer to its complement in the genomic DNA.

The use of the thermostable DNA polymerase has greatly increased the applicability of the method. Before this DNA polymerase was available, it was necessary to add a fresh aliquot of DNA polymerase to every cycle, since most enzymes are destroyed by the heating step needed to separate DNA strands (the termination step). Use of the thermostable DNA polymerase makes the process much easier, and DNA polymerase can be added at the beginning and will remain active for many cycles. Another important development for the PCR method is the thermal cycler, which is an instrument that can rapidly and accurately change the reaction temperature. This instrument is simply a very accurate heating block that can rapidly attain the temperatures needed for the three PCR steps. In addition, this instrument can be programmed to conduct a predetermined number of cycles, with the parameters of each cycle step controlled by the investigator. Finally, the availability of methods for inexpensive and rapid synthesis of oligonucleotide primers of defined sequence has helped make the PCR method very practical.

In the laboratory experiments in this chapter, we explore some of the uses of PCR technology. These and additional applications are briefly reviewed below. For a more complete discussion of the uses of the PCR, see Innis et al. (1990).

APPLICATIONS OF PCR

PRODUCTION OF PROBES

Since the PCR procedure results in the great amplification with high fidelity of the desired DNA sequence, it can be a valuable tool for the production of oligonucleotide probes needed for various procedures involving DNA hybridization. The amplified sequence (often only a portion of the DNA of interest) can be produced as a labeled fragment by inclusion of labeled nucleotides in the PCR reaction. For example, if biotinylylated or ^{32}P-labeled nucleotides are used in the PCR reaction, the amplified product will be labeled with biotin or ^{32}P. When cloning a gene from an organism, even if you do not have the information to design a perfect probe, it is often possible to use PCR to synthesize a very nearly perfect probe, since the major portion of the synthesized probe will be from the region of the genome between the primers. This probe can then be used to find the gene in that genome or even be used to search for particular genes in various other organisms.

GENE CLONING AND MANIPULATION

If a gene sequence is available, it is relatively simple to design primers based on the known sequence and amplify the desired gene directly from genomic DNA. PCR primers can be designed so that they contain 5´ "extensions" with convenient restriction endonuclease sites. Thus, after the desired target sequence has been amplified, cleavage at the added restriction sites can be used to generate a convenient fragment for cloning into the vector of choice. Indeed, PCR is one of the best methods for adding restriction sites upstream or downstream from a gene of interest for which sequence data are available. In Experiment 12-1, we amplify the gene for the SSK1 transcription factor from the budding yeast *Saccharomyces cerevisiae*. The sequence of this gene has been published, and we use primers based on the published sequence.

Sometimes, you have a gene cloned from one type of bacteria and wish to clone the same functional gene from different types of bacteria. There are a number of methods for identifying such genes. For example, we have already discussed the technique of "Southern hybridization" for detecting restriction fragments homologous to a probe sequence of interest. Another way to identify such "homologues" is to try amplifying the gene from genomic DNA of a particular organism by using primers that are based on the sequence from some other known organism. These primers may be partially "degenerate" by incorporating nucleotides that are not specific to G, C, A, or T (such as inosine, I, which can pair with U, C, or G), such that they permit the amplification of sequences that are not a perfect match with the known version of the gene. In experiment 12-3, we attempt to clone the *glnL* gene from various bacteria by using primers that match the ends of the *Escherichia coli glnL* gene. In addition, our primers have convenient restriction sites added to the 5´ ends to facilitate cloning into various vectors. The primer that anneals to the 5´ end of the *glnL* gene will have a nonnatural restriction site (for the endonuclease *Nde*I) located at the ATG sequence, which initiates translation of the message. This restriction site will allow you to clone the amplified *glnL* genes into an expression vector that has an *Nde*I site and an ATG sequence perfectly positioned downstream from very strong translational initiation signals. The resulting clones might then be useful in the overproduction of the newly cloned *glnL* gene product.

DNA MUTAGENESIS

When you are studying a protein and have its gene available, it can be useful to examine how mutations in this gene affect the function of the gene product. These types of studies are called "structure/function" studies, because they examine the effects that changes in structure (due to the mutations) have on the function of the gene product. Mutagenesis by PCR can be accomplished by using conditions that decrease the fidelity of the amplifying DNA polymerase.

Most DNA polymerases have both polymerase and "proofreading" activities, the latter being due to a $3' \rightarrow 5'$ exonuclease activity that preferentially removes mismatched nucleotides at the growing $3'$ end of the DNA chain. Most polymerases will misincorporate nucleotides ~0.05–0.1% of the time, but these misincorporated nucleotides are efficiently removed, permitting error rates of only ~1 in 10^6. The Taq (*Thermus aquaticus*) polymerase that we use in the experiments in this chapter lacks this proofreading activity and thus produces a fairly high rate of random mutations while amplifying the DNA. Other thermostable polymerases that have potent proofreading activities (and thus produce fewer mutations) are available, although they are more expensive. Such polymerases are useful for amplifying long stretches of DNA with high fidelity. The fidelity of DNA polymerases can be further decreased by increasing the concentration of Mg^{2+} in the reaction mixture. DNA polymerases usually contain two Mg^{2+} ions at the active site, and one of these is bound to the incoming deoxyribonucleotide substrate. It is not known why high Mg^{2+} concentrations lead to a loss of fidelity; one hypothesis is that under these conditions the incoming Mg^{2+}-deoxyribonucleotide binds to an enzyme that already contains two Mg^{2+} ions, thus distorting the active site. Thus to achieve various levels of random mutagenesis, PCR amplifications can be performed in the presence of increasing amounts of Mg^{2+}. Often, it is possible to obtain error frequencies of ~1% so that if the target gene is several hundred base pairs in length, virtually every copy of the amplified gene will have at least one mutation.

Other methods of decreasing the fidelity of DNA replication (and thus increasing the frequency of mutations) include using very high concentrations of all four dNTPs, or using a mixture of dNTPs in which one nucleotide is present at a much greater or lower concentration than the others.

AMPLIFICATION OF ANCIENT DNA

Modern techniques of DNA extraction permit the isolation of exceedingly tiny amounts of DNA from various museum specimens and archeological specimens. For example, dried tissues or tissues imbedded in amber may yield DNA. Unfortunately, most ancient DNA is difficult to clone directly owing to not only its very low quantity but also covalent modifications of the DNA that slowly accumulate over the eons. The PCR technique sometimes permits amplification of such DNA, permitting comparison of DNA sequences that were formed in totally different eras. Typically, primers are designed to be identical or similar to contemporary genes, and the amplified sequences will contain ancient gene sequences corresponding to that gene.

DNA SEQUENCING

PCR permits the amplification of double-stranded copies of a target sequence, as shown in Fig. 12-1. However, PCR can also be used to make many copies of a single strand of a gene by inclusion of only one primer in the reaction. This is called asymmetric PCR. Note, however, that this leads to an arithmetic increase in the replicated strand, as opposed to an exponential increase, since the replicated single strand cannot serve as a template for further rounds of replication. Thus, 30 cycles of PCR lead only to a 30-fold amplification rather than 2^{30}-fold (10^9-fold).

There are several ways the production of single strands can be used to determine the DNA sequence of a gene. In one procedure, the desired single strand is amplified in the normal way with deoxynucleotides and DNA polymerase but in the presence of a low concentration of a dideoxynucleotide. Every time the dideoxynucleotide is incorporated, the polymerization process for that chain will end, since the resulting DNA strand will have no 3′-OH group for the addition of more residues. Since the dideoxynucleotide concentration is low, the chain will terminate randomly at various positions throughout the length of the strand being synthesized. By conducting the amplification procedure in four separate tubes, each containing a different dideoxynucleotide (as well as the four normal nucleotides), a set of chains will be obtained in that tube of each possible length that can end in the particular dideoxynucleotide. The sizes of these chains can be accurately determined on denaturing urea-polyacrylamide gels. Thus, a series of bands will be obtained on the gel, which correspond to the positions in the sequence at which that particular dideoxynucleotide was incorporated. With all four tubes electrophoresing in side-by-side lanes, it is possible to deduce the sequence of the DNA. This method is a minor modification of the original enzymatic sequencing method devised by Sanger et al. (1977).

You can use asymmetric PCR to produce single strands from double-stranded DNA. The individual single strands can then be isolated (e.g., by cutting the single-stranded band out of an agarose gel) and used for sequencing by the Sanger method. In Experiment 12-4, we selectively amplify the single strands of the *E. coli glnL* gene by asymmetric PCR by using genomic DNA as the starting material. This technique has been used to sequence rapidly many different *glnL* alleles from genomic DNA, without first cloning each allele (Atkinson and Ninfa, 1992).

CHARACTERIZATION OF BACTERIAL RECOMBINANTS, GENE TYPING

Since the PCR technique permits the rapid examination of a target sequence, it may be useful for applications where one needs to verify the genotype of a sample. For example, in bacterial genetics laboratories, altered bacterial strains are frequently constructed by selecting recombinants in generalized transduction experiments by using transducing bacteriophage. To verify the genotype of recombinants, PCR is sometimes used. An example is in crosses where a deletion within *glnL* is introduced into a strain that had the wild-type gene. The deletion removes more than 0.7 kb of DNA from the *glnL* gene. Thus, by amplifying the *glnL* gene from suspected recombinants and sizing the DNA on an agarose gel, we can quickly identify strains that have either the wild-type or deleted version of the gene. The same technique is used to screen various human genes in clinical samples, such as screening for deletions in the Duchenne muscular dystrophy locus (Chamberlain et al., 1988).

FORENSIC DNA TYPING

Forensic DNA typing is used to determine issues of paternity or to determine whether biological samples containing DNA are from a suspect. The tests use PCR-amplified DNA to characterize various gene loci, exactly as in the preceding application. However, the details of the methods are different, as described below. We will conduct simplified versions of the forensic DNA typing tests in Experiment 12-5.

The great strength of using PCR methods for forensic analysis is that only very tiny amounts of starting DNA are required for the analysis. For example, a single human hair with attached hair follicle has more than enough DNA to conduct the analysis. Similarly, a few sperm, skin samples from under fingernails, or a small amount of blood can provide enough DNA for a conclusive analysis. In Experiment 12-5, we use dried blood stains on a piece of cloth (or cheek scrapings) as the sample to be analyzed. In order to extract the tiny amount of DNA in the sample, the sample will be homogenized in the presence of small beads that bind divalent metal ions very tightly (Walsh et. al, 1991). Any DNA released by homogenization of the tissue sample will be protected from degradation by the beads. After removing the beads, DNA is used directly in the PCR experiments.

The basic principle of DNA typing is as follows. The genomes of all individuals of a species have essentially the same set of genes. However, the DNA sequences of those genes are not identical in all individuals. The differences in DNA sequence of our genes is one reason why individuals within a species are recognizable from one another. For example, among humans, hair color, eye color, height, and weight are influenced by our genetic makeup. The latter two characteristics are also influenced by environmental factors. When the same gene is found to have different versions, with slightly different DNA sequences, the different versions are called **alleles**. While all humans have the same number of genes, we each carry a distinct set of alleles of those genes. Typing of DNA involves determining which alleles are present for various loci within the DNA.

The genes chosen for DNA typing are those that have been characterized and shown to have several different alleles in the human population. As an example, let us assume that for gene X there are five different alleles, and these are found to be equally distributed throughout the world population. (Actually, this is not found, since various alleles are distributed in a biased way among the different races and in people from different geographical origins.) If a forensic DNA sample has two copies of gene X allele 1 and DNA from a suspect has two copies of gene X allele 2, then the test proves the innocence of the suspect. If the suspect also had two copies of gene X allele 1, the test does not prove guilt, because 6% of the world population would also carry two copies of allele 1 of gene X. Therefore, you can see that *the test can prove innocence but can never prove guilt.* Of course, even when the clinical testing is completely unambiguous, it is nevertheless crucial for the samples to have been collected and identified properly if forensic testing is to be successful.

In order to increase the resolution of the test, many different genes are usually examined. In the typical forensic analysis, 6 different genes are examined, and these have a total of only 18 different alleles. Since humans carry two copies of each gene, they may have two different alleles for each gene. The six different genes allow for 20,412 different possible genotypes. However, since the alleles are not equally common in the human population, you have about a 1/10,000 chance of scoring identically to another person if you have the more common alleles. Thus, for these six genes there are ~500,000 people on Earth that would have the same genotype that you do. By looking at a larger number of different genes, the possibility of coincidental matches could be further reduced. For example, if there are 5 different alleles (A, B, C, D, E) for a given loci, there are 15 possible genotypes (AA, AB, AC, AD, AE, BB, BC, BD, BE, CC, CD, CE, DD, DE, EE). If five loci were examined, each with 15 different alleles, there would be 759,375 different possible genotypes. Again, the resolving power of the test becomes much lower when the actual distribution of alleles in the human population is considered. Some alleles may be very common in certain populations, rendering them less useful in distinguishing between different subjects. In Experiment 12-5, we only look at one locus by the standard test, but in a real forensic laboratory analysis, the identical protocol is used for many different loci. The locus we examine is the human leukocyte class 2 antigen locus.

The human leukocyte antigen locus encodes a set of membrane proteins that serve as receptors for peptide antigens. The receptor-antigen complex is recognized by the T-cell receptor, and this recognition leads to the activation of T cells and to the consequent immune response. There are six common alleles for the human leukocyte antigen class 2 (HLA-D) loci, referred to as alleles 1.1, 1.2, 1.3, 2, 3, and 4. The HLA-D genes are located on chromosome 6. The complete DNA sequence for each allele is known. As noted above, two different alleles may be present in a given subject. Thus, there are 21 possible genotypes for this locus involving the six common alleles. Inclusion of rare alleles raises the number of possible genotypes.

In order to determine which of these alleles are present in a given human, DNA from that person is amplified with primers that flank the HLA–D locus. The amplified DNA is made to contain biotinylylated nucleotides. The amplified DNA is then hybridized to a test strip containing immobilized probes with the six different human HLA-D alleles. The amplified allele(s) from the test subject should hybridize to one of these immobilized alleles or to two of the alleles if the person bore different alleles on each copy of chromosome 6. The nonhybridizing DNA is then washed away, leaving only one or two of the alleles on the test strip hybridized to the PCR product. The test strip is then flooded with streptavidin, which binds tightly to biotin. The streptavidin that is used in the experiment is covalently coupled to an enzyme that catalyzes a chromogenic reaction; for example, β-galactosidase, alkaline phosphatase, horseradish peroxidase, etc., may be used as the coupled enzyme. Next, the allele with annealed PCR product is detected by addition of the substrate for the chromogenic reaction.

Another type of forensic PCR analysis makes use of the fact that at certain loci, humans have a variable number of repeats of a DNA sequence. For example, at the D1S80 locus, located near the end of chromosome 1, a 16-base-pair sequence is repeated a variable number of times, ranging from 14 repeats to 40 repeats. In all, 29 different alleles have been identified, so there are 435 different possible genotypes.

The complete sequence of the flanking region is known, and we use PCR to amplify this locus from forensic samples. The length of the DNA segment that contains this locus will be different in different people, depending on which allele(s) they possess. To size the amplified DNA fragment(s), the amplified sequence can be directly compared to standards by agarose gel electrophoresis. The identity or nonidentity to the control DNA samples can then be determined.

We combine the analysis of the D1S80 locus (435 possible genotypes) with the analysis of the HLA-D locus (21 possible genotypes) to test our samples. By using these two tests in combination, there are 9135 possible genotypes. If each genotype were equally common, there would be over 27,000 people in the United States with each genotype.

REFERENCES

Atkinson, M. R., and A. J. Ninfa. 1992. Characterization of mutations in the *glnL* gene of *Escherichia coli* affecting nitrogen regulation. J. Bacteriol. **174**:4538.

Chamberlain, J. S., R. A. Gibbs, J. E. Ranier, P. N. Nguyen, and C. T. Caskey. 1988. Deletion screening of the Duchenne muscular dystrophy locus via multiplex DNA amplification. Nucleic Acids Res. **16**:11141.

Innis, M. A., D. H. Gelfand, J. J. Sinsky, and T. J. White. 1990. PCR protocols. A guide to methods and applications. Academic Press, Inc., New York.

Sanger, F., S. Nicklen, and A. R. Coulson. 1977. DNA sequencing with chain terminating inhibitors. Proc. Natl. Acad. Sci. USA **74**:5463.

Walsh, P. S., D. A. Metzger, and R. Higuchi. 1991. Chelex 100 as a medium for simple extraction of DNA for PCR-based typing from forensic material. BioTechniques **10**:506.

EXPERIMENT 12-1

AMPLIFICATION OF THE SSK1 GENE FROM YEAST GENOMIC DNA

MATERIALS

1. 10× PCR buffer:

   200 mM Tris-Cl, pH 8.5 (20 °C)

   15 mM $MgCl_2$

   250 mM KCl

   0.5% Tween 20

2. DNA primers:

   SSK1 upstream primer: GGG AAG CTT CAT ATG CTC AAT TCT GCG TTA CTG TGG AAG GTT TGG; MW 13,900

   SSK1 downstream primer: GGG GAA TTC CTA TCA TAA TGT CCT CTA CAC GGT ACA ACC; MW 11,865

3. DNA template: 1 mg/mL yeast DNA

4. Nuclease-free bovine serum albumin (BSA)

5. dNTP stock solution: purchased as 10 mM each; dilute to 1 mM each before use and store stock vials at –80 °C.

6. Taq polymerase: need 2 units/assay

PROCEDURE

1. Set up 100 μL PCR reactions containing the following:

   10 μL 10× PCR buffer

   10 ng yeast DNA template

   20 pmol each primer

   10 μg BSA

   5 μL of 1 mM dNTP working solution (50 μM final each dNTP)

   2 units Taq polymerase

   H_2O to 100 μL

2. Your instructor will put your samples into the thermal cycler and freeze your samples for you when the amplification is completed. The program to be used is as follows:

   Denaturation: 96 °C, 15 s

   Annealing: 55 °C, 30 s

   Extension: 72 °C, 90 s

   30 cycles will be run; then a final extension at 72 °C for 5 min will be performed. The samples will be frozen until the next laboratory session.

3. Characterize the PCR product by running aliquots directly on a 1% agarose gel, along with size standards (digested λ DNA).

EXPERIMENT 12-2

AMPLIFICATION OF THE *glnL* GENE OF *E. COLI* FROM WILD-TYPE AND DELETED STRAINS

MATERIALS

1. 10× bacterial DNA preparatory buffer for PCR:

    10% Triton X-100

    200 mM Tris-Cl, pH 8.5

    20 mM EDTA

2. Freshly streaked out plates with well-isolated single colonies of the wild-type strain (YMC10) and the *glnL2001* strain (RB9132)

3. Sterile wooden dowels

4. Primers:

    glnL upstream: CCC GAA TTC ATA TGG CAA CAG GCA CGC AGC CC

    glnL downstream: GGG AAT TCG AGC TCT GCG AGC GCA CGT TCA AGC ACC CAA CGG

5. 10× PCR buffer (see Experiment 12-1)

6. Taq DNA polymerase

7. Nuclease-free BSA

8. dNTP working solution (1 mM each) (see Experiment 12-1)

PROCEDURE

1. Preparation of DNA from bacterial colonies:

    Using a sterile wooden dowel, pick a well-isolated colony and suspend it into 50 µL of 1× bacterial DNA preparatory buffer for PCR.

    Heat at 95 °C for 5 min. Spin at full speed in a microfuge for 10 min. Use 5 µL of the supernatant for each PCR reaction.

2. Set up 100 µL PCR reactions containing the following:

    10 µL 10× PCR buffer

    5 µL DNA template (supernate from step 1)

20 pmol each primer

10 µg BSA

5 µL of 1 mM dNTP working solution (50 µM final each dNTP)

2 units Taq polymerase

H_2O to 100 µL

3. Perform 30 cycles of PCR using the same program as in Experiment 1, and a 72 °C extension for 5 min as in Experiment 12-1. Your instructor will freeze the samples for you when the reactions are complete.

4. Analyze the amplification product directly on a 1% agarose gel, next to size standards. The wild-type PCR product should be 1325 bps, and that from the *glnL2001* strain should be 830 bp.

EXPERIMENT 12-3

USING THE *E. COLI glnL* GENE SEQUENCE TO IDENTIFY HOMOLOGUES IN OTHER BACTERIA

MATERIALS

1. Freshly streaked rich-medium plates containing various bacterial strains. The names of the strains used will be provided by the instructor and written on the bottom of the plates.
2. Materials from Experiment 12-2

PROCEDURE

The procedure is the same as for Experiment 12-2, except that instead of using the *E. coli* strains, use the various other bacterial strains provided.

PRODUCING SINGLE-STRANDED DNA FOR EACH STRAND OF THE *E. COLI glnL* GENE

MATERIALS

The materials are identical to those used in Experiment 12-2.

PROCEDURE

The procedure is identical to that in Experiment 12-2, except that when setting up the PCR reaction, instead of using 20 pmol of each primer, use 20 pmol of one primer and either 0.4 or 0.2 pmol of the other primer.

FORENSIC DNA: DEATH IN A TENURED POSITION (WITH APOLOGIES TO AMANDA CROSS); WHO KILLED THE DEPARTMENT CHAIRMAN?

Everyone on the department staff disliked faculty meetings scheduled for late Friday afternoon in the middle of November. It was dark and cold by the time the faculty got out, and the campus and parking structures were deserted. The chairman of the department had a reason to be especially unappreciative of one such meeting. He was murdered after it.

Police found the blood-stained body on the otherwise deserted sixth floor of the parking structure. Apparently, the perpetrator used a knife to stab the victim many times. Police determined from the pattern of blood stains that a fierce struggle had occurred and that the blood stains present were possibly from both the victim and the perpetrator. None of the victim's jewelry or cash was taken, and his expensive European sports car was found at the scene, along with the keys. Robbery was therefore ruled out as a motive.

Since no one else was around at the time the crime was committed, the police considered the possibility that a discontented faculty member was responsible for the crime. They made a list of faculty members who had big research grants but had rather poor laboratory facilities. Such a situation might create hostilities between faculty and chairman. They also made a list of faculty that had a salary lower than others at the same level, since that could also engender hostility toward the chairman. Finally, they made a list of faculty members that had more extensive teaching and committee assignments than the norm. Dave and Alex were on all three lists, as were a few other faculty.

From the angle of the wounds in the chairman's body, the police concluded that they were looking for a perpetrator that was quite a bit shorter than the chairman. Yet, the perpetrator had to be fairly young and virile, since the chairman was a large and athletic man. This led the police to discount several older faculty members who were also short men and to focus their attention on Dave and Alex. Both Dave and Alex passed lie detector tests during which they professed innocence, but the police discounted this since everyone knows that persons of such high intellect can easily fool the lie detector.

Can you help prove the innocence of your professors? Samples of the victim's blood-stained white shirt are provided. These samples are believed to contain both the victim's and the perpetrator's blood. The police were able to collect some uncontaminated blood from the victim, and Dave and

Alex were required by the police to provide blood samples. To control for the condition of the blood from the crime scene, each of the control blood samples was spotted onto a clean portion of the shirt the chairman was wearing. These bits of blood-stained cloth will serve as the samples for DNA extraction and typing.

MATERIALS

1. Chelex DNA extraction agent, 5% suspension. Chelex is a very potent chelator of metal ions; it serves to protect the DNA from degradation during the extraction.

2. Forensic samples: blood-stained cloth

The experiment will also work with cheek swabs instead of blood-stained cloth. Take a sterile toothpick and vigorously scrape the inner cheek so as to remove some skin cells. Place the end of the toothpick and the associated skin cells into a microtube, and treat them the same as the cloth samples in the protocol. This eliminates most contact with human blood. (However, a new introductory story will have to be made up.)

3. AmplifiFLP D1S80 PCR amplification kit:

    PCR reaction mix

    5 mM $MgCl_2$

    Control DNA (DNA from an individual with alleles 18 and 31) 250 ng/mL

    Allelic ladder marker mix for electrophoresis

    Sterile, autoclaved H_2O

4. AmpliType HLA DQa PCR reaction kit:

    PCR reaction mix

    Control DNA (DNA from an individual with alleles 1.1 and 4) 100 ng/mL

    8 mM $MgCl_2$

    HLA probe strips

    HRP-strepavidin conjugate

    Chromogenic substrate (tetramethylbenzidine)

EXPERIMENT 12-5

0.1 M Sodium citrate, pH 5.0

20× SSPE buffer: 3.6 M NaCl, 200 mM $NaH_2PO_4 \cdot H_2O$, 20 mM EDTA, pH 7.4

20% SDS

Hybridization solution: 5X SSPE, 0.5% SDS (may require heating to 60 °C to dissolve precipitates).

Wash solution: 2.5 × SSPE, 0.1% SDS

3% Hydrogen peroxide

PROCEDURE

All procedures must be conducted while wearing gloves. All pipette tips or any other material that touches the samples (tubes, etc.) must be contained and isolated from human contact. These materials will have to be sterilized before being discarded. Biohazard bags will be provided for disposal of contaminated materials.

There must be no direct human contact at any time with the forensic samples or any object that was in contact with the forensic samples. Do not touch your skin, eyes, etc. while wearing gloves! Dispose of gloves frequently. Do not leave the room with gloves on.

The PCR amplifications that we are about to perform are designed to work with exceedingly small quantities of template DNA. Be especially aware of the possibility of cross-contamination of samples.

1. Isolation of DNA from blood-stained cloth fragments:

   a. Combine ~3 mm of square blood-stained cloth (or the toothpick end with cheek scrapings) with 1 mL of H_2O in a microtube.

   b. Incubate at room temperature for 15 min; occasionally mix gently by inversion.

   c. Spin in microfuge for 2 min at full speed; carefully remove the supernate (all but 20–30 μL) and discard. Leave the fabric in the tube.

   d. Add 5% Chelex suspension to a final volume of 200 μL.

   e. Incubate at 56 °C for 15 min.

   f. Vortex at high speed for 10 s.

    g. Incubate in a boiling water bath for 8 min.

    h. Vortex at high speed for 10 s.

    i. Spin in microfuge for 3 min at full speed.

    j. Use 20 μL of the supernate for each PCR reaction.

2. PCR amplification of the HLAα locus:

    a. PCR reaction tubes containing 50 μL of "mix" will be provided. These already contain polymerase dNTPs, etc.

    b. Add 50 μL of the 8 mM $MgCl_2$ solution to each PCR reaction tube.

    c. Add either 20 μL of control DNA, 20 μL of sample DNA, or 20 μL of H_2O (negative control). Your instructor will put your samples into the thermal cycler for you. The program will be as follows (30 cycles):

        denaturing:, 94 °C, 60 s

        annealing: 60 °C, 30 s

        extension: 72 °C, 30 s

    Your samples will be frozen for you until the next laboratory session.

3. DNA hybridization:

    a. Place one probe strip into a tray for each sample.

    b. Heat the PCR reaction tubes to 95 °C for 3–10 min. Store at 95 °C until use.

    c. Tilt the tray toward the labeled ends of the strips. Add 3.3 mL of prewarmed hybridization solution and 27 μL of enzyme conjugate HRP. Immediately add 35 μL of the amplified DNA sample (95 °C). Put lid on tray, and incubate at 55 °C for 20 min.

    d. Remove the tray from the water bath, and remove liquid.

    e. Wash with 10 mL of wash solution, and discard the wash.

    f. Add 10 mL of wash solution, and incubate 55 °C for 12 min.

    g. Remove the tray from the water bath, and discard the wash solution.

    h. Rinse with 10 mL of wash solution at room temperature for 5 min, then discard the wash solution.

4. Color development:

    a. Add 10 mL of citrate buffer to each well, rinse at room temperature 5 min and discard.

    b. Add 10 mL of color development solution to each well. The instructor will prepare this reagent immediately before use. It contains 10 mL of citrate buffer, 10 µL of 3% hydrogen peroxide, and 0.5 mL of the chromogen for each well.

    c. Develop color at room temperature for 20–30 min.

    d. Stop color development by washing with 10 mL of distilled H_2O. Repeat 2 or 3 times. Rinse for 5 min each time.

    e. Look at the strips, and record the result.

5. Amplification of D1S80 locus:

    a. PCR reaction tubes containing 20 µL of reaction mixture will be provided. Use one tube per sample.

    b. Add 10 µL of the 5 mM $MgCl_2$ solution to each tube.

    c. Add 20 µL of sample DNA, control DNA, or H_2O. Your instructor will put your samples into the thermal cycler for you. The program will be as follows (30 cycles):

        denaturing: 94 °C, 60 s

        annealing: 65 °C, 60 s

        extension: 72 °C, 60 s

    Your samples will be frozen for you until the next laboratory session.

6. Agarose gel electrophoresis of the PCR product:

    a. Prepare a 1.5% agarose gel.

    b. Mix 25 µL (1/2 of each sample) with the tracking dye and load.

    c. Stain and photograph gel as usual.

MATERIALS AND REAGENTS NEEDED FOR CHAPTER 12

✓ Tris base: Sigma T8524, 1 kg

✓ HCl

✓ Magnesium chloride ($MgCl_2$): Sigma M9272, 500 g

✓ Potassium chloride (KCl): Sigma P9541, 1 kg

✓ Tween 20: Sigma P7949, 100 mL

✓ DNA primers: these are purchased locally.

✓ BSA (acetylated): Sigma B2518, 10 mg

✓ dNTP stock solution (10 mM each): Sigma D7295, 1 vial (0.5 mL)

✓ Taq polymerase: Sigma D1806, 250 units

✓ Agarose: Sigma 99311, 100 g

✓ Boric acid: Sigma B7901, 1 kg

✓ EDTA: Sigma E5134, 100 g

✓ Ethidium bromide: Sigma E8751, 5 g

✓ DNA size markers: we make our own standards by cleaving bacteriophage λ DNA with HindIII. λ DNA is from GibcoBRL, 25250-010, 500 g. The same λ DNA is used for the ligation assay of Chapter 11.

✓ Triton X10: Sigma T8532, 100 mL

✓ Sterile wooden dowels

✓ Various bacterial strains: The E. coli glnL+ and ΔglnL strains may be obtained from A. Ninfa (aninfa@umich.edu). Alternatively, the experiment may be performed with any known bacterial deletion mutant, although in that case, the corresponding primers will have to be used and tested. Bacterial strains other than E. coli for the PCR analysis are obtained locally. Over the years, we have used Bacillus subtillis, Rhodopseudomonas capsulata, Vibrio cholera, P. aeruginosa, Caulobacter crescentus, Salmonella typhimurium, and Klebsiella aerogenes. Of these, S. typhimurium, K. aerogenes, and V. cholera gave the best results with E. coli-specific glnL primers, while B. subtillis gave the worst results. This is not unexpected, and it is useful for the students to see and think about.

✓ Yeast DNA: obtained locally from a yeast lab. Apparently, there is no commercial source. Extraction protocol available upon request (aninfa@umich.edu).

✓ Forensic DNA kits: AmpliFLP D1S80, Perkin-Elmer N808-0054, 50 test kit; Amplitype HLADQα, Perkin-Elmert N808-0056, 50 test kit

✓ Forensic samples: these are provided by the instructors. Note that the experiment will work with cheek swabs if the use of blood is objectional.

13 Using the Personal Computer and the Internet for Biochemical Research

INTRODUCTION TO THE WORLD WIDE WEB (THE INTERNET)

The most important factor in conducting successful research is the quality of the thinking involved: deciding what are the key questions, which approaches will be most informative and productive, and how to successfully carry out those approaches. This will never change. However, in contemporary research in biochemistry and in almost every other field of biomedical science, to do cutting edge work it is necessary to have some knowledge of modern computational methods and the skills to tap the wealth of information available over the Internet. A complete guide to the use of computers in biochemical research is beyond the scope of this book, but several good reference works are available (e.g., Peruski and Peruski, 1997; Abelson and Simon, 1996). In this chapter, we point you to some of the public-domain information that is available and show how you can use it to aid your research efforts. Hopefully, these exercises will not only be instructive but will stimulate you to further develop your abilities in this area.

Regardless of your discipline of study, reporting your research in the form of journal articles is essential for the work to be of any value. Thus, you need to have a good word processing program and a good graphics program. Data analysis programs such as spreadsheets, curve-fitting programs, DNA and protein sequence analysis programs, kinetic simulation programs, and others may also be required, depending on the type of work that you do. Such programs can usually be run on your microcomputer, and in most cases, they are commercially available. In this chapter, we do not endorse any particular programs; in fact, newer and better programs are constantly being developed so that anything endorsed here would probably be out of date by the time you read this. Before purchasing any programs, it may be wise to examine whether the transfer of data between the different programs is simple. Satisfying this criterion will make your life much simpler. Throughout this class, you will surely gain familiarity with some of these programs while producing the laboratory reports required, and in the set of laboratory experiments in this chapter, some of the sequence and structural analysis programs will be introduced.

In this chapter, we discuss information and programs that are available without cost over the Internet (which is also referred to as the World Wide Web). Access to the Internet may be gained by a variety of services. While all have costs, we assume that most universities provide for students, in libraries and/or in study centers, access to computers that are linked to **ethernet networks** and that can easily be connected to the Internet. The procedures for accessing the Internet and using public-domain information are essentially the same regardless of the type of computer used (i.e., Macintosh™, IBM or clone, Unix workstation). Access to Internet resources is made convenient by a **Web browser**, which is a program run on your computer. Web browsers are programs with graphical interfaces that ease your work with the Internet. The most commonly used Web browser in academic settings is **Netscape**, and this program is used in libraries, research laboratories, and even at home through telephone modems rather than ethernet connections. The presentations in this chapter, which use Netscape version 3.01, demonstrate many of the features of how Web browsers are used. Netscape is under continuous development, so that if you use a newer version, some of the specific procedures and screen displays may differ. Other Web browsers, such as Microsoft Internet Explorer for Windows™, also work well. If your local site is equipped with a different Web browser, you may need to learn a few different simple commands than those stated below, but this should not be too difficult.

Upon activating the Web browser program, you will be presented with a screen such as shown in Fig. 13-1. At the top, you will see a menu bar, as in all Windows™ and Macintosh™ programs, that permits you to access the functions of the Web browser program (Fig. 13-2). Just below the menu bar is a series of graphical "buttons" that help you direct your search (Back, Forward, Home, Reload, Images, Print, Find, and Stop), a small dialogue box (location) that displays the current Web address, and below this, another series of buttons (What's New? What's Cool? Destinations, Net Search, People, and Software). This section of the screen is specific to Netscape 3.01. The screen below this Netscape-specific section is a "window" that displays the currently addressed "**Home Page**." Figure 13-1

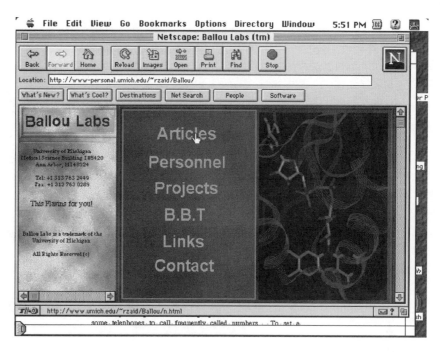

FIGURE 13-1 A (well-known) home page viewed in Netscape.

shows the home page of the "Ballou Lab" displayed by Netscape 3.01. The browser can be set to open with any Home Page as Home, but it is not polite to alter this setting unless it is your computer. If it is your computer, you may specify the Home by clicking on the item called "General Preferences" on the Options menu at the top of the screen. A window (Fig. 13-3) will open showing various settings for the Web browser, which can then be altered by using the mouse and keyboard. After setting your preferences, you can exit the "General Preferences" screen by clicking on OK or, if you change your mind, Cancel. From the Home Page, you can access virtually any other Internet site, as you will see. If you are working at any other site, you can return directly to the Home site by clicking the Home button.

WEB PAGES

Web pages are usually composed of several elements. These elements are the following: images (pictures, diagrams, graphs, etc.), text, background, and Hypertext. Each of these elements can usually be separately viewed and manipulated. You can use the mouse to point at and select an item to dis-

FIGURE 13-2 The Netscape menu bar for Macintosh™ programs.

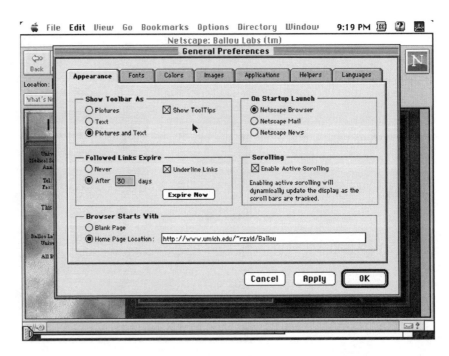

FIGURE 13-3 The General Preferences box for Netscape 3.01. This box is accessed from the Options menu.

play a dialogue box with several options (see Table 13-1). (With Windows™-based computers, point and depress the right mouse button. With Macintosh™-based computers, point and hold the button down.) Any text can be "selected," "cut," and "pasted" as in word-processing programs. Background can be selected in Windows™ by pointing and depressing the right mouse button and can then be used for other purposes. (This option is not available on Macintosh™ computers.) Selection of an image permits those options listed in Table 13-1 with "image" in the command to be carried out. Hypertext is text with special properties described below under Links and Bookmarks.

LINKS AND BOOKMARKS

As mentioned above, the Web browser program is basically a program with a graphical interface that permits you to easily connect to other sites. Just as when using a telephone, you can go to any site by specifying the electronic address. However, typing in the electronic address of each site can be tedious and is prone to error. (These addresses are often arcane; see Table 13-3). However, related sites are often linked together in logical ways, so that you can easily move from one site to another by using the mouse. Linked sites (called **Links**) are easy to identify, because the text leading to a link is usually highlighted, shown in a different color, or under-

TABLE 13-1

Options for Treating Elements on a Home Page

MENU ITEM	EXPLANATION
Back	Go back to previous location
Forward	Move forward to next location
New window	Open a new window like this
Open this link	Open the linked location
Add bookmark	Add this bookmark to your file
New window with this link	Open a new window at linked page
Save this link	Save link address to a file
Copy this link location	Copy the link address to clipboard
Open this image	Open this image by itself
Save this image	Save this image to a file
Copy this image	Copy this image to clipboard
Copy this image location	Copy image location to clipboard
Load this image	Load this image for manipulation

Not all of these items apply to each type of element.

lined (depending on the settings in Preferences). This kind of text is called **Hypertext** (written in **HTML**, which is short for "hypertext markup language"). When the mouse pointer passes over such text, the linked address will be shown in a small panel at the bottom of the screen and the arrowhead of the mouse pointer will be converted to a small pointing "hand." In Fig. 13-1, the mouse is pointing to "Articles," and the linked address is shown in the lower left box: http://www.umich.edu/~rzaid/ballou/n.html. Clicking the mouse button momentarily while pointing to the highlighted text moves you directly to that site without having to type in the electronic address. (With Windows™-based computers, clicking the left mouse button has the same effect as clicking a Macintosh™ mouse.) If you want to return to the previous site, simply click on the Back button at the top of the screen. Clicking the Forward button reverses this action.

Sometimes when "surfing" through various links, you may decide that you want to return to a site that was visited at the beginning of your session. You can get there by repeatedly clicking the Back button and retracing your path, but if many sites were visited, this is cumbersome. You can go directly to the startup site by clicking the **Home** button. Alternatively, you can point to the **Go** menu with the mouse, and a listing all of the sites visited during the current session will be shown; you can quickly choose the site you wish to revisit without having to retrace your path, simply by pointing and clicking with the mouse. The list of sites that have been visited in this session can also be displayed along with their full Web addresses by selecting Window in the menu bar and choosing the History option. A dialogue box will appear that shows the sites and addresses with options to either save a bookmark (see below) or go to that site.

Bookmarks are electronic addresses that may be quickly activated by clicking with the mouse. Bookmarks are essentially the equivalent of the programmable buttons on some telephones that are used to access frequently called numbers. To set a bookmark, you must first go to the site. Then select the Bookmark menu at the top of the screen, and the current electronic address and a short description of the site will be stored locally on your computer. To revisit a site with a Bookmark, simply open the Bookmark menu, and the list of all stored sites will appear in a small panel. Clicking on the desired site then sends you directly to that site. You can also create a bookmark from the History menu item discussed above.

Click the **Open** "button" to reach a site directly. A small dialogue box will appear with a space to type in the address you want and two buttons labeled Cancel and Open. Type in the address. When the address has been entered, click on Open, or you can disregard the new address by clicking on Cancel. Another way to open a site directly is to type the site address in the Location dialogue box located in the top portion of the Netscape screen (Fig. 13-1).

SAVING DATA AND DOWNLOADING PROGRAMS

It is a very good idea to save data often during an Internet session. Crashes may occur for many reasons outside of your control, such as a temporary loss of connection to the ethernet or to your telephone connection. Also, Internet sessions often tie up such a large amount of computer memory that occasional crashes are inevitable. When the terminal crashes, work that was not saved is lost.

In many cases, the information gained during a session will be text. Certain data files, such as amino acid or nucleotide sequences, as well as data tables such as structural coordinates, may be considered text files. The easiest way to save text data is simply "copy and paste" the information into a word-processor document file.

In a similar fashion, images may be copied and pasted into many graphics programs. Indeed, some of the better word-processing programs have built in graphics programs and may directly accept images by the copy-and-paste route. There is another way to copy images. As noted in Table 13-1, the option to copy an image is presented when the mouse button is held depressed while the mouse is pointing to an image (for a Macintosh™ computer) or when the right mouse button is depressed while pointing at the image (for a Windows™-based computer). By activating this choice, the image is copied to the clipboard, and the image can then be pasted into a word-processing or graphics file. The entire screen can be copied by simultaneously depressing the "alt" and "print screen" keys (on Windows™-based computers) or by simultaneously depressing "command" and "shift 3" keys (on Macintosh™ computers). This places the image onto the clipboard so that it can be pasted in another file. This image can then be cut and modified with a graphics program such as Paintbrush™ or Photoshop™.

Finally, in some cases, very large files or programs will be desired. In such cases, the source usually has a file transfer system in place, and the method of transfer is directed from that site. A window will appear that allows you to direct the file to the location of your choice. Large programs are often transmitted in a compressed form. An auxiliary program such as "Unstuffit" or "WinZip" is required to restore the original form of the desired file. If your computer does not have these programs, they can usually be transferred along with the file.

SEARCH ENGINES

If you do not know the address of the site you want to visit, or you are just "surfing the net" to see what information is available in a given area, you can use a program called a **search engine**. Several different engines are available, and their electronic addresses are listed in Table 13-3. (These search engines sustain themselves by selling advertising and are free sites.) Open one of the sites, and you will see a dialogue box where you may type in one or more keywords. Various tricks are available to speed searching in ways analogous to searching for references in a library system. For example, multiple keywords may be simultaneously used to more precisely define a search. The search engine screen will contain a link to a help site that provides information on how to speed your search, so you may want to visit that site to learn various tricks. After entering the keyword or words you wish to search (search parameters), click on the search button to activate the search engine. The search engine will search through the Web for sites containing the keyword(s) that you specified. A list of the sites found will be presented, and you can then scan the list and open any desired sites by clicking on the highlighted text (Link). The electronic addresses are usually also listed.

As a practice exercise, open one of the search engines in Table 13-3. Enter the name of either of the authors (use the full name) of this book, and click on the search button. The search engine will find the home pages, and likely a few other sites that contain publications from our laboratories (since these also have our names). Clicking on the links to the authors' home pages will bring you to home pages, where you can read about our research, and see pictures of what we looked like when we still had hair.

One should be aware of the reality that information obtained from the Web may not be carefully reviewed and authenticated. Therefore, you should be careful about believing information without further investigation.

For example, if the pictures that you obtained from the authors' home pages had been dated as 1997, you might think that the authors were young. Indeed, these images might have been taken from some other Web page such as that of Arnold Ziffel or even from GQ (**NOT!**).

INFORMATION ON THE INTERNET USEFUL TO BIOCHEMISTS

LITERATURE SEARCHES

It used to be true that "all good research projects start in the library." However, with Internet access, virtually all scientific literature can be searched from a connected microcomputer. In most cases, you will not be able to obtain complete publications from these searches, but you will be able to search Titles and Abstracts from research publications (this is why it is important to write good Titles and Abstracts). Even though this is a considerable limitation, you can usually narrow down your search to a few key papers that seem, from their abstracts and titles, to deal with the issue(s) under consideration. Then you need only go to the library to get those few papers. One of our exercises in this chapter will teach you how to conduct a literature search using the Internet. We use a free site maintained by the **National Center for Biotechnology Information (NCBI)** (Fig. 13-4). A powerful literature search engine known as **PubMed** is at this site. The search engine uses a database called **MEDLINE** to search the literature. In addition, you may search for DNA sequence files (**Entrez**), compare DNA sequence files (**Blast**), search for protein sequence files (Entrez), compare protein sequence files to the translated gene libraries (Blast), and obtain data files with structural coordinates for all of the macromolecular structures from the protein database (Entrez). This site has many other useful medical resources as well. This is a very important site; its electronic address is in Table 13-3 as well as in the Location box in Fig. 13-4.

To access the PubMed site, first open the NCBI site. This site is listed in Table 13-3 under **Databases** (http://www.ncbi.nlm.nih.gov/). Several links will be presented, with the most important presented as buttons across the center of the page. The first of these will be **PubMed**. Simply click on the button to go to that location (see Fig. 13-5).

The PubMed page includes a help link (located to the left, under the National Library of Medicine emblem), which is exceptionally useful in pointing out how to conduct a successful search. Unlike many help files, this file is well written and interesting. Unless you have used the program before, you will probably find the help file very useful. The PubMed page has a dialogue box in which you enter search parameters such as keyword,

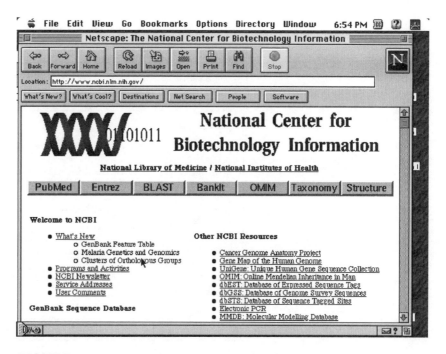

FIGURE 13-4   The NCBI home page. All of the NCBI linked pages use similar formats and procedures to access and manipulate data.

author name, journal, or year of publication (Fig. 13-5). Since you are searching all of the biomedical literature, it is essential that you limit the output appropriately. For example, searching for *"Escherichia coli"* will find thousands of papers, hardly a useful step. Searching for *"Escherichia coli* aspartate transcarbamylase active site" will yield just a few papers, dealing exactly with that issue. The program allows the use of various logical operators (AND, OR, etc.) to link the search words. The default value for a space between keywords is the AND operator.

PubMed provides the search results in a table in reverse chronological order, listing title and first author. The author name is in hypertext and links the entry to its abstract. After reading the abstract, you can decide whether to print it, and then you can return to the output table by simply clicking the Netscape Back button. Several different output formats are available via a menu box at the top of the PubMed output table screen. Experiment with the program to find the format that best serves your needs.

A wonderful feature of the PubMed site is that each of the literature entries has been linked to papers that share keywords. Whenever an output table is presented, hypertext (Related Articles) is associated with each entry allowing you to obtain another output table listing the related articles. You can also obtain the list of related articles from abstract pages; in this case, there will be a "Related Papers" button at the top of the screen. Spending a few minutes checking out the related papers can be a very good way to expand your understanding of the problem under investigation.

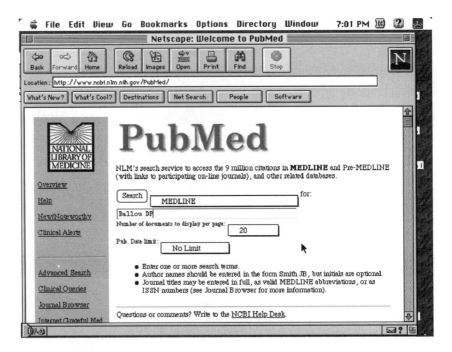

FIGURE 13-5 The PubMed Screen at the NCBI site. The name of an author has been entered in the dialogue box, and the literature database is now ready to be searched for his publications.

SCIENTIFIC JOURNALS

Many scientific journals maintain sites where you can read the table of contents and, for many, the abstracts of papers. In some cases, the complete manuscripts are available from the most recent volumes. Most of the journals available on the Web in complete form must be subscribed to by you or else your university library system must carry a subscription to gain access to them. These journal sites often also contain useful information such as instructions for formatting submissions, the address to which you might send submissions, a list of the members of the editorial board, etc. Several sites containing links to many different scientific journals are presented in Table 13-3, as well as a few sites for individual journals. It may be useful to look at two or three of these sites to familiarize yourself with their use.

DATABASES: EXTRACTING DATA

Numerous databases are found on the Web, and we discuss first those that are used most frequently in biochemistry. These are the nucleic acid sequence, protein sequence, and protein structure databases. We will be using all three in our laboratory exercise. Each database can be accessed directly from the NCBI Entrez site by clicking on the Entrez button on the NCBI home page (Fig. 13-4). The opening Entrez page will provide you with the choice of searching for nucleotides, proteins, 3-D structures, genomes, taxonomy, or literature.

The main DNA sequence database, GenBank, contains all known DNA sequences for all organisms, as well as the conceptual translation of the DNA sequences in all six possible reading frames (three on each strand) to amino acid sequences. This is updated daily with exchange of information with other major sequence databases and input from scientists all over the world. Various subsets of the DNA sequence data are also available, such as the *Escherichia coli* sequence database, the human sequence database, etc. These may be selectively searched by clicking on the appropriate links on the NCBI home page or at other sites (Table 13-3). From the Entrez page, select "nucleotides" to search for a nucleotide sequence or "proteins" to search for a protein sequence. The example below demonstrates how to obtain a protein sequence. The procedures for searching for nucleotide sequences are the same.

Figure 13-6 shows the Protein Sequence Search page of Entrez. In the center of the screen, you will find two menu boxes (Search Field and Mode), a dialogue box in which you enter your search parameters, and two buttons (Search and Reset). As in PubMed, the complete Entrez Help link is in the left column. In addition, specific help links are in hypertext for menu fields and for the dialogue box. The default settings to search "All

FIGURE 13-6 Using the Entrez link to locate the amino acid sequence of *Escherichia coli* alkaline phosphatase.

Fields" in the "Automatic" mode are sufficient for most purposes. If you happen to know the accession number (an arbitrary number assigned to sequences as they are entered in the database), you can simply enter it as the sole search parameter, and the desired sequence will be identified. Otherwise, some combination of keywords (authors, organism, year, etc.) can be searched. (Such combinations can only be used when searching All Fields.) It is desirable to limit the search by providing several simultaneous search parameters. For example, searching for "alkaline phosphatase" will result in the identification of hundreds of different alkaline phosphatase sequences from many different organisms. Searching for "*Escherichia coli* alkaline phosphatase" (Fig. 13-6) limits the output to sequences from *E. coli*. As in PubMed (and other NCBI-linked locations) a space separating terms in the dialogue box corresponds to the logical operator AND. After selecting the search parameters, begin searching by clicking the Search button.

A summary is presented listing the number of files found using your search parameters. If your parameters were not selective enough, hundreds or even thousands of hits may be obtained. In such a case, additional search parameters may be added with the dialogue box in the lower half of the summary screen (Fig. 13-7). If a reasonable number of hits were obtained, the files may be retrieved by clicking the "Retrieve Documents" button. The

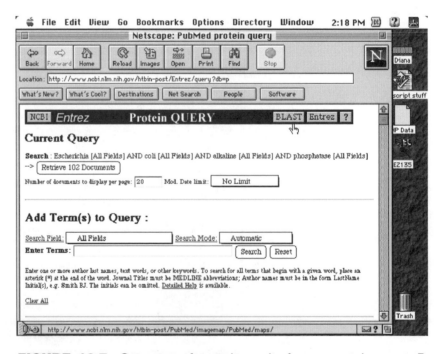

FIGURE 13-7 Summary of search results from a protein query. By scrolling down, you can see the number of "hits" obtained for each of your search parameters. In the current search, 102 different sequences were identified.

search results will then be presented in a table (Fig. 13-8) with links to the data files. Clicking on a link takes you to the sequence file, which can then be copied and pasted into a word processing document for saving. The data sets for proteins, genes, structures, and literature are linked so that once a hit is identified in any search, it is easy to access a wealth of related data.

The dialogue box at the top of the results table allows you to further hone your search. Just below this is a menu box that allows you to obtain various kinds of data for each protein identified in the search. The default setting is GenPept Report, which retrieves information including the GenBank and structure database accession numbers, literature references, and the amino acid sequence. (Other kinds of reports are also available; these are also accessible by hypertext links after each result.) The search results are then listed with a "check box" for each entry. To view the reports for entries, check the associated check boxes, and click on the Display button at the top of the table. If you only want a report for a single entry, you can click on the associated hypertext links. FASTA reports can be obtained, which contain amino acid sequences in the proper format for entry into the BLAST program for sequence alignments (see below).

The protein and DNA sequence databases are extremely useful for several different research applications. In a typical case, the investigator has identified a gene or protein of interest and has obtained sequence data. It is

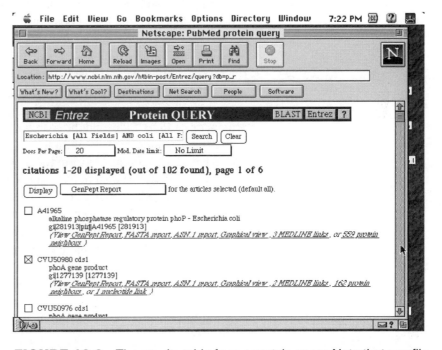

FIGURE 13-8 The results table from a protein query. Note that any file containing the combination of search parameters will be identified. For example, the first entry contains the sequence of a regulatory protein for *E. coli* alkaline phosphatase, not the alkaline phosphatase sequence.

often very useful to know whether the gene or protein is related to genes/proteins that have been studied in other organisms. This information may provide a clue to the function or mechanism of action of the unknown gene or protein. To find such information, the investigator could compare the new sequence data to the gene or protein sequence database. This task is easily accomplished from the NCBI site by clicking on the **BLAST** (Basic Local Alignment Search Tool) button. The Blast page contains a dialogue box for you to enter a "query" sequence. An easy way to enter the sequence is to copy and paste it from a text file such as the FASTA report (above). Next, the database to be searched must be specified. Protein sequence data can be searched against the conceptually translated GenBank (the DNA sequence is translated into its possible amino acid sequences for comparison to the protein sequences) using the FASTA program, while DNA sequences can be searched against the GenBank database with the FASTN program. After the proper parameters have been specified, the search will be conducted with an algorithm devised by Altschul et al. (1990), and the results will be presented in a table. The table will include the name of the related sequences, links to the sequence files, a short description of the sequence, and a similarity score. Related sequences will be presented in order of decreasing relatedness. Clicking on the links in the results table then takes you to the individual sequence files, which are usually also linked to the literature in which the sequence was presented. The alignment of the related sequences to the test sequence will be presented below the search results table, so that you may see (and save) those parts of the query sequence and related sequences that you find have good matches.

Coordinates for protein structures may also be obtained via the NCBI. However, the most user-friendly site for beginners to obtain these files (known as protein database or **PDB** files) is the NIH site known as "**Molecules R Us.**" The electronic address for this site is presented in Table 13-3 under "Databases." As at other NCBI sites, enter the name of the desired protein or its PDB code, and the search results will be presented in a table listing the file name, protein, and a link to each file. Clicking on a link takes you to a page containing the form for submitting a request (Fig. 13-9), and below it, the PDB file containing a short description of the protein, literature citations for the structure determination, a summary of the structural data, and the molecular coordinates. To save the file to a local floppy or hard drive, click on "submit request." In our laboratory exercise, you will use the **RasMol** program to view protein structures. The default option for saving files is the RasMol format (PDB Viewer), and this should be chosen. The desired file will then be transferred to your designated destination. PDB files may be quite large, in which case the transfer may take some time. Later in this chapter, you can find instructions for using RasMol to view these files.

A large number of other databases are also present on the Web. Some especially useful for the biochemist are those listing metabolic reactions and pathways. The addresses for these free sites are presented in Table 13-3.

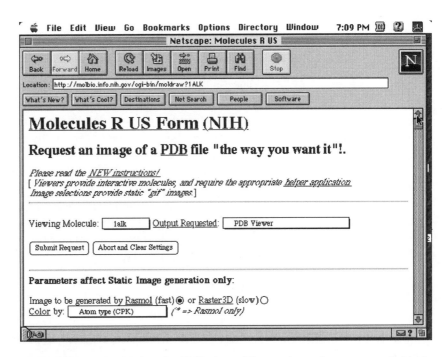

FIGURE 13-9 Molecules R Us form. The settings shown are suitable for viewing the image in RasMol. Alkaline phosphatase from *E. coli* (1alk) has been selected. Below the screen is the PDB file, which can be accessed by scrolling down.

FREE PROGRAMS

Many free programs are available through the Web. A few sites listing these programs, with links to the sites at which programs may be obtained, are presented in Table 13-3. We use the free program RasMol, which permits the viewing and analysis of protein structures on microcomputers. This spectacular program, written by Roger Sayle (Sayle and Milner-White, 1995), permits manipulation of protein structures on microcomputers. Such operations only a few years ago required considerably more powerful and expensive computer work stations. A succinct description of the program and the instructions for its use may be found in the work of Sayle and Milner-White (1995) or in the RasMol users guide, which can be obtained from a link at the program source.

Very popular free programs such as RasMol may be obtained from the original site or from several "mirror" sites. Two different versions of the program are available, one for use on Macintosh™ computers and another for use on PC machines running Windows™. The program is fairly large and takes some time to transfer, and since the program is so popular, it is sometimes hard to access sites where it is available.

To locate RasMol sites, the Yahoo search engine (Table 13-3, Search Engines) can be opened and the term RasMol can be searched. The results of this search will list sites at which different versions of RasMol (as well as their User Manuals) are available. (As with many programs, RasMol is under continuous refinement, so many versions are available. However, the latest and most powerful versions are also the versions with many "bugs" yet to be worked out, so you might wish to choose well tested versions that are sufficient for your needs). You may need to attempt repeatedly to connect to these busy sites, until eventually access is obtained. Once a suitable site is accessed (it might be a mirror site, e.g., in Massachusetts), the version of the program desired is selected with consideration of the intended type of local machine (i.e., RasMol for PC or RasMac for Macintosh™). The transfer can be activated by clicking the appropriate buttons on the screen. The program will be sent and saved to a floppy disk or other medium prescribed by you and can be serially copied to floppy disks for local distribution and use.

USING RASMOL TO VIEW MOLECULAR STRUCTURES

First, load the RasMol program onto the hard drive of your computer. For a PC running Windows™, the program may be set up with the "File New" option, as with many other programs. When you activate the program, two or three windows will appear on the screen, depending on which version of the program is used. For easiest use, these windows should be sized such that they do not overlap and all two or three can be viewed at the same time. The main window will contain the image of the protein structure; when first opening the program, this is blank with a black background. Another window contains a blinking cursor; this is the "command" window used for entering commands. The third window, when it is present, contains several buttons that permit the measurement of bond angles, atomic distances, etc. For example, when two atoms are specified and the atomic distance button is selected, the distance between the two selected atoms will be presented at the bottom of this window.

To open a PDB file, select the File menu at the top of the main screen. Choose Open, and then select the location of the PDB file. For example if a PDB file was obtained from Molecules R Us and copied onto a floppy disk in drive A (or B if appropriate), direct the program to that file (File Open, drive A, filename). In a matter of seconds, the structure of the protein will appear on the screen.

RasMol permits the structure to be displayed in several ways, providing a very valuable tool to learn about the structure. For example, the protein can be displayed as a ball-and-stick model, a ribbon diagram, or a space-filling model. These display options may be selected from the Display menu commands at the top of the main window. The image may also be saved in several different formats. Two widely used formats recognized by different graphics programs are the BMP format and the GIF format.

After saving as a BMP or GIF file, the image may be manipulated by importing the file into a graphics program.

Another powerful tool for looking at protein structures is the ability to shade different parts of the molecule individual colors. For example, if the key catalytic residues in the active site of an enzyme are known, they may be colored with a particular hue to ease their visualization (see, for example, Fig. 13-10). The procedure for doing this is as follows: point the mouse at the command window to activate that window. Next, type "select 104" to select the amino acid at position 104. Or if you want to color amino acids 104–110, you would type "select 104–110." Strike the return key to activate the selection, after which RasMol will confirm the selection by indicating the number of atoms selected. Next, type "color red" and the enter key, and the selected residues will be colored red on the screen. RasMol supports a wide variety of colors, so that different parts of the molecule can be colored differently.

Yet another powerful capability of RasMol is the ability to rotate the molecule in all directions, as well as the ability to view the whole structure or look at close-up views of selected parts of the structure. The mouse and keyboard commands for moving molecules around are found in the RasMol

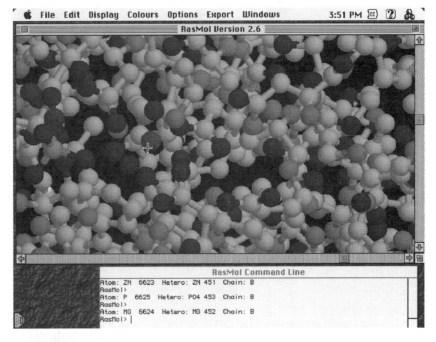

FIGURE 13-10 Active site of alkaline phosphatase displayed by RasMol. The structure is shown in ball-and-stick display, with CPK coloring (carbon grey, oxygen red, nitrogen light blue). The magnesium ion is shown in green, the zinc ions are shown in maroon, and phosphorus is shown in orange. (If you have the black and white version of this text, it will appear in various shades of gray.)

manual and in the work of Sayle and Milner-White (1995). The program is so appealing that most users cannot be bothered to read the instructions before trying it. Therefore, we note that rotations of the molecule may be accomplished by simply dragging atoms around with the mouse. A partial list of manipulations in RasMol are listed in Table 13-2.

OTHER USEFUL INFORMATION ON THE INTERNET

PROCEDURES AND TOOLS ON THE WEB

Many useful laboratory procedures are described on the Web. Several collections of links to many different biochemical and biological procedures and tools are presented in Table 13-3, Protocols, Courses, and Useful Research Tools. If you have the time, you can browse these sites to get an idea of what is available over the Web. A very useful site with many tools

TABLE 13-2

Three-Dimensional Manipulations in RasMol (abstracted from Sayle and Milner-White, 1995)

MANIPULATION	MACINTOSH™	WINDOWS™
Rotate molecule about any axis in the x,y plane	Drag cursor in the appropriate direction	Drag cursor in the appropriate direction using the left mouse button.
Zoom molecule	Drag cursor vertically using the left mouse button with the shift key depressed.	Drag cursor vertically with the shift key depressed.
Translate molecule within x,y plane.	Drag cursor with the command key depressed.	Drag cursor using the right mouse button.
Rotate molecule about the z axis.[a]	Move cursor with both the shift and command keys depressed.	Drag cursor using the right mouse button and with the shift key depressed.

[a]To make the point of rotation at the midpoint of the molecule, type "centre." To set the center of rotation at a particular residue, such as residue 30, type "centre 30."

and resources is Pedro's Biomolecular Research Tool, whose Web address is listed in Table 13-3.

COURSES

Many different courses and educational sites are present on the Web. These sites are an important source of background information, challenging homework problems, and useful literature citations. They present a way to brush up on topics that you need to know. A few good sites in this category are presented in Table 13-3.

VENDORS

Scientists need reagents, radioactive compounds, instruments, biological materials, etc., which are purchased from various vendors with funds obtained mainly from federal research grants. Essentially all of these vendors now maintain Web sites with the complete catalogue, so now the problem of obtaining current catalogues and disposing of outdated catalogues is a thing of the past. In addition, these electronic catalogs make it easier to search for chemicals and other products. For example, if you search Sigma Chemical Co. for alcohol dehydrogenase, it will give you a list of all forms of alcohol dehydrogenase that they carry. Also, such Web sites usually indicate the availability of catalogue items and allow for direct electronic ordering. A few vendor sites are listed in Table 13-3 to serve as examples.

REFERENCES

Abelson, J. N., and M. I. Simon, eds. 1996. Methods in enzymology, computer methods for macromolecules, Academic Press, New York.

Altschul, S. F., W. Gish, W. Miller, E. W. Myers, and D. J. Lipman. 1990. Basic local assignment search tool. J. Mol. Biol. **215:**403.

Peruski, L. F., and A. H. Peruski. 1997. The Internet and the new biology: tools for genomic and molecular research. A.S.M. Press, Washington, D. C.

Sayle, R. A., and E. J. Milner-White. 1995. RASMOL: biomolecular graphics for all. Trends Biochem. Sci. **20:**374.

HOMEWORK ASSIGNMENT

In several of the experiments described in this book, you have studied the enzyme alkaline phosphatase from *E. coli*. For example, you purified the enzyme and analyzed the kinetics of the reaction and its inhibition by inorganic phosphate. Now you can try to carry out some computer searches to enhance your understanding of the enzyme and develop some facility with several of the programs and procedures discussed above. These exercises are open-ended to allow you the freedom to explore these methods and to be more like dealing with a realistic research problem. No "cut-and-dried" protocols for obtaining information will be given. You will be asked to write a report about your findings just like other laboratory reports.

1. Conduct a literature search to identify the active site of the enzyme and the reaction mechanism. What metals are in the enzyme, how are they chelated, and what is their role in catalysis? Has anyone ever examined mutant forms of the enzyme containing alterations of the active site residues?

2. Extract the *E. coli* alkaline phosphatase amino acid sequence from a database. Next, compare the sequence to the translated DNA sequence database. Are there any other enzymes from other biological sources, catalyzing different phosphatase reactions, that are related to *E. coli* alkaline phosphatase? (Exclude very closely related alkaline phosphatases from *E. coli* or other bacteria.)

3. Your instructors will provide you with a copy of RasMol or RasMac, and the PDB file for *E. coli* alkaline phosphatase. Visualize the structure, and color the active site residues a distinct color. If any active site residues have been mutated and studied, color those residues a distinct color. If there are any metal ions in the structure, color the amino acids that chelate the metal ions a distinct color. Export a GIF and BMP image file to a floppy disk, and see whether you can produce a figure with your results by importing these files into a graphics program.

4. Measure the distance between the metals in the alkaline phosphatase. Do these metals occur in a sequence motif that is common to any of the other phosphatases that you have found?

5. During the action of alkaline phosphatase, a phosphoenzyme is formed. Identify the residue that is involved. How far is it from any metals in the enzyme? Does any metal have a catalytic role in activating that residue?

TABLE 13-3
Useful Web Sites

Internet Search Engines

ENGINE	ADDRESS
Yahoo	http://www.yahoo.com/
WebCrawler	http://webcrawler.cs.washington.edu/ WebCrawler/WebQuery.html
Lycos	http://lycos.cs.cmu.edu/
Harvest	http://rd.cs.colorado.edu/harvest/
JumpStation	http://www.stir.ac.uk/jsbin/js
Infoseek	http://www.infoseek.com
Altavista	http://www.altavista.com
NIKOS	http://www.rns.com/cgi-bin/nikos

Scientific Journals

JOURNAL	ADDRESS
Nature	http://www.nature.com/
Protein Science	http://www.prosci.org/
Internet Journal of Science	http://www.netsci-journal.com/
Theoretical Chemistry	http://www.elsevier.nl:80/section/chemical/ theochem/menu.htm
Electronic Protein Science	http://www.prosci.uci.edu/
ACS Publications Division (many journals)	http://pubs.acs.org/
Journal of Biological Chemistry	http://www-jbc.stanford.edu/jbc/
Journal of Bacteriology	http://asmusa.edoc.com/jb/
Genes and Development	http://207.22.83.2:443/cshl/journals/gnd/
Biochemistry On-Line (charge)	http://biochem.arach-net.com/beasley/ journals.html

Databases

DATABASE	ADDRESS
The National Center for Biotechnology Information (NCBI)[a]	http://www.ncbi.nlm.nih.gov/
NCBI Molecules-R-Us (for PDB query)[a]	http://molbio.info.nih.gov/cgi-bin/pdb
Enzyme Structures Database	http://www.biochem.ucl.ac.uk/bsm/enzymes/ index.html
The Institute for Genomic Research	http://www.tigr.org/
The European Bioinformatics Institute	http://www.ebi.ac.uk/
The National Cancer Institute 3D Structure Database	http://epnws1.ncifcrf.gov:2345/dis3d/ 3Ddatabase/dis3d.html
Carbohydrate databases	http://www.boc.chem.ruu.nl/sugabase/ databases.html
Abbreviations of chemical compounds	http://www.chemie.fu-berlin.de/cgi-bin/ abbscomp
Bacterial nomenclature	http://www.ftpt.br/cgi-bin/bdtnet/bacterianame
Microbial Strain Data Network	http://bioinfo.ernet.in/cgi-bin/asearch/msdn/ dsm/read"

TABLE 13-3 *(Continued)*
Useful Web Sites

Databases *(continued)*

DATABASE	ADDRESS
The E. coli Genome Center	http://www.genetics.wisc.edu/EcoCyc:
Encyclopedia of E. coli Genes and Metabolism	http://www.ai.sri.com/ecocyc/ecocyc.html
Biocatalysis/Biodegradation Database[a]	http://www.labmed.umn.edu/umbbd/index.html
ExPASy - Biochemical Pathways[a]	http://expasy.hcuge.ch/cgi-bin/search-biochem-index
REBASE–Restriction Enzyme Database	http://www.neb.com/rebase/

Useful Programs

PROGRAM	ADDRESS
RasMol Home Page at UMass	http://www.umass.edu/microbio/rasmol/
RasMac v2.5-ucb	http://hydrogen.cchem.berkeley.edu:8080/Rasmol/
Index of packages/kinsim[a]	http://wuarchive.wustl.edu/packages/kinsim/
Chemical Kinetics Simulation	http://www.chem.uci.edu/instruction/applets/simulation.html
Kinetics (a simulation program)	http://jcbmac.chem.brown.edu/baird/Chem22I/lectures/Kinetics/Kinetics.html
WinZip	http://www.winzip.com
Gepasi: a metabolic simulator (also good for kinetic simulations)	http://gepasi.dbs.aber.ac.uk/metab/gepasi.htm
Webcutter (Restriction Cleavage Program)	http://www.firstmarket.com/cgi-bin/firstmarket/cutup

Protocols, Courses, and Useful Research Tools

ITEM	ADDRESS
Fundamental Laboratory Approaches for Biochemistry and Biotechnology[a]	http://www.umich.edu/~aninfa/book/index.html
Pedro's BioMolecular Research Tools[a]	http://www.public.iastate.edu/~pedro/research_tools.html
Educational Computing Courses on the Web	http://lenti.med.umn.edu/~mwd/courses.html #BIOCOMPUTING
Molecular Biology on the Net	http://www.lsumc.edu/campus/micr/mirror/public_html/molbiol.html
The ESG Biology Hypertextbook Home Page	http://esg-www.mit.edu:8001/esgbio/7001main.html
Protocols on the World Wide Web	http://www.ifrn.bbsrc.ac.uk/gm/lab/docs/protocols.html
Molecular Biology Protocols	http://listeria.nwfsc.noaa.gov/protocols.html
Cell and Molecular Biology Online—Educational Resources[a]	http://www.tiac.net/users/pmgannon/teaching.html
Internet Resources for Cell Biology	http://www.gac.edu/cgi-bin/user/~cellab/phpl?links.html
Biology Links	http://cgsc.biology.yale.edu/bio.html
Chemistry Software and Information Resources	http://www.csir.org/

TABLE 13-3 (Continued)
Useful Web Sites

Protocols, Courses, and Useful Research Tools (continued)

ITEM	ADDRESS
Bio Access Guide to the Internet	http://www.bchs.uh.edu/Server/
Internet Chemistry Resources[a]	http://www.rpi.edu/dept/chem/cheminfo/chemres.html
Software for Organic Chemistry Laboratories	http://heme.gsu.edu/post_docs/koen/wsoftwar.html#OS
Organic Chemistry Resources Worldwide	http://heme.gsu.edu/post_docs/koen/worgche.html
Electrochemical Science and Technology Information Resource	http://www.cmt.anl.gov/estir/pdi.htm
MolData Chemical Data on the Web	http://pages.pomona.edu/~wsteinmetz/moldata.html
Physical Organic Chemistry	http://www.chem.qmw.ac.uk:80/iupac2/gtpoc/index.html
Swiss-Model: Automated Protein Modelling Server	http://expasy.hcuge.ch/swissmod/SWISS-MODEL.html
UCLA-DOE Structure Prediction Server	http://www-lmmb.ncifcrf.gov/~nicka/run123D.html
PredictProtein: Protein Sequence Analysis and Structure prediction[a]	http://www.embl-heidelberg.de/predictprotein/
DNA and Protein Analysis (from Stockholm University)	http://www.biokemi.su.se/~threader_server/pscan/
VRML, Visualization of PDB Files	http://ws05.pc.chemie.th-darmstadt.de/vrml/pdbvis.html
ExPASy-tools for Sequence Analysis[a]	http://expasy.hcuge.ch/tools.html
Chemistry Resources on the Web[a]	http://www.chem.rpi.edu/icr/chemres.html
Biotechnology resources[a]	http://biotech.chem.indiana.edu
Molecular Modelling for Beginners	http://swift.embl-heidelberg.de/workshop/ws.html

Vendors

Sigma Chemical Company	http://www.sigma.sial.com/safinechem.html
New England Biolabs	http://www.neb.com
Gibco BRL Life Technologies	http://www.lifetech.com
Molecular Probes	http://www.probes.com/
Invitrogen	http://www.invitrogen.com
Promega	http://www.promega.com

[a]Sites not to be missed.
These Web site addresses will be updated and expanded periodically on the Web site for "Fundamental Laboratory Approaches for Biochemistry and Biotechnology."

MATERIALS NEEDED FOR CHAPTER 13

✓ In order to save the students some time in completing their assignment, we usually provide a copy of RasMol or RasMac, along with the alk1 PDB, on a disk. We obtain these from the sites listed in Table 13-3, copy them onto the hard drives of our own computers, and then serially produce copies for the students.

Appendix 1
Safety Checklist

Before commencing with laboratory work, determine the location of each of the following safety items in the laboratory. Check them off and note their locations.

1. Exits

2. Eye wash station

3. Safety shower

4. Fire alarm

5. Fire extinguishers

6. Fume hood

7. Telephone

8. Safety glasses

9. Glass disposal

10. Plastic disposal

11. Solid chemical

12. Organic solvent

13. Acids

14. Bases

15. Radioactivity

16. Other disposal

17. Spill kits

18. First aid kit

19. MSDS folder

Appendix 2
An Example of an MSDS

1. PRODUCT IDENTIFICATION

Product name: Phenol
Formula: C_6H_5OH
Formula WT: 94.11
CAS no.: 00108-95-2
NIOSH/RTECS no.: SJ3325000
Common synonyms: Carbolic acid; hydroxybenzene; monohydroxybenzene; phenic acid; phenylic acid
Product codes: 2858,2862
    Effective: 01/22/87
    Revision no.: 04
Precautionary Labeling (Baker SAF-T-DATA™ System)

HEALTH	3 SEVERE (LIFE)
FLAMMABILITY	2 MODERATE
REACTIVITY	1 SLIGHT
CONTACT	4 EXTREME (CORROSIVE)

    Hazard ratings are 0 to 4 (0 = No hazard; 4 = Extreme hazard).

Laboratory Protective Equipment. ** CODE NOT ON FILE **

Precautionary Label Statements.
POISON DANGER—COMBUSTIBLE—CAUSES SEVERE BURNS —RAPIDLY ABSORBED THROUGH SKIN—MAY BE FATAL IF SWALLOWED, INHALED, OR ABSORBED THROUGH SKIN—EXCEPTIONAL HEALTH AND CONTACT HAZARDS—READ MATERIAL SAFETY DATA SHEET—KEEP AWAY FROM HEAT, SPARKS, FLAME. DO NOT GET IN EYES, ON SKIN, ON CLOTHING—DO NOT BREATHE DUST. KEEP IN TIGHTLY CLOSED CONTAINER. USE WITH ADEQUATE VENTILATION WASH THOROUGHLY AFTER HANDLING. IN CASE OF FIRE, SOAK WITH WATER. IN CASE OF SPILL, SWEEP UP AND REMOVE. FLUSH SPILL AREA WITH WATER.

SAF-T-DATA™ Storage color code. Red Stripe (Store Separately)

2. HAZARDOUS COMPONENTS

Component	%	CAS No.
Phenol	90–100	108-95-2

3. PHYSICAL DATA

Boiling Point: 182° C (360° F)
Vapor Pressure (MM HG): 0.35
Melting Point: 40° C (104° F)
Vapor Density (air = 1): 3.24
Specific Gravity: 1.07
Evaporation Rate: <1 (H_2O = 1) (Butyl Acetate = 1)
Solubility (H_2O): Moderate (1–10 %)
% Volatiles by Volume: 100
Appearance & Odor: Colorless crystals; characteristic odor.

4. FIRE AND EXPLOSION HAZARD DATA

Flash Point (Closed Cup): 79° C (175° F)
NFPA 704M Rating: 3-2-0
Flammable Limits: Upper —8.6%, Lower —1.5 %
Fire Extinguishing Media: Use water spray, alcohol foam, dry chemical, or carbon dioxide.
Special Fire-Fighting Procedures: Firefighters should wear proper protective equipment and self-contained breathing apparatus with full facepiece operated in positive pressure mode. Move containers from fire area if it can be done without risk. Use water to keep fire-exposed containers cool.
Unusual Fire & Explosion Hazards: Gives off heavy smoke. Gives off flammable vapors. Vapors may form explosive mixture with air. Closed containers exposed to heat might explode. Contact with strong oxidizers may cause fire.
Toxic Gases Produced: Carbon monoxide, carbon dioxide

5. HEALTH HAZARD DATA

TLV and PEL listed denote (skin).
Threshold Limit Value (TLV/TWA): 19 mg/m^3 (5 ppm)
Short-Term Exposure Limit (STEL): 38 mg/m^3 (10 ppm)
Permissable Exposure Limit (PEL): 19 mg/m^3 (5 ppm)
Toxicity: LD_{50} (ORAL-RAT)(mg/kg)–384
    LD_{50} (SKN-RAT) (mg/kg) –669
    LD_{50} (IPR-RAT)(mg/kg) –250
    LC_{50} (INHL-RAT) (mg/kg) –316
Carcinogenicity: NTP—No   IARC—No   Z LIST—No   OSHA Reg—No
Effects of Overexposure: Acute poisoning via all routes of exposure may be severe enough to be fatal. Inhalation of dust may cause headache, coughing, difficulty in breathing, chest pain, severe lung irritation, or pulmonary edema. Contact with skin or eyes may cause severe irritation or burns. Substance is readily absorbed through the skin. Ingestion may cause nausea, vomiting, gastrointestinal irritation, and burns to mouth and throat. Chronic effects of overexposure may include kidney and/or liver damage.
Target Organs: Liver, kidneys, skin
Medical Conditions Generally Aggravated by Exposure: Kidney disorders
Routes of Entry: Inhalation, absorption, eye contact, skin contact

Emergency and First Aid Procedures

- **Call a physician**.
- If swallowed, do not induce vomiting; if conscious, give water, milk, or milk of magnesia.
- If inhaled, remove to fresh air. If not breathing, give artificial repiration. If breathing is difficult, give oxygen.
- In case of contact, immediately flush eyes or skin with plenty of water for at least 15 min while removing contaminated clothing and shoes. Wash clothing before reuse.

6. REACTIVITY DATA

Stability: Stable
Hazardous Polymerization: Will not occur
Conditions to Avoid: Heat, flame, other sources of ignition, light, air
Incompatibles: Strong oxidizing agents, strong bases, alkalies, calcium hypochlorite
Decomposition Products: Carbon monoxide, carbon dioxide

7. SPILL AND DISPOSAL PROCEDURES

Steps to Be Taken in the Event of a Spill or Discharge: Wear self-contained breathing apparatus and full protective clothing. Shut off ignition sources: no flares, smoking, or flames in area. With a clean shovel, carefully place material into clean , dry container and cover. Remove from area. Flush spill area with water.
Disposal Procedure: Dispose in accordance with all applicable federal, state, and local environmental regulations.
EPA Hazardous Waste Number: U188 (Toxic Waste)

8. PROTECTIVE EQUIPMENT

Ventilation: Use general or local exhaust ventilation to meet TLV requirements.
Respiratory Protection: Respiratory protection required if airborne concentration exceeds TLV. At concentrations up to 50 ppm, a chemical cartride respirator with organic vapor cartridge is recommended. Above this level, a self-contained breathing apparatus is recommended.
Eye/Skin Protection: Safety goggles and face shield, uniform, protective suit, Viton gloves are recommended.

9. STORAGE AND HANDLING PRECAUTIONS

SAF-T-DATA™ Storage Color Code: Red Stripe (Store separately)
Special Precautions: Keep container tightly closed. Store in a cool, dry, well-ventilated, flammable-liquid-storage area or cabinet. Store in light-resistant containers.

10. TRANSPORTATION DATA AND ADDITIONAL INFORMATION

<u>**Domestic (D.O.T.)**</u>
Proper shipping name: Phenol
Hazard Class: Poison B
UN/NA: UN1671
Labels: POISON
Reportable Quantity: 1000 lbs.

<u>**International (I.M.O.)**</u>
Proper shipping name: Phenols
Hazard Class: 6.1
UN/NA: UN1671
Labels: POISON

Index